ISNM
International Series of Numerical Mathematics
Vol. 147

Free Boundary Problems

Theory and Applications

Pierluigi Colli
Claudio Verdi
Augusto Visintin
Editors

Springer Basel AG

Editors:

Pierluigi Colli
Dipartimento di Matematica
Università di Pavia
Via Ferrata, 1
27100 Pavia
Italy
e-mail: pier@imati.cnr.it

Augusto Visintin
Dipartimento di Matematica
Via Sommarive, 14
38050 Povo di Trento
Italy
e-mail: visintin@science.unitn.it

Claudio Verdi
Dipartimento di Matematica
Università di Milano
Via Saldini, 50
20133 Milano
Italy
e-mail: claudio.verdi@mat.unimi.it

2000 Mathematics Subject Classification 35R35, 35Qxx, 35Dxx

A CIP catalogue record for this book is available from the Library of Congress, Washington D.C., USA

Bibliographic information published by Die Deutsche Bibliothek
Die Deutsche Bibliothek lists this publication in the Deutsche Nationalbibliografie; detailed bibliographic data is availabl
in the Internet at <http://dnb.ddb.de>.

ISBN 978-3-0348-9613-9

Originally published by Birkhäuser Verlag,Basel in 2004
Softcover reprint of the hardcover 1st edition 2004
Printed on acid-free paper produced of chlorine-free pulp. TCF ∞

ISBN 978-3-0348-9613-9 ISBN 978-3-0348-7893-7 (eBook)
DOI 10.1007/978-3-0348-7893-7

9 8 7 6 5 4 3 2 1

www.birkhasuer-science.

Contents

Preface

This volume collects the proceedings of a conference on free boundary problems that took place in Trento (Italy) in June 2002.

Many phenomena of interest for applications are represented by differential equations which are defined in a domain whose boundary is a priori unknown, and is accordingly named a *"free boundary"*; a further quantitative condition is then provided in order to exclude indeterminacy. Free boundaries are often related to discontinuities in constitutive relations; this raises a number of interesting mathematical issues: existence of solutions in function spaces, uniqueness, regularity properties, numerical approximation procedures, and other questions have been extensively investigated.

Examples of free boundary problems include the classical Stefan problem and more general models of phase transitions: here the free boundary is represented by the moving interface between the phases. Free boundaries also occur as fronts between saturated and unsaturated regions in filtration through porous media, between plastic and elastic phases in continuous mechanics, between magnetic domains in ferromagnetism, just to mention a few cases. Relevant examples also come from reaction-diffusion, fluid dynamics, biomathematics, and so on.

Several of these problems are of interest for applications in industry and offer opportunities of collaboration among mathematicians, physicists, engineers, material scientists, biologists, and other researchers. A large community of applied scientists spread over the world has been formulating and studying these problems for many years. They have also been meeting in major conferences every three years or so; the first appointment was held at Montecatini in 1981, and was followed by Maubuissons, Irsee, Montreal, Toledo, Zakopane, Crete, Chiba. The next conference is planned for 2005 in Portugal. The proceedings of these conferences provide a comprehensive picture of the development of research on free boundary problems in the last decades.

The Trento conference saw the participation of more than 150 scientists, coming from a long list of countries. Talks included twenty plenary addresses, and seven focus sessions were devoted to selected topics: free boundary problems in polymer science, image processing, grain boundary motion, numerical aspects of free boundary problems, free boundary problems in biomathematics, modelling in crystal growth, and transitions in anisotropic materials. A poster session open to the contribution of all participants was also organized. Post-dinner entertainment included two concerts performed by professional and semi-professional musicians.

Following the tradition of the series, the contributions concern problems which are either directly related to free boundaries, or may be so in perspective. Special emphasis was put on interdisciplinarity and on issues of applicative relevance. This variety is obviously reflected in these proceedings, which are likely to meet the interests of a large spectrum of readers.

The editors thank the co-organizers G. Dziuk, A. Fasano, K.-H. Hoffmann, and J. Sprekels. Scientific assistance by the other members of the scientific committee, M. Frémond, A. Friedman, S. Luckhaus, M. Niezgódka, J. Ockendon, M. Primicerio, J.F. Rodrigues, and the collaboration of M. Paolini, are also gratefully acknowledged; R. Rossi also carefully contributed to the revision of these proceedings.

The conference was financially supported by the Istituto Nazionale di Alta Matematica (through G.N.A.M.P.A., G.N.F.M., G.N.C.S.), by three national research projects of the Italian Ministry for University, by the University of Trento, by the C.A.E.S.A.R. Institut (Bonn), and by the Weiestrass Institut (Berlin).

Pavia, Milano, Trento, May 2003 P. Colli, C. Verdi, A. Visintin

International Series of Numerical Mathematics, Vol. 147, 1–15

Structural Optimization by the Level-Set Method

Grégoire Allaire, François Jouve, and Anca-Maria Toader

Abstract. In the context of structural optimization, we describe a new numerical method based on a combination of the classical shape derivative and of the level-set method for front propagation. We have implemented this method in two and three space dimensions for models of linear or non-linear elasticity, with various objective functions and constraints on the volume or on the perimeter. The shape derivative is computed by an adjoint method. The cost of our numerical algorithm is moderate since the shape is captured on a fixed Eulerian mesh. Although this method is not specifically designed for topology optimization, it can easily handle topology changes.

1. Introduction

Shape optimization of elastic structures is a very important and popular field. The classical method of shape sensitivity (or boundary variation) has been much studied (see e.g. [13], [16], [20], [21]). It is a very general method which can handle any type of objective functions and structural models, but it has two main drawbacks: its computational cost (because of remeshing) and its tendency to fall into local minima far away from global ones. The homogenization method (see e.g. [1], [2], [5], [7], [8], [12]) is an adequate remedy to these drawbacks but it is mainly restricted to linear elasticity and particular objective functions (compliance, eigenfrequency, or compliant mechanism). Yet, another method has recently been proposed in [14], [18], [4] relying on the level-set method which has been devised by Osher and Sethian [15], [17] for numerically tracking fronts and free boundaries. The level-set method is versatile and computationally very efficient: it is by now a classical tool in many fields such as motion by mean curvature, fluid mechanics, image processing, etc.

The paper [14] studied a two-phase optimization of a membrane (modelled by a linear scalar partial differential equation), i.e. the free boundary was the interface between two constituents occupying a given domain. It combined the level-set method with the shape sensitivity analysis framework. On the other hand, the work [18] focused on structural optimization within the context of two-dimensional linear elasticity. In that setting, the free boundary was the shape of the structure, which was captured on a fixed mesh using the immersed interface method. However,

[18] did not rely on shape sensitivity analysis: rather, the structural rigidity was improved by using an ad hoc criterion based on the Von Mises equivalent stress.

In [4], we have generalized these two previous works in several aspects. We give here a brief review of our approach, based on a systematic implementation of the level-set method in which the front velocity is derived from a shape sensitivity analysis. We focus on shape optimization rather than two-phase optimization, and we replace the immersed interface method by the simpler "ersatz material" approach which amounts to filling the holes by a weak phase. This is a well-known approach in topology optimization, and it can be rigorously justified in some cases [1]. We compute a shape derivative by using an adjoint problem. Then, the shape derivative is used as the normal velocity of the free boundary which is moved during the optimization process. Front propagation is performed by solving a Hamilton–Jacobi equation for a level-set function.

Acknowledgements: This work has been supported by the grant CNRS/ICCTI n° 2002-12163 of the Centre National de la Recherche Scientifique (France) and the Instituto de Cooperação Científica e Tecnologica Internacional (Portugal).

2. Setting of the problem

We start by describing a model problem in linearized elasticity. There is no conceptual difficulty in choosing another model, and in particular a nonlinear elasticity problem. Let $\Omega \subset \mathbb{R}^d$ ($d = 2$ or 3) be a bounded open set occupied by a linear isotropic elastic material with Hooke's law A. Recall that, for any symmetric matrix ξ, A is defined by

$$A\xi = 2\mu\xi + \lambda(\mathrm{Tr}\xi)\,\mathrm{Id},$$

where μ and λ are the Lamé moduli of the material. The boundary of Ω is made up of two disjoint parts

$$\partial\Omega = \Gamma_N \cup \Gamma_D,$$

with Dirichlet boundary conditions on Γ_D, and Neumann boundary conditions on Γ_N. The two boundary parts Γ_D and Γ_N are allowed to vary in the optimization process, although it is possible to fix some portion of it (see the numerical examples below).

We denote by f (g) the vector-valued function of the volume forces (of the surface loads, respectively). The displacement field u in Ω is the solution of the linearized elasticity system

$$\begin{cases} -\mathrm{div}\,(A\,e(u)) = f & \text{in } \Omega \\ u = 0 & \text{on } \Gamma_D \\ \big(A\,e(u)\big)n = g & \text{on } \Gamma_N. \end{cases} \tag{1}$$

Since Ω is varying during the optimization process, f and g must be known for every possible configuration of Ω. We therefore introduce a working domain D (a bounded open set of $\mathbb{R}^d$) which contains all the admissible shapes Ω.

To give a precise mathematical meaning to (1), we choose $f \in L^2(D)^d$ and $g \in H^1(D)^d$ and we assume that $\Gamma_D \neq \emptyset$ (otherwise we should impose an equilibrium condition on f and g). In such case it is well known that (1) admits a unique solution in $H^1(\Omega)^d$.

The objective function is denoted by $J(\Omega)$. In this paper, we shall mostly focus on two possible choices of J (these are merely examples, and much more freedom is allowed). A first classical choice is the compliance (the work done by the load)

$$J_1(\Omega) = \int_\Omega f \cdot u \, dx + \int_{\Gamma_N} g \cdot u \, ds = \int_\Omega A\, e(u) \cdot e(u) \, dx, \tag{2}$$

which is very common in rigidity maximization. A second choice is a least square error compared to a target displacement

$$J_2(\Omega) = \left(\int_\Omega k(x) |u - u_0|^\alpha dx \right)^{1/\alpha} \tag{3}$$

which is a useful criterion for the design of compliant mechanisms [3], [19]. We assume $\alpha \geq 2$, $u_0 \in L^\alpha(D)$ and $k \in L^\infty(D)$, k being a non-negative given weighting factor. In both formulae (2) and (3), $u = u(\Omega)$ is the solution of (1). We define a set of admissible shapes which must be open sets contained in the working domain D and having a fixed volume V

$$\mathcal{U}_{ad} = \left\{ \Omega \subset D \text{ such that } |\Omega| = V \right\}. \tag{4}$$

Our model problem of shape optimization is

$$\inf_{\Omega \in \mathcal{U}_{ad}} J(\Omega). \tag{5}$$

It is well known that the minimization problem (5) is usually not well posed on the set of admissible shapes defined by (4) (i.e. it has no solution). In order to obtain existence of optimal shapes, either some smoothness or geometrical or topological constraints are required. For example, a variant of (5) with a perimeter constraint turns out to be a well-posed problem (see [6]). More precisely, if $\ell > 0$ is a positive Lagrange multiplier, the minimization problem

$$\inf_{\Omega \in \mathcal{U}_{ad}} \left(J(\Omega) + \int_{\partial\Omega} ds \right) \tag{6}$$

admits at least one optimal solution. There are other regularized variants of (5) which are well-posed and we refer to [9], [11] for such existence theories. Note that, although existence is not an issue of the present paper, we shall work with a smoother subset of (4), in order to define a proper notion of shape derivative.

3. Shape derivative

In order to apply a gradient method to the minimization problem (5), we recall a classical notion of shape derivative. This notion goes back to Hadamard, at least, and several researchers have contributed to its development (see e.g. the reference books [16], [21]). Here, we follow the approach of Murat and Simon [13], [20]. Starting from a smooth reference open set Ω, we consider domains of the type

$$\Omega_\theta = \big(\mathrm{Id} + \theta\big)(\Omega),$$

with $\theta \in W^{1,\infty}(\mathbb{R}^d, \mathbb{R}^d)$. It is well known that, for sufficiently small θ, $(\mathrm{Id} + \theta)$ is a diffeomorphism in $\mathbb{R}^d$.

Definition 3.1. The *shape derivative* of $J(\Omega)$ at Ω is defined as the Fréchet derivative $J'(\Omega)$ in $W^{1,\infty}(\mathbb{R}^d, \mathbb{R}^d)$ at $\theta = 0$ of the application $\theta \to J\big((\mathrm{Id} + \theta)(\Omega)\big)$, i.e.

$$J\big((\mathrm{Id} + \theta)(\Omega)\big) = J(\Omega) + J'(\Omega)(\theta) + o(\theta) \quad \text{with} \quad \lim_{\theta\to 0} \frac{\|o(\theta)\|}{\|\theta\|} = 0\,,$$

where $J'(\Omega)$ is a continuous linear form on $W^{1,\infty}(\mathbb{R}^d, \mathbb{R}^d)$.

A classical result states that the directional derivative $J'(\Omega)(\theta)$ depends only on the normal trace $\theta \cdot n$ on the boundary $\partial\Omega$.

Lemma 3.2. *Let Ω be a smooth bounded open set and $J(\Omega)$ a differentiable function at Ω. Its derivative satisfies*

$$J'(\Omega)(\theta_1) = J'(\Omega)(\theta_2)$$

for every $\theta_1, \theta_2 \in W^{1,\infty}(\mathbb{R}^d; \mathbb{R}^d)$ such that $\theta_2 - \theta_1 \in C^1(\mathbb{R}^d; \mathbb{R}^d)$ and

$$\theta_1 \cdot n = \theta_2 \cdot n \qquad \text{on } \partial\Omega.$$

We give two examples of shape derivatives that will be useful in the sequel.

Lemma 3.3. *Let Ω be a smooth bounded open set and $f(x) \in W^{1,1}(\mathbb{R}^d)$. Define*

$$J(\Omega) = \int_\Omega f(x)\,dx.$$

Then J is differentiable at Ω and

$$J'(\Omega)(\theta) = \int_\Omega \mathrm{div}\big(\theta(x)\,f(x)\big)\,dx = \int_{\partial\Omega} \theta(x)\cdot n(x)\,f(x)\,ds$$

for every $\theta \in W^{1,\infty}(\mathbb{R}^d; \mathbb{R}^d)$.

Lemma 3.4. *Let Ω be a smooth bounded open set and $f(x) \in W^{2,1}(\mathbb{R}^d)$. Define*

$$J(\Omega) = \int_{\partial\Omega} f(x)\,ds.$$

Then J is differentiable at Ω and

$$J'(\Omega)(\theta) = \int_{\partial\Omega} \theta\cdot n \left(\frac{\partial f}{\partial n} + Hf\right) ds,$$

for any $\theta \in W^{1,\infty}(\mathbb{R}^d; \mathbb{R}^d)$, where H is the mean curvature of $\partial\Omega$ defined by $H = \mathrm{div} n$.

Remark 3.5. In particular, Lemma 3.3 is useful in order to compute the shape derivative of a volume constraint $V(\Omega) = C$. Indeed, we have

$$V(\Omega) = \int_\Omega dx \quad \text{and} \quad V'(\Omega)(\theta) = \int_{\partial\Omega} \theta(x) \cdot n(x)\, ds.$$

Similarly, Lemma 3.4 is useful in order to compute the shape derivative of a perimeter constraint $P(\Omega) = C$. Indeed, we have

$$P(\Omega) = \int_{\partial\Omega} ds \quad \text{and} \quad P'(\Omega)(\theta) = \int_{\partial\Omega} \theta(x) \cdot n(x) H\, ds.$$

Theorem 3.6. *Let Ω be a smooth bounded open set and $\theta \in W^{1,\infty}(\mathbb{R}^d; \mathbb{R}^d)$. The shape derivative of* (2) *is*

$$J_1'(\Omega)(\theta) = \int_{\partial\Omega} \left(2 \left[\frac{\partial(g \cdot u)}{\partial n} + Hg \cdot u \right] - Ae(u) \cdot e(u) \right) \theta \cdot n\, ds. \tag{7}$$

The shape derivative of (3) *is*

$$J_2'(\Omega)(\theta) = \int_{\partial\Omega} \left(\frac{\partial(g \cdot p)}{\partial n} + Hg \cdot p - Ae(p) \cdot e(u) + \frac{C_0}{\alpha} k |u - u_0|^\alpha \right) \theta \cdot n\, ds, \tag{8}$$

where u is the solution of (1), *and p is the adjoint state, solving*

$$\begin{cases} -\mathrm{div}\,(A\, e(p)) = C_0 k(x) |u - u_0|^{\alpha - 2}(u - u_0) & \text{in } \Omega, \\ p = 0 & \text{on } \Gamma_D, \\ (Ae(p))n = 0 & \text{on } \Gamma_N, \end{cases} \tag{9}$$

where C_0 is a constant given by

$$C_0 = \left(\int_\Omega k(x) |u(x)|^\alpha dx \right)^{1/\alpha - 1}.$$

Remark 3.7. Remark that there is no adjoint state involved in (7) (indeed the minimization of (2) is a self-adjoint problem).

Proof. Although Theorem 3.6 is a classical result (see e.g. [13], [16], [20], [21]) we briefly sketch its proof for the sake of completeness. To simplify we give a short, albeit formal, proof due to Céa [10]. We consider a general objective function

$$J(\Omega) = \int_\Omega j(x, u)\, dx,$$

for which we introduce the Lagrangian, defined for $(v, q) \in \left(H^1(\mathbb{R}^d; \mathbb{R}^d)\right)^2$ by

$$\begin{aligned} \mathcal{L}(\Omega, v, q) = & \int_\Omega j(x, v)\, dx + \int_\Omega Ae(v) \cdot e(q)\, dx - \int_\Omega q \cdot f\, dx \\ & - \int_{\Gamma_N} q \cdot g\, ds - \int_{\Gamma_D} \Big(q \cdot Ae(v)n + v \cdot Ae(q)n \Big) ds. \end{aligned} \tag{10}$$

In (10), q is a Lagrange multiplier for the state equation and its boundary conditions. It is worth noticing that v and q belong to a functional space that does not depend on Ω, so that we can apply the usual differentiation rule to the Lagrangian $\mathcal{L}$. The stationarity of the Lagrangian is going to give the optimality conditions of the minimization problem. For a given Ω, we denote by (u,p) a stationary point for $\mathcal{L}$. Taking the partial derivative of $\mathcal{L}$ with respect to q in the direction $\phi \in H^1(\mathbb{R}^d;\mathbb{R}^d)$, and integrating by parts the resulting Euler condition, we obtain

$$\begin{aligned}\langle\frac{\partial\mathcal{L}}{\partial q}(\Omega,u,p),\phi\rangle = 0 = \quad & -\int_\Omega \phi\cdot\Big(\operatorname{div}(Ae(u))+f\Big)dx \\ & +\int_{\Gamma_N}\phi\cdot\big((Ae(u))n-g\big)\,ds \\ & -\int_{\Gamma_D} u\cdot Ae(\phi)n\,ds.\end{aligned} \tag{11}$$

First, choosing a test function ϕ with compact support in Ω, we deduce from (11) the state equation. Then, varying the trace function ϕ on Γ_N gives the Neumann boundary condition for u, while varying the corresponding normal stress $\big(Ae(\phi)\big)n$ on Γ_D gives the Dirichlet boundary condition for u. On the other hand, in order to find the adjoint equation, we differentiate $\mathcal{L}$ with respect to v in the direction $\phi \in H^1(\mathbb{R}^d;\mathbb{R}^d)$. This yields

$$\begin{aligned}\langle\frac{\partial\mathcal{L}}{\partial v}(\Omega,u,p),\phi\rangle = 0 \quad & = \int_\Omega j'(u)\cdot\phi\,dx + \int_\Omega Ae(\phi)\cdot e(p)\,dx \\ & -\int_{\Gamma_D}\Big(p\cdot Ae(\phi)n+\phi\cdot Ae(p)n\Big)\,ds.\end{aligned}$$

Integrating by parts we obtain

$$\begin{aligned}\langle\frac{\partial\mathcal{L}}{\partial v}(\Omega,u,p,\mu),\phi\rangle = \quad & \int_\Omega\Big(j'(u)-\operatorname{div}(Ae(p))\Big)\cdot\phi\,dx + \int_{\Gamma_N}\phi\cdot\big(Ae(p)\big)n\,ds \\ & -\int_{\Gamma_D} p\cdot Ae(\phi)n\,ds.\end{aligned}$$

Letting ϕ vary among the functions with compact support in Ω gives the adjoint state equation

$$-\operatorname{div}(Ae(p)) = -j'(u) \quad \text{in} \quad \Omega.$$

Then, varying the trace of ϕ on Γ_N yields the Neumann boundary condition

$$\big(Ae(p)\big)n = 0 \quad \text{on} \quad \Gamma_N.$$

Finally, varying the normal stress $\big(Ae(\phi)\big)n$ on Γ_D gives

$$p = 0 \quad \text{on} \quad \Gamma_D.$$

We have therefore found a well-posed boundary value problem for the adjoint state p.

The shape derivative of the objective function is obtained by differentiating

$$J(\Omega) = \mathcal{L}(\Omega, u(\Omega), p(\Omega)),$$

which, by the chain rule theorem, reduces to the partial derivative of $\mathcal{L}$ with respect to Ω in the direction θ

$$J'(\Omega)(\theta) = \frac{\partial \mathcal{L}}{\partial \Omega}(\Omega, u, p)(\theta).$$

Applying Lemma 3.3 and 3.4, we obtain

$$\begin{aligned} \frac{\partial \mathcal{L}}{\partial \Omega}(\Omega, u, p)(\theta) = & \int_{\partial\Omega} \theta \cdot n\Big(j(u) + Ae(u)\cdot e(p) - p\cdot f\Big) ds \\ & - \int_{\Gamma_N} \theta \cdot n \left(\frac{\partial (g\cdot p)}{\partial n} + H\, g\cdot p \right) ds \\ & - \int_{\Gamma_D} \theta \cdot n \left(\frac{\partial h}{\partial n} + Hh \right) ds, \end{aligned} \tag{12}$$

with $h = u \cdot Ae(p)n + p \cdot Ae(u)n$. Taking into account the boundary condition $u = p = 0$ on Γ_D which also implies

$$Ae(u)\cdot e(p) = \mu \frac{\partial u}{\partial n}\cdot\frac{\partial p}{\partial n} + (\mu+\lambda)\left(\frac{\partial u}{\partial n}\cdot n\right)\left(\frac{\partial p}{\partial n}\cdot n\right) \quad \text{on } \Gamma_D,$$

we deduce

$$\begin{aligned} \frac{\partial \mathcal{L}}{\partial \Omega}(\Omega, u, p)(\theta) = & \int_{\Gamma_N} \theta \cdot n \left(j(u) + Ae(u)\cdot e(p) - \frac{\partial (g\cdot p)}{\partial n} - H\, g\cdot p \right) ds \\ & + \int_{\Gamma_D} \theta \cdot n\Big(j(u) - Ae(u)\cdot e(p)\Big)\, ds. \end{aligned}$$

This proof is merely a formal computation (in particular we have to assume u and p to be differentiable with respect to the shape Ω) but it can be rigorously justified (see the references quoted above). □

Remark 3.8. We can generalize Theorem 3.6 to more general objective functions, including functions of the strain or stress. It is also possible to consider non homogeneous Dirichlet boundary conditions in the state equation, or even a non-linear model of elasticity.

Remark 3.9. It is possible to further restrict the class of domains by requiring some parts of the boundary Γ_{fixed} not to move: specifically, the map θ must belong to

$$T_{ad} = \left\{\theta \in W^{1,\infty}(\mathbb{R}^d, \mathbb{R}^d) \text{ such that } \theta = 0 \text{ on } \Gamma_{fixed}\right\}.$$

4. Front propagation by the level-set method

Let a bounded domain $D \subset \mathbb{R}^d$ be the working domain in which all the admissible shapes Ω are included, i.e. $\Omega \subset D$. In numerical practice, the domain D will be uniformly meshed once and for all. Then, we shall capture the shape Ω on this fixed mesh. To this purpose, we parametrize the boundary of Ω by means of a level-set function, following the idea of Osher and Sethian [15]. We define this level-set function ψ in D such that

$$\begin{cases} \psi(x) = 0 & \Leftrightarrow x \in \partial\Omega \cap D \\ \psi(x) < 0 & \Leftrightarrow x \in \Omega \\ \psi(x) > 0 & \Leftrightarrow x \in (D \setminus \Omega) \end{cases}$$

The normal n to the shape Ω is yielded by $\nabla\psi/|\nabla\psi|$ and the curvature H is given by the divergence of n (these quantities are evaluated by finite differences, since our mesh is uniformly rectangular). Remark that, although n and H are theoretically defined on $\partial\Omega$ only, the level set method allows to define easily their extension to the whole domain D.

Following the optimization process, the shape is going to evolve according to a fictitious time, which corresponds to descent stepping (we shall come back to this issue in the next section). As it is well-known, if the shape is evolving in time, then the evolution of the level-set function is governed by a simple Hamilton–Jacobi equation. To be precise, assume that the shape $\Omega(t)$ evolves in time $t \in \mathbb{R}^+$ with a normal velocity $V(t,x)$. Then

$$\psi\Big(t, x(t)\Big) = 0 \quad \text{for every } x(t) \in \partial\Omega(t).$$

Differentiating in t yields

$$\frac{\partial\psi}{\partial t} + \dot{x}(t) \cdot \nabla_x\psi = \frac{\partial\psi}{\partial t} + Vn \cdot \nabla_x\psi = 0.$$

Since $n = \nabla_x\psi/|\nabla_x\psi|$ we obtain

$$\frac{\partial\psi}{\partial t} + V|\nabla_x\psi| = 0.$$

This Hamilton Jacobi equation is posed on the whole of the box D, and not only on the boundary $\partial\Omega$, if the velocity V is known everywhere. Remark that the level-set method easily allows to compute the mean curvature $H = \text{div}n$ (which plays an important role in a perimeter penalization).

It is solved by an explicit first order upwind scheme (see e.g. [17])

$$\frac{\psi_i^{n+1} - \psi_i^n}{\Delta t} + \min(V_i^n, 0)\, g^-(D_x^+\psi_i^n, D_x^-\psi_i^n) + \max(V_i^n, 0)\, g^+(D_x^+\psi_i^n, D_x^-\psi_i^n) = 0,$$

with $D_x^+\psi_i^n = \frac{\psi_{i+1}^n - \psi_i^n}{\Delta x}$, $D_x^-\psi_i^n = \frac{\psi_i^n - \psi_{i-1}^n}{\Delta x}$, and

$$g^+(d^+, d^-) = \sqrt{\min(d^+, 0)^2 + \max(d^-, 0)^2},$$

$$g^-(d^+, d^-) = \sqrt{\max(d^+, 0)^2 + \min(d^-, 0)^2}.$$

In order to regularize the level set function (which may become too flat or too steep), we re-initialize it periodically by solving

$$\begin{cases} \frac{\partial\psi}{\partial t} + \text{sign}(\psi)\Big(|\nabla_x\psi| - 1\Big) = 0 & \text{in } D \times \mathbb{R}^+, \\ \psi(t = 0, x) = \psi_0(x) & \text{in } D, \end{cases} \tag{13}$$

which admits the signed distance to the initial interface $\{\psi_0(x) = 0\}$ as a stationary solution. In numerical practice, re-initialization is very important because the level set function often becomes too steep, which implies a bad approximation of the normal n or of the curvature H.

5. Optimization algorithm

For the minimization problem

$$\inf_{\Omega \in \mathcal{U}_{ad}} \left\{ J(\Omega) = \int_\Omega j(x,u)\,dx \right\}$$

we have computed a shape derivative

$$J'(\Omega)(\theta) = \int_{\partial\Omega} v\,\theta \cdot n\,ds,$$

where the function v is given by a result analogous to Theorem 3.6. Ignoring smoothness issues, a descent direction is found by taking

$$\theta = -v\,n.$$

Therefore, we choose the normal component $\theta \cdot n = -v$ to be the advection velocity in the Hamilton–Jacobi equation

$$\frac{\partial \psi}{\partial t} - v|\nabla\psi| = 0. \tag{14}$$

Transporting ψ by (14) is equivalent to move the boundary of Ω (the zero level set of ψ) along the descent gradient direction $-J'(\Omega)$. Finally, our algorithm is an iterative method, structured as follows:

1. Initialization of the level-set function ψ_0 corresponding to an initial guess Ω_0.
2. Iteration until convergence, for $k \geq 0$:
   (a) Computation of the state u_k and the adjoint state p_k through two problems of linear elasticity.
   (b) Deformation of the shape through the transport of the level set function: $\psi_{k+1}(x) = \psi(\Delta t_k, x)$, where $\psi(t,x)$ is the solution of (14) with velocity $v_k = v(u_k, p_k)$ and initial condition $\psi(0,x) = \psi_k(x)$. The time step Δt_k is chosen in such a way that $J(\Omega_{k+1}) \leq J(\Omega_k)$.
3. From time to time, for stability reasons, we also re-initialize the level set function ψ by solving (13).

In order to avoid an explicit meshing of the shapes Ω_k, we compute the state u_k and the adjoint state p_k on the whole working domain D. For this, we fill the empty part $D \setminus \Omega_k$ with a very weak material with Hooke's law $B = 10^{-3}A$ and we perform the elasticity analysis on a fixed rectangular mesh on D (using $Q1$ finite elements). Then, the Hooke's law in D is

$$A_k(x) = \begin{cases} A & \text{where } \psi_k(x) < 0 \\ 10^{-3}A & \text{where } \psi_k(x) > 0, \end{cases}$$

completed by a simple linear interpolation (proportional to the volume) of these values in the mesh cells where ψ_k changes sign.

Since n and H, as well as the state u and the adjoint state p, are computed everywhere in D, the shape derivative (see formulae (7) and (8)) delivers a normal velocity $-v$ which is defined throughout the domain D and not only on the free boundary $\partial\Omega$. Therefore, we do not ever have to know where the boundary $\partial\Omega$ is precisely and we apply the same numerical scheme everywhere in the working domain D.

Since the Hamilton–Jacobi equation (14) is solved by an explicit scheme, the time step Δt_k must satisfy a CFL condition. Remark that one explicit time step for (14) is much cheaper, in terms of CPU time and memory requirement, than the solution of the state equation (1) or the adjoint state equation (9). On the other hand, the usual time step Δt_k for (14) is often much smaller that the optimal descent step for the minimization of the objective function $J(\Omega)$. Therefore, for each iteration k in the above algorithm (corresponding to a single evaluation of u_k and p_k), we solve several time steps of the Hamilton–Jacobi equation (14). The number of such time steps per iteration k is monitored by the decrease of $J(\Omega_k)$.

Our algorithm never creates new holes or boundaries if the time step Δt_k does indeed satisfy a CFL condition because of the maximum principle for (14) (there is no nucleation mechanism for new holes). However the level set method is well known to handle easily topology changes, i.e. merging or cancellation of holes. Therefore, our algorithm is able to perform topology optimization if the number of holes of the initial design is sufficiently large (see Figure 1). The algorithm converges smoothly to a (local) minimum which strongly depends, of course, on the initial topology (see the differences in Figures 2 and 3). For a "good" initialization, the numerical results are very similar to those obtained by the homogenization method but the convergence is usually slower (although we have not yet tried to speed it by a quasi-Newton algorithm).

6. Numerical examples

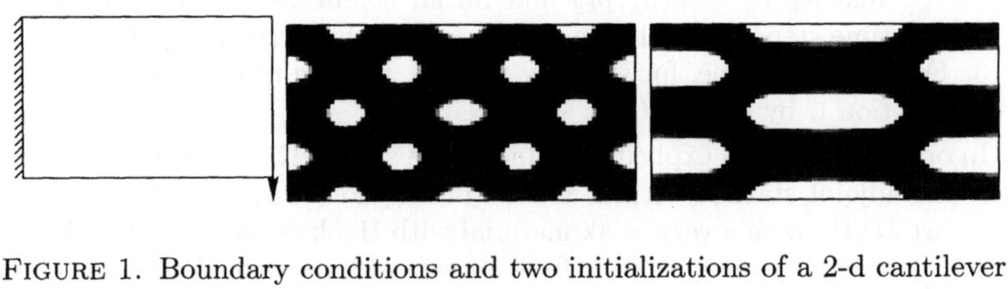

FIGURE 1. Boundary conditions and two initializations of a 2-d cantilever

We first consider a compliance problem and neglect body forces, i.e. $f \equiv 0$ in (1). The objective function is a combination of the compliance and of the weight of the structure

$$J(\Omega) = \int_{\Gamma_N} g \cdot u \, ds + \ell \int_{\Omega} dx, \tag{15}$$

where $\ell > 0$ is a Lagrange multiplier. We further impose that the Dirichlet boundary Γ_D, as well as the portion of the Neumann boundary Γ_N where the loads are applied, are kept fixed during the homogenization process. Thus, only the traction-free Neumann boundary is allowed to move. The boundary conditions and two initial configurations for a plane cantilever are displayed on Figure 1. The results are shown on Figures 2 and 3 for two different iteration numbers, showing the strong influence of the initialization. The convergence is smooth and fast (see Figure 4) and 20 explicit time steps of the Hamilton–Jacobi equation were performed at each elasticity iteration.

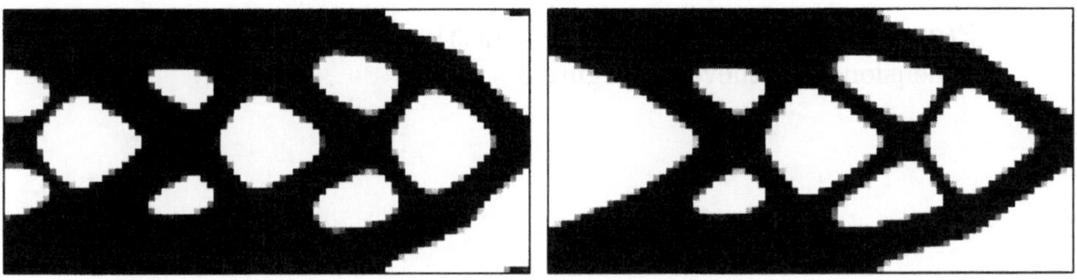

FIGURE 2. Iterations 10 and 50 of the two-dimensional cantilever initialized as in Figure 1 (middle)

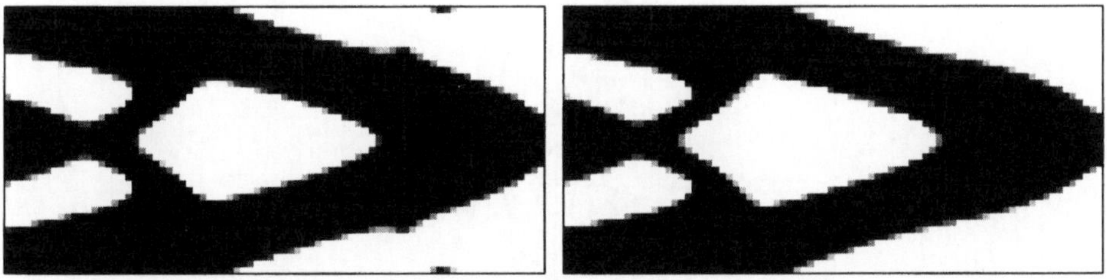

FIGURE 3. Iterations 10 and 50 of the two-dimensional cantilever initialized as in Figure 1 (right)

The advantage of the level-set method is that it can be easily extended to three space dimensions. A three-dimensional optimal electrical mast is shown in Figure 5 (see [2] for a precise definition of this test case).

Next, Figure 6 shows a numerical result for the least square objective function (3), where $k(x)$ and $u_0(x)$ have been chosen in such a way that the jaws of the mechanism close. This is a classical gripping mechanism test case which is described, e.g., in [1], [19].

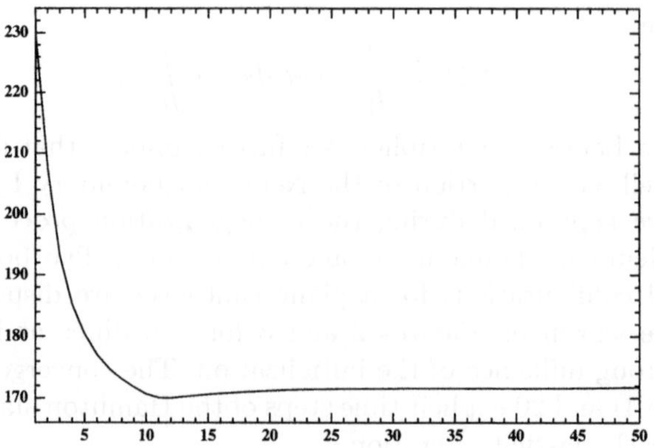

FIGURE 4. Convergence of the objective function for the two-dimensional cantilever of Figure 3

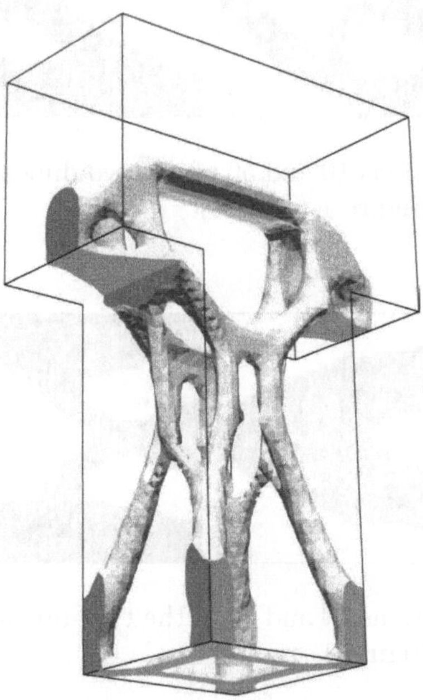

FIGURE 5. Three-dimensional electrical mast.

We come back to the compliance objective function (2) with design dependent loads and take g as a uniform pressure load on the free boundary Γ_N. Imposing 5 Dirichlet points yields a nice sea star, as shown in Figure 7.

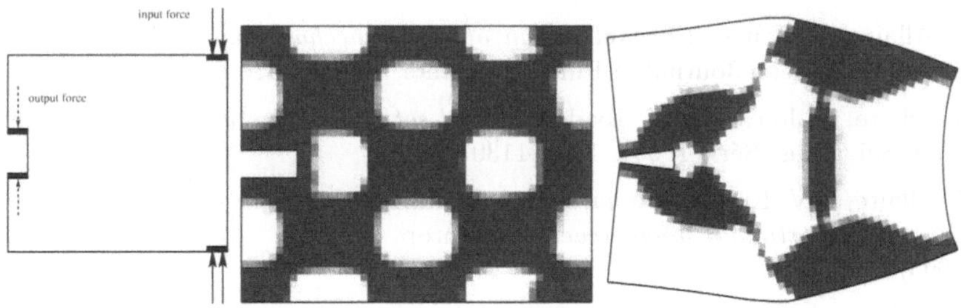

FIGURE 6. Boundary conditions, initialization, and deformed optimal shape of a plane gripping mechanism.

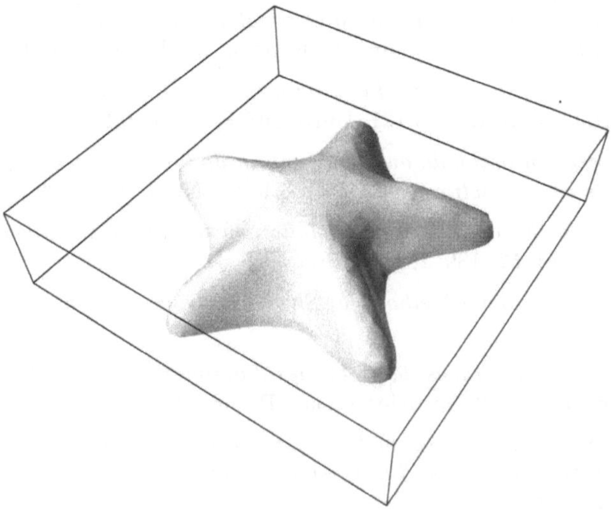

FIGURE 7. Optimal shape under a uniform pressure load with five anchor points.

Our shape optimization method can be easily extended to models of nonlinear elasticity or to multiple loads problems. This issue, as well as the analysis of the effect of first-order or second-order discretizations and of re-initialization in the numerical convergence towards an optimal shape, are the subject of a forthcoming paper.

References

[1] G. Allaire, *Shape optimization by the homogenization method*, Springer Verlag, New York, (2001).

[2] G. Allaire, E. Bonnetier, G. Francfort, F. Jouve, *Shape optimization by the homogenization method.* Nümerische Mathematik, **76**, 27–68, (1997).

[3] G. Allaire, F. Jouve, *Optimal design of micro-mechanisms by the homogenization method*, European Journal of Finite Elements, **11**, 405–416 (2002).

[4] G. Allaire, F. Jouve, A.-M. Toader, *A level-set method for shape optimization,* C. R. Acad. Sci. Paris, Série I, **334**, 1125–1130 (2002).

[5] G. Allaire, R.V. Kohn, *Optimal design for minimum weight and compliance in plane stress using extremal microstructures,* Europ. J. Mech. A/Solids, **12**, 6, 839–878, (1993).

[6] L. Ambrosio , G. Buttazzo, *An optimal design problem with perimeter penalization,* Calc. Var. **1**, 55–69 (1993).

[7] M. Bendsoe, *Methods for optimization of structural topology, shape and material,* Springer Verlag, New York, (1995).

[8] M. Bendsoe, N. Kikuchi, *Generating Optimal Topologies in Structural Design Using a Homogenization Method,* Comp. Meth. Appl. Mech. Eng., **71**, 197–224, (1988).

[9] A. Chambolle, *A density result in two-dimensional linearized elasticity and applications,* Preprint CEREMADE 121, Université Paris-Dauphine, (2001).

[10] J. Céa, *Conception optimale ou identification de formes, calcul rapide de la dérivée directionnelle de la fonction coût,* Math. Model. Num. Anal., **20**, 3, 371–402 (1986).

[11] D. Chenais, *On the existence of a solution in a domain identification problem,* J. Math. Anal. Appl. **52**, 189–289, (1975).

[12] A. Cherkaev, *Variational Methods for Structural Optimization,* Springer Verlag, New York (2000).

[13] F. Murat, S. Simon, *Etudes de problèmes d'optimal design,* Lecture Notes in Computer Science **41**, 54–62, Springer Verlag, Berlin, (1976).

[14] S. Osher, F. Santosa, *Level set methods for optimization problems involving geometry and constraints: frequencies of a two-density inhomogeneous drum,* J. Comp. Phys., **171**, 272–288, (2001).

[15] S. Osher, J.A. Sethian, *Front propagating with curvature dependent speed: algorithms based on Hamilton–Jacobi formulations,* J. Comp. Phys., **78**, 12–49, (1988).

[16] O. Pironneau, *Optimal shape design for elliptic systems,* Springer-Verlag, New York, (1984).

[17] J.A. Sethian, *Level Set Methods and fast marching methods: evolving interfaces in computational geometry, fluid mechanics, computer vision and materials science,* Cambridge University Press, (1999).

[18] J.A. Sethian, A. Wiegmann, *Structural boundary design via level set and immersed interface methods,* J. Comp. Phys., **163**, 489–528, (2000).

[19] O. Sigmund, *On the design of compliant mechanisms using topology optimization,* Mech. Struct. Mach., **25**, 493–524, (1997).

[20] J. Simon, *Differentiation with respect to the domain in boundary value problems,* Num. Funct. Anal. Optimz., **2**, 649–687, (1980).

[21] J. Sokolowski, J.P. Zolesio, *Introduction to shape optimization: shape sensitity analysis,* Springer Series in Computational Mathematics, Vol. 10, Springer, Berlin, (1992).

(Grégoire Allaire, François Jouve) Centre de Mathématiques Appliquées,
Ecole Polytechnique,
91128 Palaiseau, France
E-mail address: gregoire.allaire@polytechnique.fr,
francois.jouve@polytechnique.fr

(Anca-Maria Toader) CMAF, Faculdade de Ciências da Universidade de Lisboa,
Av. Prof. Gama Pinto 2, 1699 Lisboa, Portugal
E-mail address: amtan@lmc.fc.ul.pt

International Series of Numerical Mathematics, Vol. 147, 17–26

On a Variational Problem Arising in Image Reconstruction

Luigi Ambrosio and Simon Masnou

Abstract. We consider a variational approach to the problem of recovering missing parts in a panchromatic digital image. Representing the image by a scalar function u, we propose a model based on the relaxation of the energy

$$\int |\nabla u|(\alpha + \beta \left|\operatorname{div} \frac{\nabla u}{|\nabla u|}\right|^p), \qquad \alpha, \beta > 0, \quad p \geq 1$$

which takes into account the perimeter of the level sets of u as well as the L^p norm of the mean curvature along their boundaries. We investigate the properties of this variational model and the existence of minimizing functions in BV. We also address related issues for integral varifolds with generalized mean curvature in L^p.

1. Introduction

Several problems in digital image processing require the ability to recover missing parts of an image or to remove spurious or undesired objects. One can mention for instance the removal of scratches in old photographs and films, the recovery of pixels blocks corrupted during a binary transmission (or analogously the removal of impulse noise) or the removal of undesired advertisements, text or subtitles from a photograph.

A digital image is usually modelled as a function u from a bounded domain of $\mathbb{R}^N$ ($N = 2$ for usual snapshots, $N = 3$ for medical images or movies, $N = 4$ for moving medical images) onto $\mathbb{R}^M$ ($M = 1$ for a grey-level image, $M = 3$ for colour images). Since it is now well admitted that the essential features of any natural image are contained in its grey level representation, we shall concentrate on the panchromatic case $M = 1$.

After the work of L. Rudin and S. Osher [24], the usual representation of a panchromatic image is a sum of two components $u_1 \in BV(\mathbb{R}^N)$ and $u_2 \in L^2(\mathbb{R}^N)$. The component u_1 is supposed to describe the *geometry* of the image, i.e. its objects and their boundaries, while u_2 contains all information about *texture* and *additive noise*. The assumption that the geometry of the image can be described by a function of bounded variation sounds quite natural, for it means that there can be discontinuities in the image but supported on rectifiable curves. The need for another component not necessarily belonging to BV can be corroborated by an experimental procedure which seems to point out that, given a digital image, the subjacent "real" image may be often too oscillating to belong to BV (see [2] for the details and [11] for connected theoretic issues). The reader may refer to [3, 17] for a thorough survey of the space BV.

In recent years, a wide literature that has been published in on the recovery of missing parts in a digital image: in this framework, one can basically distinguish between two approaches and each of them corresponds in some way to the processing of one component in the decomposition above:

- the stochastic approach, which is based on the modelling of an image as a realization of a random process. This modelling is particularly adapted for texture images (thus to the processing or the component u_2 in the previous decomposition) and has motivated numerous works on texture analysis and synthesis [13, 14, 20, 27, 28],
- the deterministic approach, aiming at recovering the geometry of the image. Motivated by the seminal work of Mumford, Nitzberg and Shiota [23], several variational or PDE methods related to this approach have appeared very recently [7, 9, 10, 15, 21, 22].

Our purpose here is to discuss the mathematical aspects of a model from the latter category, which was introduced in [21]. Its underlying principle is to mimic a natural ability of the human visual system, the so-called "amodal completion" process, that has been described and studied by the Gestalt school of psychology and, particularly, G. Kanizsa [19]. In fact, our brain is able to reconstruct artificially a partially occluded object by interpolating its edges by means of curves offering a compromise between shortness and good continuation. The model introduced in [21] is based on the following claim: one indeed assumes that, in order to recover the missing part of an image or, analogously, to remove a spurious object (both can be considered as occlusions), it suffices to complete the *level lines* known outside the occlusion by means of short and not too oscillating curves that do not cross themselves or each other — and thus can be considered as level lines. Then the full image can be reconstructed, since it is well known that any Borel function is fully determined by its level sets. This constructive process, called *disocclusion* process in [21], can be written in a variational form: ideally, each interpolating level line should minimize an energy of the type

$$\int_{\{u=t\}} (\alpha + \beta|\kappa|^p)\,d\mathcal{H}^1, \quad p \geq 1, \quad \alpha, \beta > 0,$$

which penalizes the length and the oscillations. After summing over all the levels and using the change of variables formula, we get the variational criterion

$$F(u) = \int_{\tilde{A}} |\nabla u|\left(\alpha + \beta \left|\operatorname{div} \frac{\nabla u}{|\nabla u|}\right|^p\right) dx,$$

with the convention that the integrand is 0 wherever $|\nabla u| = 0$. Our problem is now to minimize F over all the functions coinciding with the original image u_0

outside the occlusion domain A. The integration domain $\tilde{A}$ in F is an open set strictly containing A; in this way, the interpolant is forced this way to verify the good continuation principle. Of course, F makes sense for smooth functions only, but it can be easily extended to L^1 by setting $F(u) = \infty$ for any $u \in L^1$. Then, as usual in the direct method of the calculus of variations, we introduce the relaxed functional associated with F, defined by

$$\overline{F}(u) = \inf\{\liminf_{h\to\infty} F(u_h),\ u_h \to u \in \mathrm{L}^1\}.$$

Besides, it is straightforward to generalize F — thus $\overline{F}$ — to higher dimensions, in which cases it is no more associated to the minimization of $\int(\alpha+\beta|\kappa|^p)d\mathcal{H}^1$ over curves but to the minimization of $\int(\alpha+\beta|\mathbf{H}|^p)d\mathcal{H}^1$ over hypersurfaces, where $\mathbf{H}$ denotes the mean curvature vector.

This paper is organized as follows. In Section 2, we state a locality result for the mean curvature vector $\mathbf{H}$ of integral $(N-1)$-varifolds when $\mathbf{H} \in \mathrm{L}^p$, $p > N-1$, $p \geq 2$ (Theorem 2.2). A direct consequence of this result is the lower semicontinuity (Theorem 2.4) of the functional

$$\int_{\partial E}(1+|\mathbf{H}_\mathrm{E}|^p)d\mathcal{H}^{\mathrm{N}-1}, \quad p > N-1,\ N \geq 3$$

with respect to the convergence in L^1 in the class of sets $E \subset \mathbb{R}^\mathrm{N}$ with $\partial E \in \mathrm{C}^2$. This result extends to higher dimensions a previous result due to G. Bellettini, G. Dal Maso and M. Paolini [6] for $N = 2$, $p > 1$.

Section 3 is devoted to the study of the disocclusion problem in dimension N. We prove the existence of an optimal solution (Theorem 3.1), the coincidence between F and the associated relaxed functional $\overline{F}$ for smooth functions (Theorem 3.2) and give some results on the regularity of the optimal solution in the particular case $N = 2$ (Corollary 3.4). For the sake of simplicity and with absolutely no loss of generality, we shall assume that $\alpha = \beta = 1$ in the sequel.

2. Locality of the mean curvature and applications

The main result of this section is stated in Theorem 2.2, whose proof is based on the quadratic decay of the tilt-excess established by R. Schätzle in [25]. We recall his result below (the notation is the standard one in geometric measure theory, see [3, 16, 26]).

Theorem 2.1 (Quadratic tilt-excess decay [25, Thm. 5.1]). *Let $U = \underline{\underline{\mathbf{v}}}(M,\theta)$ be an integral $N-1$-varifold in an open set $\Omega \subseteq \mathbb{R}^\mathrm{N}$ with weight μ_U and mean curvature $\mathbf{H}_\mathrm{U} \in \mathrm{L}^p_{\mathrm{loc}}(\mu_U)$, $p > N-1$, $p \geq 2$. Then for μ_U-almost all $x \in \operatorname{spt}\mu_U$, the tilt-excess*

$$\operatorname{tiltex}_\mu(x,\rho) := \rho^{1-N}\int_{B^N_\rho(x)} \|T_yM - T_xM\|^2\, d\mu_U(y)$$

decays quadratically, that is

$$\operatorname{tiltex}_\mu(x,\rho) = O_x(\rho^2).$$

Theorem 2.2 (Locality of the mean curvature). *Let $U = \underline{\underline{\mathbf{v}}}(M,\theta_U)$, $V = \underline{\underline{\mathbf{v}}}(N,\theta_V)$ be integral $(N-1)$-varifolds in $\Omega \subseteq \mathrm{I\!R}^{\mathrm{N}}$. If $\mathbf{H}_{\mathrm{U}} \in \mathrm{L}^p_{\mathrm{loc}}(\mu_U)$ and $\mathbf{H}_{\mathrm{V}} \in \mathrm{L}^p_{\mathrm{loc}}(\mu_V)$ for some $p > N-1$, $p \geq 2$ then*

$$\mathbf{H}_{\mathrm{U}}(x) = \mathbf{H}_{\mathrm{V}}(x)$$

for $\mathcal{H}^{\mathrm{N}-1}$-almost all $x \in M \cap N$.

Proof. (see [4]) It is based on the relationship between the first variation of a varifold, the approximate tangent space and the mean curvature, combined with Theorem 2.1 and finally on a result by Brakke on the orthogonality of the mean curvature vector [8]. □

Theorem 2.2 is a key point in the proof of the lower semicontinuity of the mean curvature's L^p norm, stated in Theorem 2.4 below, which is a generalization to the higher dimensional case of a previous result obtained in dimension two by G. Bellettini, G. Dal Maso and M. Paolini in [6]. We recall their result below.

Theorem 2.3 ([6, Thm 7.1]). *Let Ω be an open subset of $\mathrm{I\!R}^2$, let $p > 1$ and let E be an open bounded subset of $\mathrm{I\!R}^2$ such that $\partial E \cap \Omega \in \mathrm{C}^2$. Then*

$$\int_{\partial E \cap \Omega} (1 + |\kappa|^p) d\mathcal{H}^1 \leq \liminf_{h\to\infty} \int_{\partial E_h \cap \Omega} (1 + |\kappa_h|^p) d\mathcal{H}^1$$

for any sequence $(E_h)_{h\in\mathrm{I\!N}}$ of bounded open sets such that $\partial E_h \cap \Omega \in \mathrm{C}^2$ and $E_h \to E$ in $\mathrm{L}^1(\Omega)$ as $h \to \infty$.

Theorem 2.4 (Lower semicontinuity of the mean curvature's L^p norm).
Let Ω be an open subset of $\mathrm{I\!R}^{\mathrm{N}}$, $N \geq 3$, and $p > N-1$. Let $\{E_h\}_{h\in\mathrm{I\!N}} \subset \mathrm{I\!R}^{\mathrm{N}}$ be converging in $L^1(\Omega)$ to E, with $\partial E_h \cap \Omega \in \mathrm{C}^2$ and $\partial E \cap \Omega \in \mathrm{C}^2$. Then

$$\int_{\partial E \cap \Omega} (1 + |\mathbf{H}_{\mathrm{E}}|^p) d\mathcal{H}^{\mathrm{N}-1} \leq \liminf_{h\to\infty} \int_{\partial E_h \cap \Omega} (1 + |\mathbf{H}_{\mathrm{E_h}}|^p) d\mathcal{H}^{\mathrm{N}-1},$$

where $\mathbf{H}_{\mathrm{E}}$ (resp. $\mathbf{H}_{\mathrm{E_h}}$) denotes the mean curvature vector on $\partial E \cap \Omega$ (resp. $\partial E_h \cap \Omega$).

Proof. Due to the lower semicontinuity of the perimeter, it is clearly sufficient to prove the part of the claim that involving the curvature. We can assume that the right hand side of the inequality to prove is finite, otherwise the result is trivial. In addition, possibly taking a subsequence, we can assume without loss of generality that

$$\sup_{h\in\mathrm{I\!N}} \int_{\partial E_h \cap \Omega} (1 + |\mathbf{H}_{\mathrm{E_h}}|^p) d\mathcal{H}^{\mathrm{N}-1} \leq C < \infty.$$

Let $V_h = \underline{\underline{\mathbf{v}}}(\partial E_h \cap \Omega, 1)$ be the unit-density rectifiable $(N-1)$-varifolds associated with the sets $E_h \cap \Omega$ and let $\mu_{V_h} = \mathcal{H}^{\mathrm{N}-1} \llcorner \partial E_h \cap \Omega$ be the corresponding weights. By the divergence theorem, the first variation of the varifolds V_h in Ω can be written as

$$\delta V_h(X) = -\int_\Omega X \cdot \mathbf{H}_{\mathrm{E_h}} d\mu_{V_h}, \quad \forall X \in \mathrm{C}^1_c(\Omega, \mathrm{I\!R}^{\mathrm{N}})$$

hence the L^p norms of δV_h with respect to μ_{V_h} are uniformly bounded.

By Allard's compactness theorem (see [1] or Theorem 42.7 in [26]), we obtain, possibly passing to a subsequence, that there exists a limit integral $(N-1)$-varifold V in Ω such that $V_h \rightharpoonup V$ and $V = \underline{\underline{\mathbf{v}}}(M, \theta_V)$, with M a countably $\mathcal{H}^{\mathrm{N}}$-rectifiable set and θ_V a positive integer-valued and locally $\mathcal{H}^{\mathrm{N}}$-integrable function on Ω. As $\delta V_h = \mathbf{H}_{\mathbf{E_h}} d\mu_{V_h} \rightharpoonup \delta V$, a well-known lower semicontinuity theorem (see for instance Example 2.36 in [3]) yields that $\delta V = \mathbf{H}_{\mathbf{V}}\mu_V$ with $\mathbf{H}_{\mathbf{V}} \in L^p(\mu_V)$ and

$$\int_M \theta_V |\mathbf{H}_{\mathbf{V}}|^p d\mathcal{H}^{\mathrm{N}-1} \le \liminf_{h\to\infty} \int_{\partial E_h \cap \Omega} |\mathbf{H}_{\mathbf{E_h}}|^p d\mathcal{H}^{\mathrm{N}-1}. \tag{1}$$

Notice that so far we have just used the fact that $p > 1$.

Now we show that $\mathcal{H}^{\mathrm{N}-1}$-almost all points in $\partial E \cap \Omega$ belong to M and that $\mathbf{H}_{\mathbf{V}}$ coincides with the classical mean curvature $\mathbf{H}_{\mathbf{E}}$ for $\mathcal{H}^{\mathrm{N}-1}$-almost every point of $\partial E \cap \Omega$ whenever $p > N-1$. Let $x \in \partial E \cap \Omega$. Since E has finite perimeter, for all $r > 0$ except possibly for a countable set, $\mu_V(\partial B_r^N(x)) = 0$, hence $\mu_V(B_r^N(x)) = \lim_{h\to\infty} \mu_{V_h}(B_r^N(x)) = \lim_{h\to\infty} \mathcal{H}^{\mathrm{N}-1}(B_r^N(x) \cap \partial E_h \cap \Omega) \ge \mathcal{H}^{\mathrm{N}-1}(B_r^N(x) \cap \partial E \cap \Omega)$, using the lower semicontinuity of the perimeter. It follows that x is a point where the lower $(N-1)$-dimensional density of μ_V is strictly positive. As $\mu_V = \theta_V \mathcal{H}^{\mathrm{N}-1} \llcorner M$, $\mathcal{H}^{\mathrm{N}-1}$-almost every point with this property belongs to M.

Let $U := \underline{\underline{\mathbf{v}}}(\partial E \cap \Omega, 1)$ be the unit-density rectifiable $(N-1)$-varifold in Ω associated with $\partial E \cap \Omega$. By the divergence theorem we have

$$\mathbf{H}_{\mathbf{U}}(x) = \mathbf{H}_{\mathbf{E}}(x)$$

for $\mathcal{H}^{\mathrm{N}-1}$-almost all $x \in \partial E \cap \Omega$. Since $\partial E \cap \Omega \in \mathrm{C}^2$, it is easily seen that $\mathbf{H}_{\mathbf{U}} \in \mathrm{L}^p_{\mathrm{loc}}(\mu_U)$. By Theorem 2.2 we obtain

$$\mathbf{H}_{\mathbf{E}}(x) = \mathbf{H}_{\mathbf{U}}(x) = \mathbf{H}_{\mathbf{V}}(x)$$

for $\mathcal{H}^{\mathrm{N}-1}$-almost all $x \in M \cap \partial E \cap \Omega$ and therefore for $\mathcal{H}^{\mathrm{N}-1}$-almost all $x \in \partial E \cap \Omega$.

Plugging this into (1) and using that $\theta_V(x) \ge 1$ for $\mathcal{H}^{\mathrm{N}-1}$-almost every $x \in M$, we finally obtain

$$\int_{\partial E \cap \Omega} |\mathbf{H}_{\mathbf{E}}|^p d\mathcal{H}^{\mathrm{N}-1} \le \int_{\partial E \cap \Omega} |\mathbf{H}_{\mathbf{V}}|^p \theta_V d\mathcal{H}^{\mathrm{N}-1} \le \liminf_{h\to\infty} \int_{\partial E_h} |\mathbf{H}_{\mathbf{E_h}}|^p d\mathcal{H}^{\mathrm{N}-1}$$

and the theorem follows. □

Remark 2.5. The varifold arguments we use require the technical assumption $p \ge 2$, which prevents the result by Bellettini *et al* from being a particular case of Theorem 2.4 whenever $N = 2$ and $1 < p < 2$.

Remark 2.6. Whenever $\mathcal{H}^{\mathrm{N}-1}(\partial E \cap \Omega) = \lim_{h\to\infty} \mathcal{H}^{\mathrm{N}-1}(\partial E_h \cap \Omega)$, the lower semicontinuity is true for any $p \ge 1$. This is an easy consequence of (1) and of Reshetnyak continuity Theorem (see for instance [3, Thm. 2.39]), which implies that $V \equiv \underline{\underline{\mathbf{v}}}(\partial E \cap \Omega, 1)$.

3. Analysis of the disocclusion problem

Let Ω be a bounded open set in $\mathbb{R}^N$ ($N \geq 2$) with Lipschitz boundary, representing the image domain, let $A \subset\subset \Omega$ be an open, connected set with Lipschitz boundary representing the occlusion and let $u_0 \in \mathrm{BV}(\Omega \setminus \bar{A})$ be the original image. By Theorem 3.87 in [3], for any $\lambda \in \mathbb{R}$ the function given by $u|_{\Omega\setminus\bar{A}} = u_0$ and $u|_A = \lambda$ belongs to $\mathrm{BV}(\Omega)$. This ensures that the occlusion can always be filled in.

Let $\mathcal{O}(\Omega)$ denote the family of open subsets of Ω. We consider the functional mapping $\mathrm{L}^1(\Omega) \times \mathcal{O}(\Omega)$ onto $[0,\infty]$ and defined for every $(u,B) \in \mathrm{L}^1(\Omega) \times \mathcal{O}(\Omega)$ by

$$F_p(u,B) := \begin{cases} \displaystyle\int_B |\nabla u|\left(1 + \left|\operatorname{div}\frac{\nabla u}{|\nabla u|}\right|^p\right) dx & \text{if } u \in C^2(B) \\ +\infty & \text{otherwise} \end{cases},$$

with the convention that the integrand is 0 wherever $|\nabla u| = 0$.

The relaxed functional associated with F is defined for every $(u,B) \in \mathrm{L}^1(\Omega)\times \mathcal{O}(\Omega)$ by

$$\overline{F_p}(u,B) := \inf\left\{ \liminf_{h\to\infty} F_p(u_h,B) : u_h \xrightarrow{\mathrm{L}^1(B)} u \right\}.$$

Since $F_p(u,B) \geq \int_B |\nabla u|\, dx$ whenever $u \in C^2(B)$, the lower semicontinuity of the total variation yields

$$\overline{F_p}(u,B) \geq |Du|(B) \qquad \forall (u,B) \in \mathrm{L}^1(\Omega) \times \mathcal{O}(\Omega). \tag{2}$$

In the sequel, we assume that there exist an open set $\tilde{A} \subset \Omega$ such that $\tilde{A} \supset\supset A$ and a function $u \in \mathrm{L}^1(\Omega)$ such that $u = u_0$ on $\Omega\setminus\bar{A}$ and $\overline{F_p}(u,\tilde{A}) < \infty$. This could be considered as a mild regularity and compatibility condition between the image and the occlusion.

Theorem 3.1. *The problem*

$$\mathrm{Min}\,\{\overline{F_p}(u,\tilde{A}) : u = u_0 \text{ on } \Omega \setminus \bar{A}\} \tag{3}$$

has at least one solution $u \in \mathrm{BV}(\Omega)$.

Proof. Let $(v_h)_{h\in\mathbb{N}} \subset \mathrm{L}^1(\Omega)$ be a minimizing sequence. Without loss of generality we may assume that $\sup_{h\in\mathbb{N}} \overline{F_p}(v_h,\tilde{A}) < \infty$. Then (2) yields $\sup_{h\in\mathbb{N}} |Dv_h|(\tilde{A}) < \infty$, and therefore $\sup_{h\in\mathbb{N}} |Dv_h|(\Omega) < \infty$ because $v_h = u_0$ on $\Omega\setminus\bar{A}$. Since the values of v_h are fixed on $\Omega\setminus\bar{A}$, the generalized Poincaré inequality given in Theorem 5.11.1 of [29] ensures that the $L^1(\Omega)$ norms of v_h are uniformly bounded. Hence $\sup_{h\in\mathbb{N}} \|v_h\|_{\mathrm{BV}(\Omega)} < \infty$ and there exists a subsequence, still denoted by $(v_h)_{h\in\mathbb{N}}$, converging in $\mathrm{L}^1(\Omega)$ to a function $u \in \mathrm{BV}(\Omega)$. Obviously, $u = u_0$ on $\Omega\setminus\bar{A}$. From the lower semicontinuity of $\overline{F_p}$ we finally obtain that

$$\overline{F_p}(u,\tilde{A}) \leq \liminf_{h\to\infty} \overline{F_p}(v_h,\tilde{A}) = \inf\{\overline{F_p}(v,\tilde{A}) : v = u_0 \text{ on } \Omega\setminus\bar{A}\},$$

and the theorem ensues. □

Now we can show that the relaxed functional coincides with F_p on C^2 functions. The proof is based on the geometric lower semicontinuity results of the previous section and on the identity

$$F_p(u,B) = \int_{\mathbb{R}} \int_{\partial\{u\geq t\}} (1+|\mathbf{H}_{\{u\geq t\}}|^p)\, d\mathcal{H}^{N-1}\, dt \qquad \forall u \in C^2(B). \tag{4}$$

The above identity is a straightforward consequence of the coarea formula and of the equality $\mathbf{H} = -\mathrm{div}(\frac{\nabla u}{|\nabla u|})\frac{\nabla u}{|\nabla u|}$, wherever $|\nabla u| > 0$, see [3].

Theorem 3.2. *Let $B \subset \Omega$ be an open set and assume that $N \geq 2$ and $p > N-1$. The functional $F_p(\cdot, B)$ is lower semicontinuous on $L^1(\Omega) \cap C^2(B)$ with respect to the L^1 topology. In particular*

$$F_p(u,B) = \overline{F_p}(u,B) \qquad \forall u \in C^2(B).$$

Proof. Let $(u_h)_{h\in\mathbb{N}} \subset L^1(\Omega) \cap C^2(B)$ be converging in $L^1(B)$ to $u \in C^2(B)$ and set $L := \liminf_{h\to\infty} F_p(u_h, \tilde{A})$, assuming with no loss of generality that $L < \infty$, that the lim inf is a limit and that u_h converge a.e. to u. By the dominated convergence theorem, $\chi_{\{u_h\geq t\}} \to \chi_{\{u\geq t\}}$ in $L^1(B)$ whenever $\{u = t\}$ is Lebesgue negligible, hence for a.e. $t \in \mathbb{R}$. In addition, by Morse Theorem, for almost every $t \in \mathbb{R}$, $\{u_h \geq t\}$, $h \in \mathbb{N}$, and $\{u \geq t\}$ have smooth boundaries. Therefore, by applying either Theorem 2.3 or Theorem 2.4 we obtain that

$$\int_{\partial\{u\geq t\}\cap B} (1+|\mathbf{H}_{\{u\geq t\}}|^p) d\mathcal{H}^{N-1} \leq \liminf_{h\to\infty} \int_{\partial\{u_h\geq t\}\cap B} (1+|\mathbf{H}_{\{u_h\geq t\}}|^p) d\mathcal{H}^{N-1}$$

for a.e. $t \in \mathbb{R}$. Integrating over $\mathbb{R}$ and using Fatou's lemma, (4) yields

$$F_p(u,B) \leq \liminf_{h\to\infty} F_p(u_h, B).$$

□

In the two-dimensional case we can say something more about the structure of the solutions. The following theorem is easily deduced from the proofs of Theorems 4.1 and 7.1 in [6]. In particular, it suffices to replace ∂E by $\partial^* E$ in the last part of the proof of Theorem 7.1, page 292.

Theorem 3.3 ([6]). *Let $\Omega \subset \mathbb{R}^2$ be an open set and $p > 1$. Let E be a Borel set such that there exists a sequence $\{E_h\}_{h\in\mathbb{N}}$ of bounded open sets of class $\mathrm{C}^2(\Omega)$ converging to E in $\mathrm{L}^1(\Omega)$ and satisfying*

$$\sup_{h\in\mathbb{N}} \int_{\partial E_h \cap \Omega} (1+|\kappa|^p)\, d\mathcal{H}^1 < \infty.$$

Then E has finite perimeter in Ω and there exists a locally finite family $\Gamma = \{\gamma_i\}_{i\in I}$ of regular curves of class $\mathrm{W}^{2,p}$ such that

1. *$\partial^* E \cap \Omega \subset \bigcup_{i\in I}(\gamma_i)$;*
2. *Γ is without crossings, i.e. $\frac{d\gamma_i(t_1)}{dt} // \frac{d\gamma_j(t_2)}{dt}$ whenever $\gamma_i(t_1) = \gamma_j(t_2) \in \Omega$ and t_1, $t_2 \in [0,1]$.*

Corollary 3.4. *Let $B \subset \Omega \subset \mathbb{R}^2$ be an open set and $u \in \mathrm{L}^1(\Omega)$ such that $\overline{F_p}(u,B) < \infty$, with $p > 1$. Then, for almost every $t \in \mathbb{R}$, there exists a locally finite family $\Gamma^t = \{\gamma_i^t\}_{i\in I_t}$ of regular curves of class $\mathrm{W}^{2,p}$ such that $\partial^*\{u > t\} \cap B \subset \bigcup_{i\in I_t}(\gamma_i^t)$ and Γ^t is without crossings.*

Proof. Let $\{u_h\}_{h\in\mathbb{N}} \subset C^2(B)$ be converging to u in $\mathrm{L}^1(\Omega)$ and a.e. and satisfying $L := \lim_{h\to\infty} F_p(u_h, B) = \overline{F_p}(u,B) < \infty$. Using Fatou's Lemma and (4) we get

$$\int_{\mathbb{R}} \liminf_{h\to\infty} \int_{\partial\{u_h>t\}\cap B} (1 + |\mathbf{H}_{\{u_h>t\}}|^p)\, d\mathcal{H}^{N-1} \leq \liminf_{h\to\infty} \int_{\mathbb{R}} \int_{\partial\{u_h>t\}\cap B} (1 + |\mathbf{H}_{\{u_h>t\}}|^p)\, d\mathcal{H}^{N-1} = L < \infty,$$

thus $\liminf_{h\to\infty} \int_{\partial\{u_h>t\}\cap B} (1 + |\mathbf{H}_{\{u_h>t\}}|^p) d\mathcal{H}^{N-1}$ is finite for almost every $t \in \mathbb{R}$. The conclusion follows by applying Theorem 3.3 for almost every $t \in \mathbb{R}$, possibly passing to a subsequence (depending on t). □

An obvious consequence of this result is that *the same regularity holds* for the boundaries $\partial^*\{u > t\} \cap \tilde{A}$ (for almost every $t \in \mathbb{R}$) of any solution of the disocclusion problem in $\mathbb{R}^2$. In addition, since this regularity holds in $\tilde{A}$ and not only within the occlusion A, it gives a necessary condition for the existence of a solution: namely, the level lines of the initial function u_0 must satisfy this regularity property, at least near the boundary of A. As a consequence, the only way — essentially — for these level lines to intersect on ∂A is to form a cusp point.

Remark 3.5. This regularity result cannot be extended to higher dimensions, due to the fact that controlling the mean curvature does not necessarily yield the regularity of a hypersurface. By Allard's regularity theorem (see [1] or Theorem 23.1 in [26]), a $(N-1)$-varifold with density 1 and generalized mean curvature in L^p, $p > N-1$, is supported on a set that can be locally represented as the graph of a $\mathrm{C}^{1,1-\frac{N-1}{p}}$ function (J. Duggan [12] showed that a $\mathrm{W}^{2,p}$ regularity actually holds). Unfortunately, such regularity does not apply to the multiple density case anymore. An example is given in [8] of a varifold V with *bounded* mean curvature whose support contains a set A of strictly positive measure with the property that, if $a \in A$, then $\operatorname{spt} V$ does not correspond to the graph of even a multiple-valued function in any neighborhood of a. Thus, controlling the mean curvature only is not sufficient.

On the other hand, it has been shown by J. Hutchinson [18] that if the *second fundamental form* of a varifold V is in L^p, $p > N-1$, then V is locally supported on the graph of a multiple-valued $\mathrm{C}^{1,1-\frac{N-1}{p}}$ function.

In our disocclusion problem, we can neither ensure that the varifolds supported on the sets $\partial\{u_h > t\}$ converge to unit-density varifolds, nor that the second fundamental form is uniformly bounded in L^p, except in the particular case $N = 2$ where the mean curvature coincides with the second fundamental form. This explains why our regularity result is stated only for $N = 2$.

Acknowledgements

The work of the first author is part of the European Research Training Network "Homogenization and Multiple Scales" under contract HPRN-2000-00109.

The second author would like to thank Reiner Schätzle and Carlo Mantegazza for the fruitful discussions and their useful comments on this work.

References

[1] Allard, W.K. (1972), *On the first variation of a varifold,* Ann. of Math. **95** (2), 417–491.

[2] Alvarez, L., Gousseau, Y. and Morel, J.-M. (1999), *The size of objects in natural and artificial images*, Adv. Imaging Electr. Phys. **111**, 167–242.

[3] Ambrosio, L., Fusco, N. and Pallara, D. (2000), *Functions of Bounded Variation and Free Discontinuity Problems*, Oxford University Press.

[4] Ambrosio, L. and Masnou, S. (2003), *A direct variational approach to a problem arising in image reconstruction*, Interfaces and Free Boundaries, To appear.

[5] Ballester, C., Bertalmio, M., Caselles, V., Sapiro, G. and Verdera, J. (2001), *Filling-in by joint interpolation of vector fields and gray levels*, IEEE Trans. On Image Processing.

[6] Bellettini, G., Dal Maso G. and Paolini, M. (1993), *Semicontinuity and relaxation properties of a curvature depending functional in 2D*, Ann. Scuola Norm. Sup. Pisa Cl. Sci (4) **20**, no 2, 247–297.

[7] Bertalmio, M., Bertozzi, A. L. and Sapiro, G. (2001), *Navier–Stokes, fluid dynamics and image and video inpainting*, IMA preprint.

[8] Brakke, K. (1978), *The motion of a surface by its mean curvature*, Princeton University Press.

[9] Chan, T.F., Kang, S.H. and Shen, J. (2002), *Euler's elastica and curvature based inpainting*, SIAM J. Appl. Math., **63** (2), 564–592.

[10] Chan, T.F. and Shen, J. (2001), *Mathematical models for local deterministic inpaintings*, SIAM J. Appl. Math., **62** (3), 1019–1043.

[11] Cohen, A., Dahmen, W, Daubechies, I and DeVore, R. (2002), *Harmonic analysis of the space* BV, submitted to Revista Mathematica Iberoamericana.

[12] Duggan, J.P. (1986), $W^{2,p}$ *regularity for varifolds with mean curvature*, Comm. Partial Diff. Equations, **11**, 903–926.

[13] Efros, A. and Freeman, W. (2001), *Image quilting for texture synthesis and transfer*, Proc. of SIGGRAPH'01, Los Angeles, California, 341–346.

[14] Efros, A. and Leung, T. (1999), *Texture synthesis by non-parametric sampling*, Int. Conf. on Comp. Vision, Vol. 2, 1033–1038.

[15] Esedoglu, S. and Shen, J. (2002), *Digital inpainting based on the Mumford–Shah–Euler image model*, European J. Appl. Math., **13**, 353–370.

[16] Federer, H. (1969), *Geometric Measure Theory*, Springer Verlag.

[17] Giusti, E. (1994), *Minimal surfaces and functions of bounded variations*, Birkhäuser.

[18] Hutchinson, J.E. (1986), $C^{1,\alpha}$ *multiple function regularity and tangent cone behaviour for varifold with second fundamental form in* L^p, in: Geometric measure theory and the calculus of variations, W.K. Allard and F.J. Almgren eds., Proc. Symp. Pure Math., **44**, Am. Math. Soc., 281–306.

[19] Kanizsa, G. (1979), *Organisation in Vision*, New-York: Praeger.

[20] Liang, L., Liu, C., Xu, Y, Guo, B. and Shum, H.-Y. (2001) *Real-time texture synthesis by patch-based sampling*, Microsoft Res. Tech. Rep., MSR-TR-2001-40, March 2001.

[21] Masnou, S. and Morel, J.-M. (1998), *Level lines based disocclusion*, Proc. ICIP'98 IEEE Int. Conf. on Image Processing, Chicago, USA, 3, 259–263.

[22] Masnou, S. (2002), *Disocclusion: a variational approach using level lines*, IEEE Trans. on Image Processing, **11** (2), 68–76.

[23] Nitzberg, M., Mumford, D. and Shiota, T. (1993), *Filtering, Segmentation and Depth, Lecture Notes in Computer Science*, Vol. 662, Springer-Verlag, Berlin.

[24] Rudin, L. and Osher, S (1994), *Total variation based image restoration with free local constraints*, Proc. of the IEEE ICIP'94, Austin, Texas, Vol 1, 31–35.

[25] Schätzle, R. (2000), *Quadratic tilt-excess decay and strong maximum principle for varifolds*, Preprint.

[26] Simon, L. (1983), *Lectures on geometric measure theory*, Proc. Centre for Math. Anal., Australian Nat. Univ., **3**.

[27] Wei, L.-Y., and Levoy, M (2000), *Fast texture synthesis using tree-structured vector quantization*, Proc. of SIGGRAPH 2000, 479–488.

[28] Zhu, S., Wu, Y. and Mumford, D. (1998), *Filters, random fields and maximum entropy (FRAME) — towards a unified theory for texture modeling*, Int. Journal of Comp. Vision, **27** (2), 107–126.

[29] Ziemer, W.P. (1989), *Weakly differentiable functions*, Springer Verlag.

(L. Ambrosio) Scuola Normale Superiore,
Piazza dei Cavalieri 7,
56126 Pisa, Italy
E-mail address: `l.ambrosio@sns.it`

(S. Masnou) Laboratoire Jacques-Louis Lions,
B.C. 187,
Université Pierre et Marie Curie,
75252 Paris Cedex 05, France
E-mail address: `masnou@ann.jussieu.fr`

International Series of Numerical Mathematics, Vol. 147, 27–51

Ill-Posed Hele–Shaw Flows

Daniele Andreucci, Giovanni Caruso, and Emmanuele DiBenedetto

1. Introduction

The Hele–Shaw flow, introduced at the end of the 19th Century in [14], is a model approximation of the motion of two immiscible fluids confined between two parallel planes, separated by a small distance $0 < h \ll L$, where L is a characteristic transversal length. Such a device is called a Hele–Shaw Cell. Its physical interest rests on its formal analogy with Darcy's law. This makes it suitable for designing simulations of the flow of immiscible fluids in a porous medium.

One such experiment consists in sucking out through a nozzle a blob of a viscous fluid, say for example oil, surrounded by a less viscous fluid, say for example water or air (see Figure 1). The blob recedes and abruptly ruptures into disordered fingers. Reversing the role of the two fluids, one can pump a bubble of air into oil, attempting to remove the oil "at infinity" out of the Hele–Shaw cell (see Figure 2). As before, the bubble advances and ruptures abruptly into disordered fingers. A further experiment consists in injecting a "finger" of water into oil within an infinite strip of finite width (see Figures 3,4). While the finger advances smoothly, experimentally it always seems to choose a "preferred" configuration among all the admissible ones. Receding blobs and advancing bubbles rupture in a "too disordered" way. Advancing fingers evolve in a "too ordered" fashion.

The mathematical interest of the Hele–Shaw flows is in understanding these phenomena. The common feature is that all of them are ill posed in the sense of Hadamard [13]. A direct relation to such an ill-posedness stems from their distributional formulation. Let u be the pressure of the more viscous fluid. The set occupied by this fluid is $[u > 0]$ whereas the interior of set $[u = 0]$ is the set occupied by the less viscous fluid. It turns out that all the aforementioned experimental settings can be translated into the distributional formulation,

$$\frac{\partial}{\partial t} H(u) + \Delta u \ni 0 \quad \text{in } \mathcal{D}'(\text{Hele–Shaw cell}), \tag{1}$$

where $H(u)$ is the Heaviside function of u. The natural initial datum associated with (1) is $H\big(u(\cdot, 0)\big)$, i.e., the geometrical configuration of the set occupied by the more viscous fluid at time $t = 0$. The formulation (1) represents a "forced backward" evolution of a process which is not reversible. This is the origin of the mathematical and physical ill-posedness of these problems. It turns out that solutions can be identified only for a very limited class of initial configurations of the two fluids. The main mathematical problem consists in identifying such configurations.

In this note we review these issues and point out the open problems. In § 2 we derive the Hele–Shaw approximation starting from the Navier–Stokes equations in the Hele–Shaw cell and we set up the basic formalism of free boundaries. In § 2.1 we give a first pointwise mathematical formulation of each of the three experiments described above and discuss their structures and differences. A common distributional formulation is derived in § 4. The problem of receding blobs is examined in § 5 and the initial configurations yielding solutions are characterized as the coincidence sets of suitable variational inequalities. The problem of expanding bubbles hinges upon identifying the initial air bubble that would remove the oil out of the cell "at infinity". Surprisingly such initial bubbles are *homeoids*, i.e., sets for which a uniform distribution of masses on their boundary would generate constant Newtonian potential inside their cavity. These results and such a connection with potential theory form § 6. Receding blobs and expanding bubbles are reasonably understood [8, 9], while the problem of advancing fingers is not, and we concentrate on it in the remaining sections.

The "experimental" finger (see Figure 3), chooses each time a symmetric configuration with an asymptotic width which is exactly one-half of the width of the channel. The physical mechanism by which such a selection occurs is not yet well understood.

We first review recent results of [3]. The problem is transformed into a variational inequality in an infinite strip and a precise asymptotic behavior is identified for a solution to exists. This asymptotic expansion implies uniqueness of solutions (modulo a choice of a coordinate system) and allows one to identify a peculiar quantity, which we call the *area function*, which might be used as a selection criterion of the *"physical"* solution among the admissible ones. Other possible criteria seem to be related to a motion by mean curvature of the free boundary and we discuss them in § 10. While supported by experimental evidence (Pitts [25]), a convincing mathematical explanation still eludes the researchers.

In this note we have collected some unpublished remarks on the Hele–Shaw flow. Some of them pave the way to a unification of the theory, see sections 3 and 4, and other ones, such as (18′), point to a novel way of formulating the problem of advancing fingers by means of a parameter $\delta \in (0,1)$, weighting the contribution of the time-evolutions and of the travelling-wave structure of solutions.

2. Hele–Shaw Cell [14, 19]

A viscous incompressible fluid of kinematic viscosity μ, velocity $\mathbf{v}$ and pressure p is confined within two parallel and slightly separated, horizontal planes π_i, $i = 1, 2$, at mutual distance $0 < h \ll L$, where L is a characteristic transversal length. Introduce Cartesian coordinates (x, y, z) in such a way that π_1 and π_2 are identified with the planes $\{z = 0\}$ and $\{z = h\}$ respectively. The fluid is Newtonian and locally governed by the Navier–Stokes equations. It is assumed that there are no external forcing terms, including gravity which is neglected, and that the velocity is small ($|\mathbf{v}| \ll 1$) and close to the steady state $(\mathbf{v}_t \approx 0)$. Thus $\mathbf{v}$ satisfies the steady-state Stokes system[1]

$$\begin{aligned} -\mu\Delta\mathbf{v} = \nabla p; \quad \operatorname{div}\mathbf{v} = 0 \quad &\text{for } 0 < z < h; \\ \mathbf{v}(x,y,0;t) = \mathbf{v}(x,y,h;0) = 0.& \end{aligned}$$

Since $0 < h \ll 1$ the vertical component v_3 of the velocity is negligible with respect to the horizontal components v_1 and v_2. Moreover the pressure p is independent of the vertical variable z. Since the term $\Delta\mathbf{v}$ accounts for the internal friction of the fluid ([19]) and the motion takes place in ideal horizontal layers, the components $v_{i,xx}$ and $v_{i,yy}$ of the friction along the horizontal variables (x,y) are negligible with respect to the component $v_{i,zz}$ along the vertical variable. Combining these remarks, the Stokes system reduces to

$$\begin{aligned} &-\mu\, v_{1,zz} = p_x; \quad -\mu\, v_{2,zz} = p_y; \quad v_{1,x} + v_{2,y} = 0 \quad \text{for } 0 < z < h; \\ &v_i(x,y,0;t) = v_i(x,y,h;t) = 0 \qquad i = 1,2. \end{aligned}$$

Multiply these by ζ and integrate in $d\zeta$ over $(0,z)$ for $z \in (0,h)$ and then integrate in dz over $(0,h)$ to obtain,

$$\mathbf{V} = -\frac{1}{12}\frac{h^2}{\mu}\nabla p \quad \text{where} \quad V_i = \frac{1}{h}\int_0^h v_i dz, \quad i = 1,2.$$

This is formally Darcy's law for a fluid of viscosity μ, pressure p and fictitious porosity $h^2/12$. Incompressibility implies $\operatorname{div}\mathbf{V} = 0$ and $\Delta p = 0$ within the fluid. This derivation is local in the interior of the fluid. A Hele–Shaw cell is the system of two slightly separated planes filled with viscous fluids whose dynamics has been so approximated. Because of the formal analogy with Darcy's law, Hele–Shaw cells serve to simulate experimentally flows in porous media ([14, 19]).

2.1. Immiscible Fluids in a Hele–Shaw Cell

Two viscous, immiscible, incompressible fluids of kinematic viscosities μ_j, velocities $\mathbf{v}_j$ and pressures p_j, $j = 1,2$ are confined within a Hele–Shaw cell. A similar derivation in the interior of each fluid yields

$$\begin{aligned} \mathbf{V}_j &= \nabla\varphi_j \\ \Delta\varphi_j &= 0 \end{aligned} \quad \text{in the fluids, where} \quad \varphi_j = -\frac{1}{12}\frac{h^2}{\mu_j}p_j, \quad j = 1,2. \tag{2}$$

The functions φ_j are the velocity potentials of the two fluids and each of them is defined within its corresponding fluid. Let $\Omega(t)$ be the domain occupied by the fluid of viscosity μ_2 at each time t. Then the fluid of viscosity μ_1 occupies the complement of $\Omega(t)$ within the Hele–Shaw cell. The interface $\Gamma(t)$ separating the

[1] Δ, ∇ and div are meant with respect to the space variables.

two fluids at time t is in local instantaneous equilibrium. Assuming that the surface tension is negligible amounts to,[2]

$$p_1 = p_2 \quad \text{on the interface } \Gamma(t). \quad \big(\text{ but not in general } \varphi_1 = \varphi_2\big) \tag{3}$$

If the motion evolves during some time interval $(0, T)$, for some $T > 0$, finite or infinite, set

$$\Gamma_T = \bigcup_{t \in (0,T)} \Gamma(t), \qquad \Omega_T = \bigcup_{t \in (0,T)} \Omega(t).$$

Assume Γ_T is smooth and has a local intrinsic representation as $\Phi(x, y; t) = 0$ in proximity of each of its points $P_o = (x_o, y_o)$ at time t_o, for a function Φ, smooth in the neighborhood of (P_o, t_o), and such that $|\nabla\Phi| \neq 0$ on Γ. A material point $P(t) = \big(x(t), y(t)\big)$ moving along its Lagrangian path starting from P_o at $t = t_o$, remains on Γ since the fluids are immiscible. Thus $\Phi\big(P(t); t\big) = 0$ for t close to t_o. Computing the total derivative gives $\nabla_{(x,y)}\Phi \cdot \dot{P} + \Phi_t = 0$, i.e.,

$$\mathbf{N} \cdot \dot{P} = -N_t \quad \text{where} \quad \mathbf{N} = \frac{(\Phi_x, \Phi_y)}{|\nabla_{(x,y;t)}\Phi|} \quad \text{and} \quad N_t = \frac{\Phi_t}{|\nabla_{(x,y;t)}\Phi|}. \tag{4}$$

The function Φ is chosen in such a way that $(\mathbf{N}, N_t)$ points toward the fluid of viscosity μ_1. The material point P belongs to both fluids so that $\dot{P} = V_j$, $j = 1, 2$. Combining this with (2), we get

$$\mathbf{V}_1 \cdot \mathbf{N} = \mathbf{V}_2 \cdot \mathbf{N} = -N_t, \quad \text{i.e.,} \quad \frac{h^2}{12\mu_1}\nabla p_1 \cdot \mathbf{N} = \frac{h^2}{12\mu_2}\nabla p_2 \cdot \mathbf{N} = N_t. \tag{5}$$

3. Receding Blobs, Expanding Bubbles and Advancing Fingers

Assume that the Hele–Shaw cell occupies the whole $I\!R^2$ and that the viscosity of one of the two fluids is negligible with respect to the other, say for example $\mu_1 \ll \mu_2$ (for example air or water versus oil). Then $\mu_1 \approx 0$ and from the definition of φ_1 in (2) one has $p_1 = 0$. By (3) one has $p_2 = 0$ on $\Gamma(t)$ for all t, and the motion is determined by $\Omega(t)$ and the pressure p_2 within it for all $t \in (0, T)$. Setting $u = \varphi_2$, the problem consists in finding a family of sets $(0, T) \to \Omega(t) \subset I\!R^2$ with smooth boundary, and a measurable function u defined in $I\!R^2 \times (0, T)$, such that

$$(x, y) \to u(x, y, t) \in C^2\left(\Omega(t)\right) \cap C^1\left(\overline{\Omega}(t)\right) \quad \text{for all } t \in (0, T)$$

and u satisfies

$$\Delta u = 0 \text{ in } \Omega(t); \quad u = 0, \quad \nabla u \cdot \mathbf{N} = -N_t \text{ on } \Gamma(t) \quad \forall t \in (0, T). \tag{6}$$

The fluid is receding if $\Omega(t_2) \subset \Omega(t_1)$ for $t_1 < t_2$ and it is expanding if the converse inclusion holds.

[2]Instantaneous equilibrium implies $p_1 - p_2 = \tau\kappa$, where τ is the surface tension and κ is the curvature of $\Gamma(t)$.

3.1. Receding Blobs

For all times the fluid occupies a bounded domain in $\mathbb{R}^2$ and surrounds a "core" or sink G with smooth boundary ∂G, through which fluid is removed (see Figure 1).

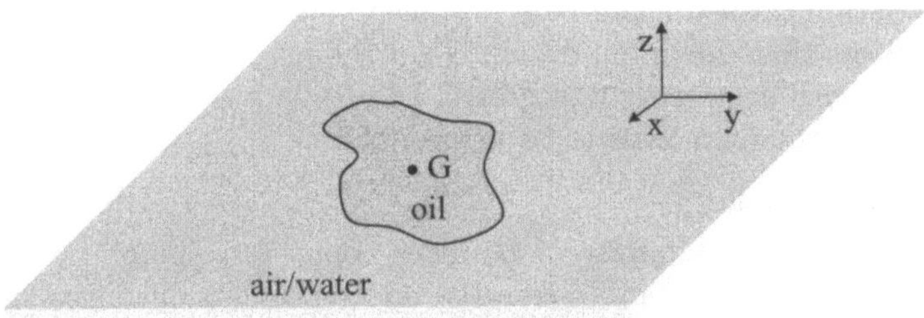

FIGURE 1. Bubble of oil into water

The suction occurs at constant rate $Q > 0$ and causes the fluid blob to recede. Denoting by ν the unit normal to ∂G directed towards the fluid,

$$\nabla u \cdot \nu = -Q \quad \text{on } \partial G \quad \text{(pressure forced to decrease along } \partial G). \tag{7}$$

It is experimentally observed that the suction process ends with the abrupt formation of irregular fingers in finite time. Repeating the experiments with the same initial configuration $\Omega(0)$ of the blob ("same" within the limits of experimental errors), it is observed that the extinction time and the configuration of the rupturing fingers are dramatically different. The problem is physically ill-posed in this sense. It is then natural to wonder whether there are initial configurations $\Omega(0)$ of the blob for which the suction process ends smoothly at some finite time T, i.e., with no abrupt rupture into irregular fingers. More generally, if one prescribes a final configuration of fingers, is there an initial blob $\Omega(0)$ for which the suction process ends abruptly by rupturing into that finger configuration?

The boundary of $\Omega(t)$ consists of two portions. $\Gamma(t)$, the first portion, is the boundary separating the more viscous fluid from the less viscous one, and the second portion is ∂G. Since u is harmonic in $\Omega(t)$, vanishes on $\Gamma(t)$ and has exterior positive flux on ∂G, by the maximum principle we get $[u > 0]$ in the fluid and $\Omega(t) = [u(t) > 0]$. Assume that this problem has a smooth solution u, positive within the fluid, null on the interface Γ and such that $|\nabla u| \neq 0$ on Γ. Then, taking $\Phi = -u$ in (4) and in the third relation in (6), we have

$$-|\nabla u|^2 = u_t \quad \text{on } \Gamma(t) \text{ for all } t \in (0, T). \tag{8}$$

Since u_t is harmonic in $\Omega(t)$, it has zero flux on ∂G and it is negative on Γ, by the maximum principle $u_t \leq 0$ in $\Omega(t)$ for all $t \in (0, T)$. Therefore the blob is receding in the sense that $\Omega(t_1) \supset \Omega(t_2)$ for $t_1 < t_2$. The mathematical formulation is

$$\begin{cases} \Delta u = 0 \text{ in } \Omega(t); \quad [u(t) > 0] = \Omega(t) \quad \forall t \in (0,T); \\ u = 0 \quad \text{and} \quad \nabla u \cdot \mathbf{N} = -N_t \quad \text{on } \Gamma(t) \quad \forall t \in (0,T); \\ \nabla u \cdot \nu = -Q \text{ on } \partial G \text{ for some fixed } Q > 0 \text{ and } \forall t \in (0,T). \end{cases} \tag{9}$$

The initial datum is the domain $\Omega(0)$ occupied by the fluid, as the suction process starts. We stress that the initial datum is not the pressure $u(0)$ at time $t = 0$, but the set where such a pressure is negative, i.e., $[u(0) > 0] = \Omega(0)$. The extinction time T is one of the unknowns to be determined.

3.2. Expanding Bubbles

Let $(0,T) \rightarrow B(t)$ be a family of bounded, open sets in $I\!R^N$. For $t = 0$ the set $B(0)$ is bounded, open and connected and represents an initial bubble of air surrounded by fluid which entirely fills the remaining part of $I\!R^N$ (see Figure 2).[3] The domain $\Omega(t)$ occupied by the fluid is $I\!R^N - B(t)$ and $\Gamma(t) = \partial B(t)$. If the

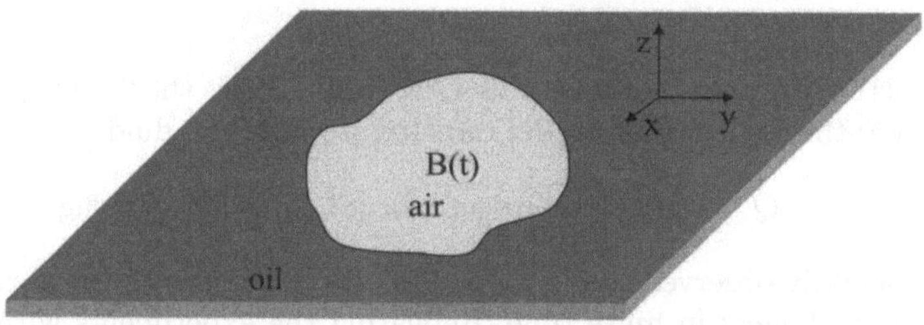

FIGURE 2. Bubble of air into oil

bubble $B(t)$ expands it pushes the fluid to $|x| \rightarrow \infty$. It is observed that the process ends abruptly with the formation of disordered fingers. In this case as well, the same bubble $B(0)$ ("same" within the experimental errors) may generate, in real world experiments, different extinction times and different finger configurations. A natural question is whether there exist bubbles $B(0)$ for which the whole fluid is pushed smoothly away to infinity.

The potential $u = \varphi_2$ is harmonic within the fluid and the fluid "outflows at infinity" essentially like a Newtonian potential, i.e.,

$$\nabla u(x;t) = -Q\,\nabla F_N(x) + O(|x|^{-N}) \quad \text{as } |x| \rightarrow \infty \text{ for all } t \in (0,T), \tag{10}$$

where F_N is the fundamental solution of Laplace's equation in $I\!R^N$, $N \geq 2$ and Q is a positive given number. Thus the fluid tends to flow out to infinity. Since $u = 0$ on $\Gamma(t)$ and is harmonic in $\Omega(t)$, by the strong maximum principle $u > 0$ within

[3]Although the problem arises physically in $I\!R^2$ we set it in $I\!R^N$ in view of its connection with N-dimensional homeoids. In this context $x = (x_1, \ldots, x_N)$. More generally, one has a bubble of fluid of smaller viscosity μ_1 which has been neglected.

the fluid and $[u(t) > 0] = \Omega(t)$ for all $t \in (0, T)$. If the problem has a sufficiently smooth solution with smooth free boundary $\Gamma(t)$, then $|\nabla u| \neq 0$ on Γ and one might take $\Phi = u$ in (4) to find a relation identical to (8) in N space variables, and conclude that $u_t \leq 0$. Thus the fluid is receding towards infinity and the air bubble $B(t)$ is expanding. We have

$$\begin{cases} \Delta u = 0 \text{ in } \Omega(t) = I\!R^N - B(t); \quad [u(t) > 0] = \Omega(t) \quad \forall t \in (0,T); \\ u = 0 \quad \text{and} \quad \nabla u \cdot \mathbf{N} = -N_t \quad \text{on } \Gamma(t) \quad \forall t \in (0,T); \\ \nabla u = -Q\,\nabla F_N(x) + O(|x|^{-N}) \text{ as } |x| \to \infty \quad \forall t \in (0,T). \end{cases} \tag{11}$$

The initial datum is $B(0)$, i.e. the domain occupied by the air bubble at time $t = 0$. Once again, the rupturing time T is one of the unknowns to be determined.

3.3. Advancing Fingers

Here the Hele–Shaw cell is an infinite strip $S = I\!R \times (-1, 1)$ and the fluid of smaller viscosity μ_1 (water for example) penetrates the more viscous fluid (oil for example), in the form of a finger, whose asymptotic, vertical section, as $x \to -\infty$, is an interval $(-\lambda, \lambda) \subset (-1, 1)$ for some $\lambda \in (0, 1)$. The shape of the finger and λ depend on t. It is assumed that, at each time t, the penetrating fluid advances with constant velocity $U\mathbf{i}$. Therefore its velocity potential is $\varphi_1 = Ux + const$, where U might depend on time. In this Subsection, we do not stipulate the assumption $p_1 = 0$; let us note that even in the former cases it was not strictly necessary.

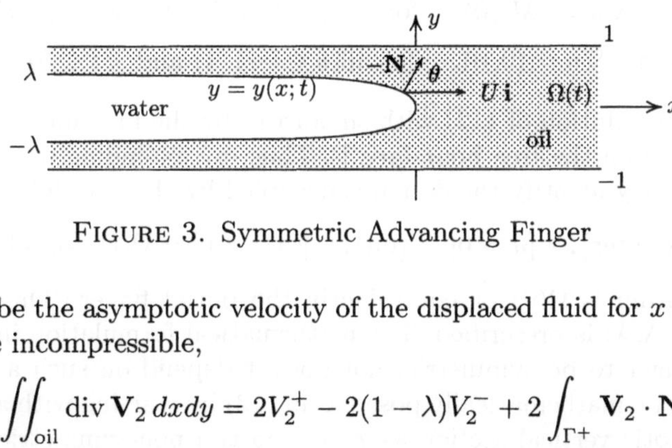

FIGURE 3. Symmetric Advancing Finger

Let $V_2^{\pm}\mathbf{i}$ be the asymptotic velocity of the displaced fluid for $x \to \pm\infty$. Since both fluids are incompressible,

$$0 = \iint_{\text{oil}} \operatorname{div} \mathbf{V}_2 \, dxdy = 2V_2^+ - 2(1-\lambda)V_2^- + 2\int_{\Gamma+} \mathbf{V}_2 \cdot \mathbf{N} d\sigma$$

$$0 = \iint_{\text{water}} \operatorname{div} \mathbf{V}_1 \, dxdy = -2\lambda U - 2\int_{\Gamma+} \mathbf{V}_1 \cdot \mathbf{N} d\sigma.$$

Using the first relation in (5), this implies $V_2^+ = U\lambda + (1-\lambda)V_2^-$. By the second of (5), there holds as $x \to -\infty$ on $\Gamma(t)$

$$\varphi_2 \longrightarrow \frac{\mu_1}{\mu_2}\varphi_1, \quad \text{i.e.,} \quad V_2^- = \frac{\mu_1}{\mu_2}U, \quad \text{and} \quad V_2^+ = U\lambda + \frac{\mu_1}{\mu_2}(1-\lambda)U.$$

The velocity potential $\varphi_1 = Ux + \text{const}$ is only defined in the region occupied by water. We regard it as defined on the whole strip S by extending it to be the same form. Then, motivated by (3), we set

$$\left(1-\frac{\mu_1}{\mu_2}\right) u = -\frac{1}{12}\frac{h^2}{\mu_2}(p_2-p_1) = \varphi_2 - \frac{\mu_1}{\mu_2}\varphi_1 \quad \text{on the set occupied by oil.}$$

The function u is harmonic on such a set and $u = 0$ on Γ. Moreover,

$$\lim_{x\to\infty} \nabla u = \lambda U\,\mathbf{i} \qquad \text{and} \qquad \lim_{x\to-\infty} \nabla u = 0.$$

From the first of (5),

$$\nabla u \cdot \mathbf{N} = -N_t \quad \text{on } \Gamma(t) \quad \forall t \in (0,T).$$

Since the edges of the cell are sealed, $\mathbf{V}_2 \cdot \mathbf{j} = 0$ for $y = \pm 1$, i.e., $u_y(x, \pm 1) = 0$ for all $x \in I\!R$. Collecting these remarks, the problem reduces to finding:

1. a domain $\Omega \subset S \times I\!R^+$ whose boundary is a space-time surface Γ and whose t-sections $\Gamma(t)$ are in the form of fingers;
2. a function $u(x, y, t)$ defined in $\overline{\Omega}$ and such that,

$$\begin{cases} \Delta u = 0 \quad \text{in } \Omega(t) \quad \forall t \in (0,T); \\ u = 0 \text{ and } \nabla u \cdot \mathbf{N} = -N_t \text{ on } \Gamma(t),\ \forall t \in (0,T); \\ u_y(x, \pm 1, t) = 0 \quad \text{for all } x \in I\!R \text{ and all } t \in (0,T); \\ \lim\limits_{x\to\infty} \nabla u = (\lambda U, 0) \quad \text{for all } y \in (-1,1) \text{ and } \forall t \in (0,T); \\ \lim\limits_{x\to-\infty} \nabla u = (0,0). \quad \text{for all } y \in (-1,1) \text{ and } \forall t \in (0,T). \end{cases} \tag{12}$$

If u is a smooth solution of (12) with smooth Γ, by the maximum principle $u > 0$ on Ω and $u = 0$ on the free boundary Γ. Thus, by extending u to be null inside the finger, we may identify the domains occupied by the two fluids as

$$[u = 0] = \{\text{water}\}, \quad [u > 0] = \{\text{oil}\}, \quad [u(t) > 0] = \Omega(t) \text{ for all } t \in (0,T).$$

The initial datum is $\Omega(0)$ or equivalently the initial finger. Thus in particular the interval $(-\lambda, \lambda)$ is prescribed. The mathematical formulation in (12) does not require the finger to be symmetric, nor does it depend on such a property. The problem could be mathematically posed as in (12) by starting with an initial finger whose asymptotic vertical section as $x \to -\infty$ is a non-symmetric interval as in Figure 4, or even a measurable set $E \subset (-1,1)$ of given measure $\mu(E) = 2\lambda$.

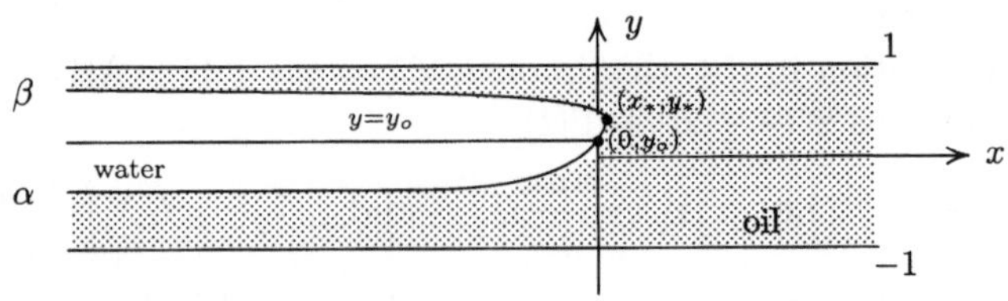

FIGURE 4. Non Symmetric Advancing Finger

Denote by $y = f(x,t)$ the local representation of the curve $\Gamma(t)$, and let $\mathbf{t}$ be its unit tangent. The unit vector $-(\mathbf{N}, N_t) = -(N_x, N_y, N_t)$ points inside the more viscous fluid. From the first of (5), Figure 3 and the definition of u, we compute

$$\nabla u \cdot \mathbf{N} = UN_x \quad \text{and} \quad N_t = -UN_x \text{ on } \Gamma(t), \quad \forall t \in (0,T). \tag{13}$$

Therefore the third equality in (12) can be equivalently replaced by (13).

4. Distributional Formulations

These pointwise formulations differ from each other only on the requirements on the fluid set $\Omega(t)$ and on the boundary conditions. However, they locally have the same structure. Namely, the fluid is identified as the set where $u > 0$, and u vanishes on the interface $\Gamma(t)$ separating the two fluids. For this reason, the partial differential equation and the fluid set where it is required to hold are linked by the same distributional formulation which we derive next. We define

$$H(u) = \begin{cases} 1 & \text{if } u > 0 \\ [0,1] & \text{if } u = 0. \end{cases} \tag{14}$$

If the time T, finite or infinite, has been determined, choose any

$$\varphi \in C_o^\infty\left(\{\text{Hele–Shaw cell}\} \times (0,T)\right),$$

and compute, in the sense of distributions,[4]

$$\begin{aligned} \langle \Delta u, \varphi \rangle &= \langle u, \Delta\varphi \rangle = \iint_{\Omega_T} u \Delta\varphi dxdt = \iint_{\Gamma_T} u \nabla\varphi \cdot \mathbf{N}\, d\sigma - \\ &\iint_{\Omega_T} \nabla u \nabla\varphi\, dxdt = -\iint_{\Gamma_T} \nabla u \cdot \mathbf{N}\varphi\, d\sigma + \iint_{\Omega_T} \Delta u\, \varphi\, dxdt \\ &= \iint_{\Gamma_T} N_t\, \varphi\, d\sigma = \iint_{\Omega_T} \varphi_t\, dxdt = \langle H(u), \varphi_t \rangle = -\left\langle \frac{\partial}{\partial t} H(u), \varphi \right\rangle. \end{aligned} \tag{15}$$

Therefore,

$$\frac{\partial}{\partial t} H(u) + \Delta u = 0 \quad \text{in} \quad \mathcal{D}'\left(\{\text{Hele–Shaw cell}\} \times (0,T)\right).$$

This includes the partial differential equation in (9), (11), (12), as well as the conditions on $\Gamma(t)$ represented by the second relations of (9), (11), (12). The remaining conditions still hold and lead to the following distributional formulation of these problems.

---

[4]The Hele–Shaw cell could be the whole $\mathbb{R}^2$ as in § 3.1, or the whole $\mathbb{R}^N$ as in § 3.2, or the infinite two-dimensional strip S as in § 3.3. The distribution pairings and calculations are formally the same for all these occurrences. x will denote a N-dimensional variable as well.

4.1. Receding Blobs

Although the physical problem arises in $I\!R^2$ we formulate it in $I\!R^N$ since its mathematical structure does not depend on the dimension.

$$\begin{cases} u \in L^2\big(0,T; W^{1,2}(I\!R^N) \cap W^{2,2}([u(t)>0])\big); \\ \dfrac{\partial}{\partial t} H(u) + \Delta u \ni 0 \quad \text{in } \mathcal{D}'\big(I\!R^N \times (0,T)\big); \\ \nabla u \cdot \nu = -Q \ \text{ on } \partial G \ \text{ for a.e. } t \in (0,T). \end{cases} \tag{16}$$

Here Q is a given positive number and ν is the unit normal to ∂G pointing towards the fluid. The inclusion in (16) is meant in the sense of the graphs.

4.2. Expanding Bubbles

$$\begin{cases} u \in L^2\big(0,T; W^{1,2}_{\text{loc}}(I\!R^N) \cap W^{2,2}([u(t)>0])\big); \\ \dfrac{\partial}{\partial t} H(u) + \Delta u \ni 0 \quad \text{in } \mathcal{D}'\big(I\!R^N \times (0,T)\big); \\ \nabla u = -Q\nabla F_N(x) + O\left(|x|^{-N}\right) \ \text{ as } |x| \to \infty \ \text{ a.e. } t \in (0,T). \end{cases} \tag{17}$$

4.3. Advancing Fingers

The Hele–Shaw cell is a 2-dimensional strip S and we will formulate the problem of advancing fingers in such a 2-dimensional setting. For $n \in I\!N$ let S_n denote the truncated half strip $(-\infty, n) \times (-1,1)$. Then we look for a function $u : S_T \to I\!R^+$ such that,

$$\begin{cases} u \in L^2\big(0,T; W^{1,2}(S_n) \cap W^{2,2}([u(t)>0] \cap S_n)\big) \quad \forall n \in I\!N; \\ \dfrac{\partial}{\partial t} H(u) + \Delta u \ni 0 \quad \text{in } \mathcal{D}'\big(S \times (0,T)\big); \\ u_y(x, \pm 1, t) = 0 \quad \text{for all } x \in I\!R \text{ and all } t \in (0,T); \\ \displaystyle\lim_{x \to \infty} \nabla u = (\lambda U, 0) \quad \text{for all } y \in (-1,1) \text{ and } \forall\, t \in (0,T); \\ \displaystyle\lim_{x \to -\infty} \nabla u = (0,0) \quad \text{for all } y \in (-1,1) \text{ and } \forall\, t \in (0,T). \end{cases} \tag{18}$$

From (13) and the distributional computation (15),

$$\frac{\partial}{\partial t} H(u) = -U \frac{\partial}{\partial x} H(u) \quad \text{in } \mathcal{D}'\big(S \times (0,T)\big). \tag{19}$$

This implies that the second of (18) can be equivalently rewritten as

$$\delta \frac{\partial}{\partial t} H(u) - (1-\delta) U \frac{\partial}{\partial x} H(u) + \Delta u \ni 0 \tag{18$'$}$$

in $\mathcal{D}'\big(S \times (0,T)\big)$, for all $0 \le \delta \le 1$.

5. Receding Blobs

Let u be a classical solution of (16). Since the fluid is receding, setting

$$v(t) = \int_t^T u(\tau)d\tau, \qquad \text{one has} \qquad H(u) = H(v). \tag{20}$$

Therefore (16) is transformed in terms of v as,

$$\Delta v \in H(v) - \xi \qquad \text{where } \xi \in H\big(u(T)\big). \tag{20'}$$

The inclusion is in the sense of the graphs and (15) is required to hold for all $t \in (0,T)$. The time T is the extinction time so that $u(\cdot,T) \equiv 0$, smoothly or abruptly. The selection,

$$\xi(\cdot) \in H\big(u(\cdot,T)\big) = H(0) = [0,1],$$

can be interpreted as a description the extinction profile and it is called the *terminal phase*. Integrating (20′) in dx over $\Omega(t)$ gives,

$$Q(T-t)|\partial G| = \int_{\Omega(t)} \{H(v) - \xi\}dx \qquad \text{and} \qquad QT|\partial G| = \int_{\Omega(0)} \{1-\xi\}dx.$$

This relation links the initial datum $\Omega(0)$, the extinction time T and the extinction profile ξ. Given an initial blob $\Omega = \Omega(0)$ and a measurable selection $\xi(\cdot) \subset [0,1]$ in the sense of the graphs, we seek a positive time T and a non-negative function v such that

$$\begin{cases} v \in C\big([0,T] : W^{2,2}(\Omega) \cap W_o^{1,2}(\Omega)\big); \\ \Delta v \in H(v) - \xi \;\; \text{a.e. in } \Omega, \; \forall t \in (0,T); \\ \nabla v \cdot \nu = -Q(T-t) \;\; \text{on } \partial G, \; \forall t \in (0,T); \\ v(x,T) = 0 \;\; \text{and} \;\; QT|\partial G| = \displaystyle\int_\Omega (1-\xi)dx. \end{cases} \tag{21}$$

Since the fluid recedes $[v(t) > 0] = \Omega(t) \subset \Omega$ for all $t \in (0,T)$. The flux condition on ∂G is meant weakly although it is shown in [8] that ∇v is Hölder continuous in $\overline{\Omega} \times (0,T)$, so that such a condition is actually classical.

As for ξ, we may assign it in the following way. Select a measurable set $\Sigma \subset \Omega$ to serve as a description of the geometrical configuration of the final fingers and put $\xi = \chi_\Sigma$. The set Σ could be the finite or infinite union of disjoint subsets of $\Omega - \overline{G}$, each being a finger. If $\Sigma = \emptyset$, then $\xi \equiv 0$ and the extinction occurs smoothly without abrupt formation of fingers.

5.1. Relating the Initial Blob to the Final Finger

Given a final finger configuration ξ, we wonder whether there exists an initial blob Ω and a time T such that the suction process ends at time T with the formation of the terminal phase ξ. Such initial configurations $\Omega(0)$, if existing, are confined within a rather small class of sets, which we will now identify.[5]

[5]One can expect the initial configurations $\Omega(0)$ to be in a narrow class by virtue an heuristical argument of analogy with the backward heat equation. If in (16) $H(u) = u$, one would have the backward heat equation, and fixing a "final" datum $u(\cdot,T)$, one would ask for the "initial"

Lemma 5.1. *If* $\{\Omega, T, \xi, v\}$ *is a weak solution of* (21), *then* $\nabla v(\cdot, 0) \cdot \mathbf{N} = 0$ *on* $\partial\Omega$, *where* $\mathbf{N}$ *is the outward unit normal to* $\partial\Omega$.

Proof. Since $v(\cdot, 0) > 0$ in Ω and it vanishes on $\partial\Omega$, one has $\nabla v(\cdot, 0) \cdot \mathbf{N} \leq 0$ on $\partial\Omega$. Integrating the equation in (21) for $t = 0$ gives

$$0 \geq \int_{\partial\Omega} \nabla v(x, 0) \cdot \mathbf{N} d\sigma \geq 0.$$

□

From this we deduce

Corollary 5.2. *If* $\{\Omega, T, \xi, v\}$ *is a weak solution of* (21), *then* $V = v(\cdot, 0)$ *is the unique solution of the variational inequality,*

$$\begin{cases} \left.\begin{cases} \Delta V \leq 1 - \xi \\ V \geq 0 \\ V(\Delta V - 1 + \xi) = 0, \end{cases}\right\} & \text{a.e. in } I\!R^N - \overline{G} \\ \nabla V \cdot \nu = -QT & \text{on } \partial G. \end{cases} \tag{22}$$

Moreover, $\Omega = [V > 0]$.

The Corollary asserts that if (21) has a solution, then the initial blob must be concentrated on the support of the solution of (22). The solution V determines its own support from (22). Thus $\Omega(0)$ cannot be prescribed in an arbitrary manner.

The following Theorem, proved in [8], establishes the converse.

Theorem 5.3. ([8]) *Fix* $\xi = \chi_\Sigma$ *and* $\lambda > 0$ *and let* $\Omega(\xi, \lambda)$ *be the open set* $[V_{\xi,\lambda} > 0]$ *where* $V_{\xi,\lambda}$ *is the unique solution of* (22) *for* ξ *and for* $QT = \lambda$. *Then there exists a unique solution* $\{\Omega, T, \xi, v\}$ *of* (21) *with* $\Omega = \Omega(\xi, \lambda)$ *and* $T = \lambda/Q$.

This characterizes completely the class of initial domains for which (21) admits a solution. Given any "terminal finger" Σ and the corresponding solution of (22) for $\xi = \chi_\Sigma$ and some $\lambda > 0$, the fluid recedes to the finger Σ in time $T = \lambda/Q$ if and only if the initial blob is confined in the complement of the coincidence set of the variational inequality (22) for this choice of parameters.

If $\Sigma = \emptyset$, i.e., if $\xi \equiv 0$, then the variational inequality (22) gives a way of choosing the initial blob for which there is smooth extinction, in time T, with no finger formation. We stress that the smooth extinction and the smoothness of the initial blob $\Omega(0)$ are not related. Only the relations in (22) with $\xi = 0$ prescribe the structure of the initial blob $\Omega(0)$ in such a way that smooth extinction may occur. In [8] we give examples showing that any finger configuration, however irregular, can be achieved starting from an initial blob $\Omega(0)$ whose boundary is nearly spherical and analytic.

---

datum $u(\cdot, 0)$ yielding the prescribed configuration $u(\cdot, T)$. Such a datum, if any, would have to analytic in Ω.

6. Expanding Bubbles

Let Σ be the final configuration of the set occupied by the fluid; equivalently, $B(T) = I\!R^N - \Sigma$ is the final configuration of the bubble at time T. Such extinction/rupture time T might be finite or infinite irrespective of the regularity of Σ. For example if $\Sigma = \{|x| > R\}$, the fluid might be smoothly displaced along of a sphere $\{|x| = r(t)\}$, and it might abruptly disappear at some finite time T when $r(T) = R$. In the other extreme case $\Sigma = \emptyset$, the bubble smoothly displaces the fluids for all times and it invades the whole $I\!R^N$ as $t \to \infty$. The problems (17) for finite or infinite T have a different nature.

6.1. Expanding Bubbles Bursting in Finite Time T

Having fixed a final configuration Σ, we let $\xi = \chi_\Sigma$ and we keep on denoting by $\Omega = \Omega(0) = I\!R^N - B(0)$ the initial fluid set for the initial bubble $B(0)$. Defining v and $H(v)$ as in (20), the new unknown v satisfies (20′) in $\mathcal{D}'(I\!R^N)$ for the choice $\xi = \chi_\Sigma$. Integrate this equation over a ball B_ρ of radius ρ centered at the origin, use the asymptotic condition in (17), and let $\rho \to \infty$ and $t \to 0$ to obtain,

$$QT = \int_\Omega (1-\xi)dx = |\Omega - \Sigma| = |I\!R^N - \Sigma| - |B(0)|. \tag{23}$$

The initial bubble $B(0)$ is confined within a ball $\{|x| < R\}$ centered at the origin. Then, for a burst or extinction in finite time the measure of $I\!R^N - \Sigma$ must be finite. Summarizing, given a final configuration $\xi = \chi_\Sigma$ and an extinction time T one seeks an initial configuration Ω of the fluid, or equivalently an initial bubble $B(0) = I\!R^N - \Omega$, and a non-negative function v such that

$$\begin{cases} v \in C\big([0,T]; W^{2,2}_{\text{loc}}(I\!R^N)\big),\ v \geq 0; \\ \Delta v \in H(v) - \xi \ \text{ a.e. in } I\!R^N,\ \forall t \in (0,T); \\ \nabla v = -Q(T-t)\nabla F_N(x) + O\big(|x|^{-N}\big) \ \text{ as } |x| \to \infty,\ \forall t \in (0,T); \\ v(x,T) = 0 \ \text{ and } \ QT = \displaystyle\int_\Omega (1-\xi)dx = |\Omega - \Sigma|. \end{cases} \tag{24}$$

If v is a weak solution, $B(t) = [v = 0]$ for all $t \in (0,T)$. Moreover,

$$\Omega = [V > 0] \ \text{ where } V = v(\cdot, 0), \qquad \text{and} \qquad \lambda = QT = |\Omega - \Sigma|.$$

Arguing as before, V is a solution to the variational inequality,

$$\begin{cases} \left.\begin{cases} \Delta V \leq 1 - \xi \\ V \geq 0 \\ V(\Delta V - 1 + \xi) = 0, \end{cases}\right\} & \text{a.e. in } I\!R^N \\ \nabla V\cdot = -\lambda \nabla F_N(x) + O\big(|x|^{-N}\big) & \text{as } |x| \to \infty. \end{cases} \tag{25}$$

Theorem 6.1. [9] *For every fixed final configuration Σ and $\lambda > 0$, there exists a unique solution of the variational inequality* (25). *As a consequence, there exists a unique weak solution of* (23) *with $T = \lambda/Q$.*

Thus, given a final fluid configuration Σ and $\lambda > 0$, the expanding bubble problem (23) has a solution with $QT = \lambda|\Omega - \Sigma|$ if and only if the initial bubble $B(0)$ is the coincidence set of the variational obstacle problem (24). In such a case, $\Omega = I\!R^N - B(0)$. The proof in [9] makes use of potential representations in terms of the fundamental solution $F_N(\cdot)$. The proof also shows that

$$v(x,t) = \int_{\Omega-\Sigma} \{F_N(y-x) - F_N(x)\}dy. \tag{26}$$

Moreover $B(t) = [v = 0]$ for all $t \in (0,T)$, and,

$$|B(t)| = |B(0)| + Qt \quad \text{for all } t \in (0,T). \tag{27}$$

6.2. Extinction in Infinite Time

The same transformation as in (20) has compactness problems; also (23) is not well defined. However, we can carry out the same procedure to construct approximating solutions. Fix an asymptotic fluid configuration Σ such that $|I\!R^N - \Sigma| = \infty$. Then, for $n \in I\!N$ set

$$\Sigma_n = \Sigma \cup \{|x| > n\}, \quad \xi_n = \chi_{\Sigma_n}, \quad \lambda_n = |I\!R^N - \Sigma_n| - \mu, \tag{28}$$

where μ is a fixed positive parameter to be chosen. For these choices, one solves the approximating problem,

$$\begin{cases} v_n \in C\big([0,T]; W^{2,2}_{\text{loc}}(I\!R^N)\big),\ v_n \geq 0; \\ \Delta v_n \in H(v_n) - \xi_n \ \text{ a.e. in } I\!R^N,\ \forall t \in (0,T_n); \\ \nabla v_n = -Q(T_n - t)\nabla F_N(x) + O\big(|x|^{-N}\big) \ \text{ as } |x| \to \infty,\ \forall t \in (0,T); \\ v_n(x,T_n) = 0, \quad T_n = \lambda_n/Q, \quad \Omega_n = I\!R^N - \Sigma_n - B_n(0); \\ QT_n = \displaystyle\int_{\Omega_n} (1-\xi_n)dx = |\Omega_n - \Sigma_n|. \end{cases} \tag{29}$$

This is a family of variational inequalities (parametrized with $t \in (0,T_n)$), with coincidence sets $B_n(t) = [v_n(t) = 0]$. Moreover, $\Omega_n = [v_n(\cdot,0) > 0]$. For each $n \in I\!N$, this problem has a unique solution v_n, which in addition satisfies,

$$v_n(x,t) = \int_{I\!R^N - \Sigma_n - B_n(t)} \{F_N(y-x) - F_N(x)\}dy \quad \forall t \in (0,T_n). \tag{30}$$

Also,

$$|B_n(t)| = \mu + Qt \qquad \forall t \in (0,T_n). \tag{31}$$

We briefly comment on the positive parameter μ. Comparing (31) with (27) and keeping (23) in mind, such a parameter prescribes the measure of the initial bubble $B_n(0)$ and keeps it fixed as $n \to \infty$. Notice that only the volume and not the shape of the initial bubble are specified. Thus when $n \to \infty$ in (29) one expects to get a solution of the expanding bubble problem with infinite extinction time and for a prescribed volume of the initial bubble. Such a solution will then depend on an arbitrary positive parameter μ. This dependence is not surprising since $T = \infty$ and it corresponds to an arbitrary choice of the initial time, as it appears from (27) and (31) for $n \to \infty$. Letting $n \to \infty$ in (29) hinges upon some a priori estimates on $\{v_n\}$, which are derived in [9] under the assumption that the fingers in Σ are not too "thick", in the sense

$$\int_1^\infty \frac{|\Sigma \cap \{|x| = \rho\}|}{|\{|x| = \rho\}|} d\rho < \infty.$$

Assuming such a "non-thickness" condition of the terminal phase Σ, the sequence $\{v_n\}$ is equi-bounded in $C^{1,\alpha}_{\rm loc}(I\!R^N)$ and a strong pointwise limit can be taken in (29). Referring to [9] for the technical arguments, the resulting limiting problem is

$$\begin{cases} v \in C\big(I\!R^+; W^{2,2}_{\rm loc}(I\!R^N)\big),\ v \geq 0; \\ \Delta v \in H(v) - \xi \ \text{ a.e. in } I\!R^N,\ \forall\, t \in I\!R^+; \\ v(x,t) \leq C|x|^2 \ \text{ as } \ |x| \to \infty,\ \forall\, t \in I\!R^+, \end{cases} \tag{32}$$

for a positive constant C independent of t. The last two of (29) lose meaning as $T_n \to \infty$, whereas the asymptotic behavior as $|x| \to \infty$ yields the last of (32). This is established in Theorem 4.5 of [9], page 39, by a suitable expansion of F_N in terms of harmonic polynomials as $|x| \to \infty$. This limiting problem is a family of variational inequalities parametrized by $t \in I\!R^+$ and for each t the coincidence set is $B(t) = [v(t) = 0]$. By a classical stability result of coincidence sets of variational inequalities, the Hausdorff distance between $B_n(t)$ and $B(t)$ tends to zero as $n \to \infty$, for all $t > 0$. It follows from this fact and from (31) that

$$|B(t)| = \mu + Qt \qquad \text{for all } \ t > 0. \tag{33}$$

The variational inequality (32) for $t = 0$ completely characterizes the initial bubbles $B(0)$, of measure μ, yielding the terminal finger configuration Σ in infinite time. These are the coincidence sets of the solution of (32) for $t = 0$.

6.3. Smooth Extinction in Infinite Time

An interesting characterization of the initial bubbles $B = B(0)$ can be given in the case when the fluid is displaced to $|x| \to \infty$ smoothly at all times. Thus, there are no terminal fingers and $\Sigma = \emptyset$ and $\xi \equiv 0$. Accordingly, $I\!R^N - \Sigma_n$ is the ball $\{|x| < n\}$ and (30), for $t = 0$, takes the form,

$$V(x) = \lim_{n \to \infty} \int_{\{|y|<n\}-B} \{F_N(y-x) - F_N(y)\} dy \quad (V = v(\cdot, 0)).$$

Write this for $x \in B$ so that

$$\lim_{n \to \infty} \int_{\{|y|<n\}-B} \{F_N(y-x) - F_N(y)\} dy = 0. \tag{34}$$

Coming back now to (29), we write it for $t = 0$ and $V_n = v_n(\cdot, 0)$. In the resulting problem, we perform the change of variables $x \to x/\delta$ for a parameter $\delta > 1$ and we set

$$V_{n,\delta} = \delta^2 V_n(x/\delta), \quad T_{n,\delta} = \delta T_n, \quad B_{n,\delta} = \delta B_n(0).$$

Then,

$$V_{n,\delta}(x) = \int_{\{|y|<n\delta\}-\delta B_n(0)} \{F_N(y-x) - F_N(y)\} dy = 0.$$

Computing this for $x \in B(0)$ and letting $n \to \infty$ gives

$$\lim_{n\to\infty} \int_{\{|y|<\delta n\}-\delta B} \{F_N(y-x) - F_N(y)\}dy = 0.$$

Thus, taking the difference with (34),

$$\int_{\delta B - B} \{F_N(y-x) - F_N(y)\}dy = 0, \quad \forall \delta > 1. \tag{35}$$

This is a geometric condition on the initial bubble $B = B(0)$ for a smooth extinction to occur in infinite time. It says that the Newtonian potential generated by a uniform distribution of masses on the "shell" $\delta B - B$ is constant inside the cavity B. Such domains are called *homeoids* (see Figure 5). A theorem by Newton

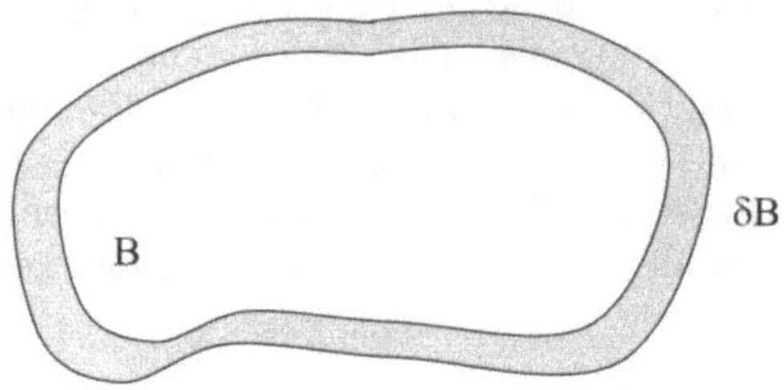

FIGURE 5. Constructing Homeoids

asserts that ellipsoids are homeoids ([17]).

The converse is also true, that is homeoids are ellipsoids. The proof for $N = 2$ is due to Kellogg [17] and in dimension $N = 3$ the proof is given in Dive [10]. The proof is calculational, in the sense that makes extensive use of potential representations. A different approach is taken in [4], where a proof is developed by exploiting a structure result for solutions to variational inequalities. It is shown in [9] by an indirect argument that the statement is indeed true in every dimension N. Thus

Theorem 6.2. *Every N-dimensional homeoid is an ellipsoid and viceversa.*

Returning to the Hele–Shaw problem of expanding bubbles, we can conclude that if the initial bubble is a homeoid, then the fluid is removed smoothly in infinite time. Viceversa, if the extinction occurs smoothly in infinite time, then the process must have started from a homeoid. Equivalently,

Theorem 6.3. *A classical solution of the Hele–Shaw problem in $\mathbb{R}^N$ of displacing a fluid by an advancing bubble and with empty final phase exists if and only if the initial bubble is an ellipsoid.*

7. Advancing Fingers and the Selection Problem

We look for a travelling wave solution of the form $u(x - Ut)$ for the problem of advancing fingers in its classical formulation (12)–(13). Relabelling by x the new variables, (12) reduces to finding a domain $\Omega \subset S$ whose boundary is a curve Γ in the form of a finger and a function u defined in $\overline{\Omega}$ and such that

$$\begin{cases} \Delta u = 0 \text{ in } \Omega, \text{ and } u = 0 \text{ on } \Gamma; \\ \nabla u \cdot \mathbf{N} = -U \cos\theta \text{ on } \Gamma; \\ u_y(x, \pm 1) = 0 \quad \text{for all } x \in I\!R; \\ \lim_{x \to \infty} \nabla u = (\lambda U, 0) \\ \lim_{x \to -\infty} \nabla u = (0,0). \end{cases} \quad \text{for all } y \in (-1,1) \tag{36}$$

Experiments consist in injecting water into an oil-filled Hele–Shaw cell from one of its "sides", say at $x = -\infty$. It is observed that after an initial formation of several incipient and invading fingers, the penetrating fluid reaches a steady state and takes the form of a single, symmetric finger as in Figure 3. Thus the formulation (36) can also be regarded as a steady state configuration of (12). The weak formulation of (36) is derived from (18) by either seeking traveling wave solutions of (18) or by imposing (19) and seeking steady state solutions. In either case, its weak formulation takes the form,

$$\begin{cases} u \in W^{1,2}(S_n) \cap W^{2,2}([u > 0] \cap S_n); \\ U \dfrac{\partial}{\partial x} H(u) - \Delta u \ni 0 \quad \text{in } \mathcal{D}'(S); \\ u_y(x, \pm 1) = 0 \quad \text{for all } x \in I\!R; \\ \lim_{x \to \infty} \nabla u = (\lambda U, 0) \\ \lim_{x \to -\infty} \nabla u = (0,0). \end{cases} \quad \text{for all } y \in (-1,1). \tag{37}$$

Assuming that the finger is symmetric, Saffman and Taylor [29] produced an implicit family of solutions of (36) parameterized by $\lambda \in (0,1)$, where 2λ is the asymptotic width of the finger as $x \to -\infty$. The equation of the curve limiting the finger can be computed explicitly as

$$x(y) = \frac{1-\lambda}{\pi} \ln \left\{ \frac{1}{2} \left(1 + \cos \frac{\pi y}{\lambda} \right) \right\}. \tag{38}$$

More generally, one might prescribe a generic interval of width 2λ, not necessarily symmetric about $y = 0$, with horizontal bisecting line $y = y_o \neq 0$, say for example,

$$(\alpha, \beta) \subset (-1,1), \quad (\beta - \alpha) = 2\lambda; \quad \alpha = y_o - \lambda, \quad \beta = y_o + \lambda, \tag{39}$$

and solve (36) by requiring the asymptotic cross section of the finger to be the interval (α, β). If $\beta \neq -\alpha$ the finger would not be symmetric and its nose would be located at some $(x_*, y_*) \neq (0,0)$ (see Figure 4). Using λ and y_o as parameters,

Saffman and Taylor [29] computed explicitly such profiles as

$$x(y) = \frac{1-\lambda}{\pi} \ln\left\{\frac{1}{2}\left(1+\cos\frac{\pi(y-y_o)}{\lambda}\right)\right\} + \frac{y_o}{\pi} \ln\left\{\frac{1+\sin\dfrac{\pi}{2}\dfrac{y-y_o}{\lambda}}{1-\sin\dfrac{\pi}{2}\dfrac{y-y_o}{\lambda}}\right\}. \tag{40}$$

In this formula, the origin of the x-axis is taken in such a way that $(0, y_o)$ is on the finger, even though not necessarily coinciding with its nose. For $y_o = 0$, this reduces to (38). Saffman and Taylor computed solutions for non-symmetric fingers as the limiting case of a deformed and elongated bubble moving in the Hele–Shaw cell. However, asymmetric fingers are not observed in experiments. The fingers occurring in experiments are symmetric and, for a wide range of the asymptotic speed U, their observed asymptotic width is $\frac{1}{2}$ of the width of the channel. In our setting $\lambda = \frac{1}{2}$ (see [29, 30]). Thus it appears that out of the families (38) and (40), Nature selects the symmetric profiles with the parameter $\lambda = \frac{1}{2}$.

The ill-posedness of this problem is of different nature than for receding blobs (16) and expanding bubbles (17). In the context of advancing fingers, families of solutions can be found in an explicit form but one does not know which one is the "good" solution. The physical principle that would favor the solution corresponding to $\lambda = \frac{1}{2}$ is not yet fully understood.

The original approach of Saffman and Taylor was based on conformal mappings and separations of variables in the conformally transformed problem. Most of the investigations on this problems have followed the same pattern or are of a numerical nature. The wide range of analytical and numerical attempts to the problem ([1,4,10,13,14,17,19]) confirms that the symmetric solution $\lambda = \frac{1}{2}$ plays some special role. Nevertheless, to our knowledge a clear and simple principle explaining the ill-posedness of the problem still eludes the efforts of the investigators.

A different approach is adopted in [11], where attention is drawn to a selection problem posed in a semi-infinite domain, essentially obtained assuming that the displaced fluid has a second free boundary on the right, open to the atmosphere. It is shown in [11] that the only admissible translating solution to this problem corresponds to the $\lambda = 1/2$ Saffman–Taylor curve. The author argues that the case of the right bounded strip is the one physically relevant, and that an infinite strip does not represent the experimental setting of a very long body of fluid.

We have approached the problem from a more theoretical point of view along the lines of [8, 9]. While far from having solved the problem, we have identified elements of the asymptotic behavior of the solution as $x \to \infty$ which seem to have physical significance and which are maximized for $\lambda = \frac{1}{2}$.

8. Transforming the Problem into a Variational Inequality

To investigate how the asymptotic shape of the finger as $x \to -\infty$ affects the solution, prescribe any measurable set $E \subset (-1, 1)$ with $|E| = 2\lambda$ for some $\lambda \in (0, 1)$ and attempt to solve (37), by requiring the domain occupied by the water to have asymptotic x-section E as $x \to -\infty$. We also ask the set occupied by the less viscous fluid to be *finger-like*, i.e., for all $\eta \in (-1,1)$ the line $y = \eta$ intersects the boundary of the set $[u > 0]$ in at most one point. Any collection of fingers is a finger-like domain. Nonetheless, the notion is more general as it is independent of the geometry and the regularity of the free boundary $[u = 0]$. In particular, it may include any disordered penetration of water into oil, provided the penetration occurs in the direction of the x-axis. Assuming a smooth solution exists, set

$$v(x,y) = \frac{1}{U}\int_{-\infty}^{x} u(s,y)ds, \qquad \text{and} \qquad A = (-1,1) - E. \tag{41}$$

The finger-like structure of the free boundary implies that v is positive or zero whenever u is positive or zero. Thus, by formally integrating the equation in (37) over $(-\infty, x)$ we obtain the problem examined in ([3])

$$\begin{cases} v \in W^{2,2}(S_n);\ v \geq 0 \ \text{ in } \ S_n \ \text{ for all } n \in I\!N; \\ \Delta v \in H(v) - \chi_{A \times I\!R} \ \text{ in } \ \mathcal{D}'(S); \\ |E| = 2\lambda > 0 \ \text{ with } \ 0 < \lambda < 1; \\ v_y(x, \pm 1) = 0 \ \text{ for all } \ x \in I\!R; \\ \lim\limits_{x \to -\infty} \sup\limits_{|y| < 1} v(x,y) = 0; \\ \sup\limits_{|y|<1} v(x,y) \leq Cx^2 \ \text{ for } \ x \gg 1, \end{cases} \tag{42}$$

for some positive constant C. The problem has a formal analogy with (32), i.e. a bubble expanding smoothly into oil, for all times with no rupture. In particular the last of (42) exhibits the same upper bound on the growth of the solution as the last of (32). The basic difference is the form of ξ. In (32) it represents the characteristic function of the terminal set occupied by the fluid. In (42) it is the characteristic function of the complement of the smallest horizontal strip containing the finger. While the set $A \times I\!R$ is occupied by the fluid, it is not the whole fluid set. All the requirements in (42) are essential. In [3] counterexamples are given showing that if any one of them is removed (for example positivity, or $\lambda = 0$), the resulting problem while admitting a solution does not translate the physical information contained in (36)–(37). In particular the requirement $|E| > 0$ allows one to generate solutions with a "finger".

Theorem 8.1. *There exists a solution to* (42). *The solution is unique up to horizontal translations.*

Construction of solutions hinges upon a priori estimates and upon identifying the finger through the abscissa x_* of its "nose". This is defined as

$$x_* = \inf \left\{ x \in I\!R \mid w(x,y) > 0, \ \text{ for all } \ y \in (-1,1) \right\}.$$

It turns out that the requirement $|E| > 0$ is essential to identify x_*. It is shown in [3] that no solution exists if $|E| = 0$. Thus the Hele–Shaw problem of advancing fingers has a solution if and only if $|E| > 0$. However, the physical problem is invariant by an horizontal shift of the x-axis. Equivalently, it is independent of the abscissa of the "nose" of the finger, provided one can be found. Thus, uniqueness up-to-translations is the only correct notion of uniqueness.

9. Asymptotic Behavior and the Area Function

Uniqueness requires a precise description of the asymptotic behavior of the solutions of (42) as $x \to \infty$, in terms of the notion of area function, which we describe next. Choose a coordinate system so that the nose of the finger has coordinates (x_*, y_*). The notion of area function is best understood in the context of the Saffmann–Taylor solutions (40) originating from an asymptotic cross section $E = (\alpha, \beta)$ as in (39). The area function as a function of x is the measure of the shaded region in Figure 6 when computed at the abscissa x_* of the nose of the finger.

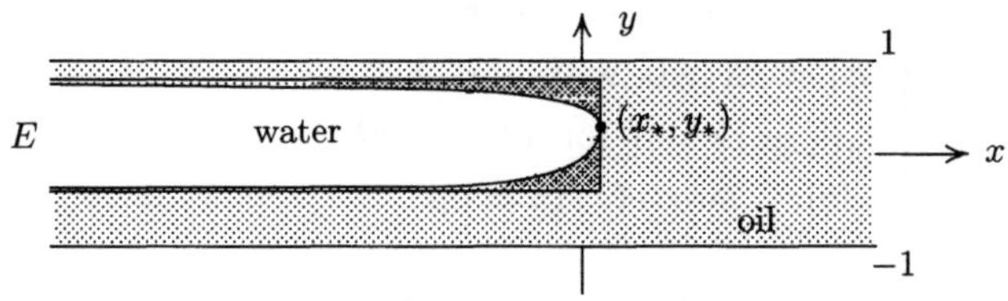

FIGURE 6. The Area Function

For a general asymptotic cross section $E \subset (-1, 1)$, the area function is defined by

$$\begin{cases} a(x) \stackrel{\text{def}}{=} \displaystyle\int_E \int_{-\infty}^{x} \chi_{[w>0]}(s, y)dsdy & \text{if } x \leq x_*; \\ a(x_*) + (x - x_*)|E|, & \text{if } x > x_*. \end{cases} \tag{43}$$

Such a function depends on the geometrical configuration of the finger only, and not on the values of the solution w. It can be determined from the knowledge of w from $-\infty$ up to the tip of the finger. The values of w for $x > x_*$ are immaterial for the determination of the area function. A physical interpretation of the area function can be drawn from the formula

$$a(x) = \int_{-1}^{1} w_x(x, y)dy = \int_{-1}^{1} u(x, y)dy \quad \text{for all } x \in I\!R,$$

which is established in Corollary 7.4 of [3]. Interpreting $u(x, y)dy$ as an elemental force acting in the x-direction, this formula suggests that $a(x)$ is the thrust exerted by the fluid across the x-section in the direction of the advancing finger. Set also

$$b(x) = \int_{-\infty}^{x} a(s)ds; \quad g(y) = g(-1) - \frac{\lambda}{2}(1 + y)^2 + \int_{-1}^{y} (y - \eta)\chi_E(\eta)d\eta, \tag{44}$$

where $g(-1)$ is a constant chosen in such a way that $g(\cdot)$ has zero average over $(-1, 1)$.

Theorem 9.1. *Let w be a solution of* (42). *Then, for $x \gg 1$,*

$$w(x,y) = \frac{1}{2}\{\lambda(x-x_*)^2 + a(x_*)(x-x_*) + b(x_*)\} + g(y) + O(1). \tag{45}$$

Analogously, let $u = w_x$ be a solution of (37). *Then for $x \gg 1$,*

$$u(x,y) = \lambda(x-x_*) + \frac{1}{2}a(x_*) + O(1). \tag{46}$$

While negligible with respect to $\lambda(x-x_*)$, the term containing $a(x_*)$ seems to be essential in determining a precise asymptotic expansion of w or $u = w_x$ as $x \to \infty$. The value $a(x_*)$ is the thrust exerted by the fluid across the vertical section of the Hele–Shaw channel exactly at the nose of the finger.

The asymptotic expansion (46) enables one to compute $a(x_*)$ in the case of the Saffman–Taylor solutions (38) and (40). For the symmetric fingers (38), one computes

$$a(0)_{\text{sym}} = \frac{4U}{\pi}\lambda(1-\lambda)\ln 2 \quad (\text{ since } x_* = 0). \tag{47}$$

Such an area is maximized for $\lambda = \frac{1}{2}$. For non-symmetric fingers, the computation is better carried out in terms of the parameter y_o introduced in (39), and it gives

$$a(x_*)_{\text{non-sym}} = a(0)_{\text{sym}}\{1 + f(\lambda, y_o)\}, \tag{48}$$

where

$$(\ln 4) f(\lambda, y_o) = \ln\left\{\left(1 + \frac{y_o}{1-\lambda}\right)^{1+\frac{y_o}{1-\lambda}}\left(1 - \frac{y_o}{1-\lambda}\right)^{1-\frac{y_o}{1-\lambda}}\right\}. \tag{48'}$$

For a fixed value of $\lambda \in (0,1)$, i.e., for a fixed asymptotic width of the finger, regardless of its geometry, the function $y_o \to f(\lambda, y_o)$ is even, it vanishes at $y_o = 0$ and is positive otherwise. Therefore it has a minimum at $y_o = 0$, i.e., for $\alpha = -\beta$. This implies that the corresponding finger is symmetric.

Thus, among all the non-symmetric fingers of given asymptotic width, the symmetric ones are those with the minimal area. On the other hand, among all the symmetric fingers, the one of largest area occurs for $\lambda = \frac{1}{2}$. The notion of area function for symmetric fingers and its extremal property for $\lambda = \frac{1}{2}$ had already been noticed by Saffman and Taylor in [28]. However, no connection had been drawn with the asymptotic behaviour of solutions nor with its physical interpretation as the thrust of the fluid at the nose of the finger. [6]

10. Motion by Mean Curvature as a Selection Criterion

Let $\{\Gamma_t\}$ be a family of planar, regular curves parameterized by the time variable, say for example $X = X(y,t)$ for $y \in (y_*, y_* + \lambda)$, for some $y_* \in I\!R$ and $\lambda > 0$. The

[6] We would like to thank B. Matkowsky for pointing out reference [28] to us.

family $\{\Gamma_t\}$ moves by mean curvature if there exists a constant k such that

$$X_t = k\frac{X_{yy}}{1+X_y^2}, \qquad \text{within its domain of definition.} \tag{49}$$

If the family is of the travelling wave type, i.e., $\{X = x(y) + Ut\}$ for some given constant U, then (49) takes the form

$$\frac{d}{dy}\arctan x' = \gamma, \qquad \text{where} \qquad \gamma = \frac{U}{k}.$$

By integration,

$$\begin{aligned} x(y) &= \int_{y_*}^{y} \tan\{\gamma(\eta - y_*) + \zeta_*)d\eta + x_* \\ &= -\frac{1}{\gamma}\ln\left|\frac{\cos\{\gamma(y-y_*)+\zeta_*)}{\cos\zeta_*}\right| + x_* \end{aligned} \tag{50}$$

where $x_* = x(y_*)$ and $\zeta_* = \arctan\{x'(y_*)\}$. If any of the curves in (50) has to represent a finger for (36)–(37) with nose at (x_*, y_*), we must have $x'(y_*) = 0$ and $x'(y) \to -\infty$ as $y \to y_* + \lambda$. This implies that $\zeta_* = 0$ and $\gamma\lambda = -\pi/2$, i.e. $k = -2U\lambda/\pi$. Thus, a curve of the sort of (50) moving by mean curvature would represent a finger for (36)–(37) only if it has the form,

$$x(y) = \frac{2\lambda}{\pi}\ln\left|\cos\frac{\pi}{2\lambda}(y-y_*)\right| + x_*.$$

In particular $x(y)$ is symmetric about $y = y_*$. This implies that none of the non-symmetric Saffman–Taylor solution (40) represents a motion by mean curvature.

Up to a suitable change of coordinates we take $(x_*, y_*) = (0,0)$. This gives

$$x(y) = \frac{2\,\lambda}{\pi}\ln\cos\frac{\pi}{2\,\lambda}\,y \qquad\qquad y \in (-\lambda, \lambda).$$

Among the symmetric curves in (38), only the one corresponding to $\lambda = \frac{1}{2}$ moves by mean curvature, since

$$x(y) = \frac{1-\lambda}{\pi}\ln\left\{\frac{1}{2}\left(1+\cos\frac{\pi\, y}{\lambda}\right)\right\}\Big|_{\lambda=\frac{1}{2}} = \frac{2\lambda}{\pi}\ln\cos\frac{\pi}{2\lambda}y.$$

From (49) it follows that the family of curves $\{x = x(y) + Ut\}$ moves by mean curvature if

$$R\cos\theta = k/U, \tag{51}$$

where R is the radius of curvature to the curve Γ_o and θ is the angle between the normal to Γ_o and the positive direction of the x-axis. Requiring a curve to satisfy (51) is equivalent to imposing that it moves by mean curvature.

It was observed by Pitts [25] as a reading of experimental data that condition (51) forces a curve to be a finger in the setting described in § 3.3. Pitts, however, does not seem to have connected it to a motion by mean curvature.

It is observed that for small values of U, curves satisfying (51) fit the experimental curves better than those of Saffmann–Taylor ([25] and [22] pages 462 and 465). In fact, for small U, the Saffman–Taylor solutions do not agree with the experimental data ([29] page 326). This seems to suggest a selection mechanism of solutions based on motion by mean curvature of the free boundary.

10.1. A Further Remark on the Motion by Mean Curvature

While supported by the experimental data of Pitts, such a mechanism needs to be mathematically better understood. Indeed all the symmetric Saffman–Taylor curves (38) move by mean curvature in the following sense. Consider the family of travelling wave curves

$$X_\lambda(y,t) = \frac{\lambda}{1-\lambda}\,x(y) + Ut \qquad \lambda \in (0,1)$$

where $x = x(y)$ is the symmetric Saffman–Taylor solution (38). A straightforward computation that uses the explicit form of (38) gives

$$X_{\lambda t} = k\,\frac{X_{\lambda yy}}{1+X_{\lambda y}^2} \qquad \Longleftrightarrow \qquad \frac{U}{k} = \frac{\dfrac{\lambda}{1-\lambda}x_{yy}}{1+\left(\dfrac{\lambda}{1-\lambda}\right)^2 x_y^2} = -\frac{\pi}{2\lambda}.$$

Note that, for any given $\lambda \in (0,1)$, this is precisely the condition for $X_\lambda(y,t)$ to move by mean curvature.

References

[1] A. P. Aldushin, B. J. Matkowsky, *Instabilities, Fingering and the Saffman–Taylor Problem in Filtration Combustion,* Comb. Sci. & Tech., **133** (1998), 293–344.

[2] A. P. Aldushin, B. J. Matkowsky, *Extremum principles for selection in the Saffman-Taylor finger and TaylorSaffman bubble problems,* Physics of Fluids, **11** n. 6 (1999), 1287–1296.

[3] D. Andreucci, G. Caruso and E. DiBenedetto, *Fingering in a Hele–Shaw Cell as an Obstacle-like Problem,* Indiana. Univ. Math. Journal, **49** n. 4 (2000), 1287–1321.

[4] D. Andreucci, M. A. Herrero and J. J. L. Velázquez, *The Classical One-phase Stefan Problem: a Catalogue of Interface Behaviours,* Surveys on Mathematics for Industry, **9** (2001), 247–337.

[5] C. Baiocchi, *Su un Problema a Frontiera Libera Connesso a Questioni di Idraulica,* Ann. Mat. Pura Appl., **92** (1972), 107–127.

[6] G. Caruso, *Il Problema di Frontiera Libera in una Cella di Hele–Shaw,* Tesi di Laurea, Dip. di Mat., Univ. di Roma Tor Vergata (1998).

[7] R. Combescot, T. Dombre, V. Hakim. Y. Pomeau and A. Pumir, *Shape Selection of Saffman–Taylor Fingers,* Phys. Rev. Lett., **56** n. 19 (1986), 2036–2039.

[8] E. DiBenedetto and A. Friedman, *The Ill-posed Hele-Shaw Model and the Stefan Problem for supercooled water,* Trans. AMS, **282** n. 1 (1984), 183–204.

[9] E. DiBenedetto and A. Friedman, *Bubble Growth in Porous Media,* Indiana. Univ. Math. Journal, **35** n. 3 (1986), 573–606.

[10] P. Dive, *Attraction des Ellipsoides Homogènes et Reciproque d'un Théorème de Newton,* Bull. Soc. Math. France, **59** (1931), 128–140.

[11] M. J. Feigenbaum, *Pattern Selection: Determined by Symmetry and Modifiable by Long-Range Effects,* preprint.

[12] A. Friedman, *Variational Principles and Free-Boundary Problems,* John Wiley, New York (1982).

[13] J. Hadamard, *Sur les Problèmes aux Derivées Particles et Leur Signification Physique,* Bull. Univ. Princeton, **13** (1902), 49–52.

[14] H. J. S. Hele Shaw, *On the Motion of a Viscous Fluid between two Parallel Plates,* Nature, **58** (1898), 34–36.

[15] D.C. Hong and J. S. Langer, *Analytic Theory of the Selection Mechanism in the Saffman–Taylor Problem,* Phys. Rev. Lett., **56**(19) (1986), 2032–2035.

[16] S. D. Howison, *Bubble Growth in Porous Media and Hele–Shaw Cells,* Proc. Roy. Soc. Edinburgh Sect. A, **102** n. 1-2 (1986), 141–148.

[17] O. D. Kellogg, *Foundations of Potential Theory,* Dover, New York (1953).

[18] O. A. Ladyzhenskaya, N. N. Ural'tseva, *Linear and Quasilinear Elliptic Equations,* Academic Press, New York (1968).

[19] H. Lamb, *Hydrodynamics,* Oxford Univ. Press (1937).

[20] R. Lattes and J. J. Lions, *The Method of Quasi-reversibility and Applications to Partial Differential Equations,* American Elsevier (1969).

[21] M. M. Lavrentiev, *Some Improperly Posed Problems of Mathematical Physics,* Springer (1967).

[22] J. W. McLean and P.G. Saffman, *The Effect of Surface Tension on the Shape of Fingers in a Hele–Shaw Cell,* J. Fluid Mech., **102** (1981), 455–469.

[23] J. Ockendon, S. Howison, A. Lacey and A. Movchan, *Applied Partial Differential Equations,* Oxford University Press, (1999).

[24] L. Payne, *Improperly Posed problems in Partial Differential Equations,* Regional Conference Series in Appl. Maths., **22** (1975).

[25] B. Pitts, *Penetration of Fluid into a Hele–Shaw Cell: the Saffman–Taylor Experiment,* J. Fluid Mech., **97** n. 1 (1980), 53–64.

[26] S. Richardson, *Hele–Shaw Flows With a Free Boundary Produced by the Injection of Fluid Into a Narrow Channel,* J. Fluid Mech., **66** (1972), 609–618.

[27] S. Richardson, *Some Hele–Shaw Flows With Time Dependent Free Boundaries,* J. Fluid Mech., **102** (1981), 263–278.

[28] P. G. Saffman and G. I. Taylor, *Cavity Flows of Viscous Liquids in Narrow Spaces,* Proc. of the Second Symposium on Naval Hydrodynamics, 1958, Washington, D.C.

[29] P. G. Saffman and G. I. Taylor, *The Penetration of a Fluid Into a Porous Medium or Hele–Shaw Cell Containing a More Viscous Fluid,* Proc. Roy. Soc. Edinburgh Sect. A, **245** (1958), 312–329.

[30] P. G. Saffman, G. I. Taylor, *A Note on the Motion of Bubbles in a Hele–Shaw Cell and Porous Medium,* Quart. J. Mech. Appl. Math., **12** (1959), 265–279.

[31] B. I. Shraiman, *Velocity Selection and the Saffman–Taylor Problem,* Phys. Rev. Lett., **56** n. 19 (1986), 2028–2031.

[32] R. E. Showalter, *The Final Value Problem for Evolution Equations,* Journal of Math. Anal. and Appl., **47** (1974), 563–572.

[33] R. E. Showalter, *Quasi Reversibility of First and Second Order Parabolic Evolution Equations. Improperly Posed Boundary Value Problems,* (Conf. Univ. New Mexico Albuquerque N. M. 1974), Res. Notes in Math., **1**, Pitman, London (1975), 76–84.

[34] G. Talenti, *Sui Problemi Mal Posti,* Boll. U.M.I. (5), **19**, A (1978), 1–29.

[35] A. Tikhonov and V. Arsenin, *Méthodes de Resolution de Problèmes Mal Posés,* Ed. Mir, Moscow (1976).

[36] S. Tanveer, *Analytic Theory for the Selection of a Symmetric Saffman–Taylor Finger in a Hele–Shaw Cell,* Phys. Fluids, **30** n. 6 (1987), 1589–1603.

[37] J. M. Vanden-Broeck, *Fingers in a Hele–Shaw Cell with Surface Tension,* Phys. Fluids, **26** (1983), 2033–2034.

[38] X. Xie and S. Tanveer, *Rigorous results in steady finger selection in viscous fingering,* Archive Rat. Mech. Anal., **166** (2003), 219–286.

[39] Jian-Jun Xu, *Interfacial Wave Theory of Pattern Formation: Selection of Dendritic Growth and Viscous Fingering in Hele–Shaw Flow,* Springer-Verlag, Berlin (1998).

(D. Andreucci) Dipartimento di Metodi e Modelli Matematici,
Università di Roma La Sapienza,
Via A. Scarpa 16,
00161 Rome, Italy
E-mail address: `andreucci@dmmm.uniroma1.it`

(G. Caruso) ITC–CNR,
Via C. Marx 43,
Rome, Italy
E-mail address: `g.caruso@ing.uniroma2.it`

(E. DiBenedetto) Dept. of Mathematics,
Stevenson Center, Vanderbilt University,
Nashville, TN 37240, U.S.A.
Partially supported by DMS 0100660
E-mail address: `em.diben@Vanderbilt.edu`

International Series of Numerical Mathematics, Vol. 147, 53–63

Finite Element Methods for Surface Diffusion

Eberhard Bänsch, Pedro Morin, and Ricardo H. Nochetto

Abstract. Surface diffusion is a (4th order highly nonlinear) geometric driven motion of a surface with normal velocity proportional to the surface Laplacian of the mean curvature. We present a novel variational formulation for the parametric case, develop a finite element method, and propose a Schur complement approach to solve the resulting linear systems. We also introduce a new graph formulation and state an optimal a priori error estimate. We conclude with several significant simulations, some with pinch-off in finite time.
Keywords: Surface diffusion, fourth-order parabolic problem, finite elements, a priori error estimates, Schur complement, smoothing effect, pinch-off.

AMS subject classification: 35K55, 65M12, 65M15, 65M60, 65Z05.

1. Surface Diffusion and Mixed Formulation

The overall goal of this work is to devise efficient numerical tools for simulating morphological changes in stressed epitaxial films and thereby study their complicated nonlinear dynamics. To model the misfit between the crystalline structure of the substrate and the epitaxial film, the film may be thought of as subject to mechanical stresses. This causes a plastic deformation of the free surface of the film. This morphological instability of the free surface may eventually lead to crack formation and fracture, an issue of paramount importance in Materials Science; see for instance [1, 4, 14] and also the list of references in [2].

The dynamics of the free surface Γ in $\mathbb{R}^d$ is governed by

$$V = -\Delta_S(\kappa + \varepsilon), \tag{1}$$

which is a 4th order highly nonlinear PDE. Hereafter, $d \in \{2, 3\}$, V and κ are the (scalar) normal velocity and mean curvature of Γ, respectively, ∇_S is the surface gradient, $\Delta_S = \operatorname{div}_S \nabla_S$ the Laplace–Beltrami operator and ε is the elastic energy density of the bulk $\Omega(t)$ enclosed by $\Gamma(t)$. We will take $\varepsilon = 0$ throughout this paper.

A number of issues arise, going from existence, well posedness, and regularity to effective algorithms for (1). The chief mathematical and numerical difficulties stem from the 4th order nonlinear operator $\Delta_S \kappa$ and the fact that one cannot work directly with the curvature vector $\vec{\kappa}$ as in [8]. In contrast to [5, 10], we have been able to derive from a suitable *semi-implicit* time discretization a variational finite element formulation for surfaces in $\mathbb{R}^d$ which involves the four unknowns, namely the scalar curvature κ, the curvature vector $\vec{\kappa}$, the normal velocity $\vec{V}$, and the (scalar) normal velocity V, as follows:

$$\vec{\kappa} = \Delta_S \vec{X}, \quad \kappa = \vec{\kappa} \cdot \vec{\nu}, \quad V = -\Delta_S \kappa, \quad \vec{V} = V\vec{\nu}, \tag{2}$$

where $\vec{\nu}$ denotes the unit normal vector to Γ, pointing outward of the bulk enclosed by Γ. In view of this *mixed formulation*, which features only second order operators, our finite element method solely requires continuity of the discrete functions; in particular we do not need C^1 elements to handle curvature but can accommodate any polynomial degree. A Schur complement approach, with a symmetric and positive definite operator, is used to reduce the system (2) to the single unknown V. This allows for an efficient solution technique via preconditioned CG.

This paper is organized as follows. We introduce the time discretization and natural variational formulation in §2, and present its finite element discretization in §3. We discuss the ensuing algebraic problem along with a Schur complement approach to its solution in §4. We present in §5 the graph formulation together with an error estimate. We document the performance of parametric and graph methods in §6 via several simulations which exhibit singularities in finite time.

2. Time Discretization and Variational Formulation

We now consider a *semi-implicit* time discretization of first order with time-step τ, representing the next surface Γ^{n+1} in terms of the current surface Γ^n as follows:

$$\vec{X}^{n+1} = \vec{X}^n + \tau\vec{V}. \tag{3}$$

This time discretization implies that all *geometric* quantities such as $\vec{\nu}$ are evaluated on the current surface Γ^n. Consequently, letting Δ_S denote now the Laplace–Beltrami operator on $\Gamma = \Gamma^n$, the only modification to (2) is as follows:

$$\vec{\kappa} - \tau\Delta_S\vec{V} = \Delta_S\vec{X}^n. \tag{4}$$

To derive a weak formulation to (2) and (4), we simply multiply them by test functions and integrate. If $\mathcal{V}(\Gamma) := H^1(\Gamma)$ and $\mathcal{X}(\Gamma)$ is the subspace of $\mathcal{V}(\Gamma)$ of functions with mean value zero, we thus seek $\vec{V} \in \vec{\mathcal{V}}(\Gamma), \kappa \in \mathcal{V}(\Gamma), \vec{\kappa} \in \vec{\mathcal{X}}(\Gamma)$, and $V \in \mathcal{X}(\Gamma)$ such that

$$\langle \vec{\kappa}, \varphi \rangle + \tau \left\langle \nabla_S \vec{V}, \nabla \varphi \right\rangle = -\left\langle \nabla_S \vec{X}, \nabla_S \varphi \right\rangle \qquad \forall\, \varphi \in \vec{\mathcal{X}}(\Gamma), \tag{5}$$

$$\langle \kappa, \phi \rangle - \langle \vec{\kappa} \cdot \vec{\nu}, \phi \rangle = 0 \qquad \forall\, \phi \in \mathcal{V}(\Gamma), \tag{6}$$

$$\langle V, \phi \rangle - \langle \nabla_S \kappa, \nabla_S \phi \rangle = 0 \qquad \forall\, \phi \in \mathcal{X}(\Gamma), \tag{7}$$

$$\left\langle \vec{V}, \varphi \right\rangle - \langle V, \varphi \cdot \vec{\nu} \rangle = 0 \qquad \forall\, \varphi \in \vec{\mathcal{V}}(\Gamma). \tag{8}$$

Hereafter the symbol $\langle \cdot, \cdot \rangle$ stands for the L^2 scalar product over the current surface $\Gamma = \Gamma^n$, which is described by the vector function $\vec{X} = \vec{X}^n$; we suppress the superscripts n and $n+1$ since no confusion arises. Since $V \in \mathcal{X}(\Gamma)$ has mean value zero, we realize that volume is preserved in the sense that $\int_\Gamma V = 0$.

3. Finite Element Discretization

Let $\mathcal{T}$ be a regular but possibly graded mesh of triangular finite elements over the surface Γ, which is assumed to be polyhedral from now on. Let h denote the local meshsize of $\mathcal{T}$. Let $T \in \mathcal{T}$ be a typical triangle and let $\vec{\nu}_T = (\nu_T^k)_{k=1}^d$ be the unit normal to T pointing outwards. We denote by $\vec{\nu}$ the outward unit normal to Γ, which satisfies $\vec{\nu}|_T = \vec{\nu}_T$ for all $T \in \mathcal{T}$, and is thus discontinuous across ∂T.

Let $\{\phi_i\}_{i=1}^I$ be the set of canonical basis functions of the finite element space $\mathcal{V}_h(\Gamma)$ of piecewise linear functions over Γ; this is a *conforming* approximation of $\mathcal{V}(\Gamma)$. If $\vec{Id} \in \mathbb{R}^{d\times d}$ is the identity matrix, consider the following matrix entries

$$M_{ij} := \langle \phi_i, \phi_j \rangle, \qquad A_{ij} := \langle \nabla_S \phi_i, \nabla_S \phi_j \rangle, \tag{9}$$

$$\vec{M}_{ij} := M_{ij}\vec{Id}, \qquad \vec{N}_{ij} := \left\langle \phi_i, \phi_j \nu^k \right\rangle_{k=1}^d, \tag{10}$$

and corresponding mass and stiffness matrices

$$M := (M_{ij})_{i.j=1}^I, \qquad \vec{M} := (\vec{M}_{ij})_{i.j=1}^I, \qquad \vec{N} := (\vec{N}_{ij})_{i.j=1}^I, \tag{11}$$

$$A := (A_{ij})_{i.j=1}^I, \qquad \vec{A} := (\vec{A}_{ij})_{i.j=1}^I. \tag{12}$$

We point out that $\vec{M}, \vec{A}$ and $\vec{N}$ possess matrix-valued entries and therefore the matrix-vector product is understood in the following sense

$$\vec{M}\vec{\boldsymbol{V}} = \Big(\sum_{j=1}^I \vec{M}_{ij}\vec{V}_j\Big)_{i=1}^I,$$

because each component of $\vec{\boldsymbol{V}} = (\vec{V}_i)_{i=1}^I$, as well as of $\vec{M}\vec{\boldsymbol{V}}$, is itself a vector in $\mathbb{R}^d$.

The vector of nodal values of a finite element function is written in bold face, namely $\boldsymbol{V} = (V_i)_{i=1}^I \in \mathbb{V} := \mathbb{R}^I$ is equivalent to $V = \sum_{i=1}^I V_i\phi_i \in \mathcal{V}_h(\Gamma)$. Let $\mathcal{X}_h(\Gamma)$ be the subspace of $\mathcal{V}_h(\Gamma)$ of functions with mean value zero, and let $\mathbb{X} = \{\boldsymbol{V} \in \mathbb{V} : \ \boldsymbol{V}\cdot M\boldsymbol{1} = 0\}$ be the corresponding subspace of $\boldsymbol{V}$ with $\boldsymbol{1} := (1)_{i=1}^I$, so that

$$V = \sum_{i=1}^I V_i\phi_i \in \mathcal{X}_h(\Gamma) \quad \Leftrightarrow \quad \boldsymbol{V} = (V_i)_{i=1}^I \in \mathbb{X}. \tag{13}$$

Upon expanding the functions V, κ, $\vec{V}$, $\vec{\kappa}$ in terms of the basis functions $\{\phi_i\}_{i=1}^I$ and testing against the latter ones, a discretization of system (5)–(8) can be written in matrix form as follows: *find* $\vec{\boldsymbol{V}} \in \vec{\mathbb{V}}, \vec{\kappa} \in \vec{\mathbb{V}}, \vec{\boldsymbol{K}} \in \vec{\mathbb{X}}, \boldsymbol{V} \in \mathbb{X}$ *such that*

$$\begin{bmatrix} \tau\vec{A} & 0 & \vec{M} & 0 \\ 0 & -A & 0 & M \\ \vec{M} & 0 & 0 & -\vec{N} \\ 0 & M & -\vec{N}^T & 0 \end{bmatrix} \begin{bmatrix} \vec{\boldsymbol{V}} \\ \boldsymbol{K} \\ \vec{\boldsymbol{K}} \\ \boldsymbol{V} \end{bmatrix} = \begin{bmatrix} -\vec{A}\vec{\boldsymbol{X}} \\ \boldsymbol{0} \\ \vec{\boldsymbol{0}} \\ \boldsymbol{0} \end{bmatrix}. \tag{14}$$

We discuss the solvability of (14) and propose an algorithm for its solution in §4.

4. Schur Complement Approach

Consider the following vector equation with a (possibly singular) square block A:

$$\begin{bmatrix} A & B \\ C & D \end{bmatrix} \begin{bmatrix} \boldsymbol{U} \\ \boldsymbol{Q} \end{bmatrix} = \begin{bmatrix} \boldsymbol{F} \\ \boldsymbol{G} \end{bmatrix}.$$

Let A be symmetric with (nontrivial) kernel $\ker(A)$. Then the range $\mathbb{Y}$ of A is the orthogonal complement of $\ker(A)$. Let $S : \mathbb{Y} \to \mathbb{Y}$ be the *right* inverse of A: $AS = Id$ on $\mathbb{Y}$. If P denotes the orthogonal projection onto $\ker(A)$, we have

$$SA\boldsymbol{V} = \boldsymbol{V} - P\boldsymbol{V} = (Id - P)\boldsymbol{V} \quad \forall\, \boldsymbol{V} \in \mathbb{R}^I. \tag{15}$$

The *Schur complement* equation for $\boldsymbol{Q}$ then reads

$$(-CSB + D)\boldsymbol{Q} + CP\boldsymbol{U} = \boldsymbol{G} - CS\boldsymbol{F}. \tag{16}$$

The solvability of this system depends on the structure of the two terms on the left hand-side of (16). We intend to apply this splitting to (14), which involves dealing with the block containing $\vec{A}$ and A on the diagonal.

Since the kernel of A in (12) is $\mathbb{Z} = \operatorname{span}\{\boldsymbol{1}\}$, with $\boldsymbol{1} = (1)_{i=1}^I$, the range of A is $\mathbb{Y} = \mathbb{Z}^\perp$. Then the spaces $\mathbb{X}$, defined in (13), and $\mathbb{Y}$ are related by

$$\boldsymbol{V} \in \mathbb{X} \qquad \Leftrightarrow \qquad M\boldsymbol{V} \in \mathbb{Y}. \tag{17}$$

Let $S : \mathbb{Y} \to \mathbb{Y}$ be the right inverse of A and let $P : \mathbb{V} \to \mathbb{Z}$ be the orthogonal projection into $\mathbb{Z}$, thereby satisfying (15) with $P = \frac{\boldsymbol{1}\otimes\boldsymbol{1}}{\boldsymbol{1}\cdot\boldsymbol{1}}$.

Applying (16) to (14) with vectors $\boldsymbol{U} = [\vec{\boldsymbol{V}}, \boldsymbol{K}]^T$ and $\boldsymbol{Q} = [\vec{\boldsymbol{K}}, \boldsymbol{V}]^T$, we obtain the symmetric system

$$\begin{bmatrix} \frac{1}{\tau}\vec{M}\vec{S}\vec{M} & \vec{N} \\ \vec{N}^T & -MSM \end{bmatrix} \begin{bmatrix} \vec{\boldsymbol{K}} \\ \boldsymbol{V} \end{bmatrix} = \begin{bmatrix} -\frac{1}{\tau}\vec{M}\vec{S}\vec{A}\vec{\boldsymbol{X}} + \vec{M}\vec{P}\vec{\boldsymbol{V}} \\ MP\boldsymbol{K} \end{bmatrix}. \tag{18}$$

We observe that $\vec{S}\vec{A}\vec{\boldsymbol{X}}$ makes sense because $\vec{A}\vec{\boldsymbol{X}} \in \vec{\mathbb{Y}}$; this could be viewed as a compatibility condition. To investigate the solvability of (18), we note that both components of $\boldsymbol{Q}$ satisfy $\vec{\boldsymbol{K}} \in \vec{\mathbb{X}}$ and $\boldsymbol{V} \in \mathbb{X}$ or, in view of (17), $\vec{M}\vec{\boldsymbol{K}} \in \vec{\mathbb{Y}}$, and $M\boldsymbol{V} \in \mathbb{Y}$. Since the upper left block of (18), $\vec{M}\vec{S}\vec{M} : \vec{\mathbb{X}} \to \vec{M}\vec{\mathbb{Y}}$, is nonsingular with inverse $\vec{M}^{-1}\vec{A}\vec{M}^{-1}$, we can apply (16) again to get

$$\left(\tau\vec{N}^T\vec{M}^{-1}\vec{A}\vec{M}^{-1}\vec{N} + MSM\right)\boldsymbol{V} + MP\boldsymbol{K} = -\vec{N}^T\vec{M}^{-1}\vec{A}\vec{\boldsymbol{X}}. \tag{19}$$

Let $\Pi := Id - \frac{M\mathbf{1}\otimes M\mathbf{1}}{M\mathbf{1}\cdot M\mathbf{1}}$ be the orthogonal projection onto $\mathbb{X}$. Since $MP\boldsymbol{K} \in \operatorname{span}\{M\mathbf{1}\} = \mathbb{X}^\perp$, applying Π to (19) yields the final form of the *Schur complement*:

$$\Pi\Big(\tau\vec{N}^T\vec{M}^{-1}\vec{A}\vec{M}^{-1}\vec{N} + MSM\Big)\Pi\boldsymbol{V} = -\Pi\vec{N}^T\vec{M}^{-1}\vec{A}\vec{\boldsymbol{X}}, \tag{20}$$

because $\Pi\boldsymbol{V} = \boldsymbol{V}$. The ensuing matrix of this *reduced* system is nonsingular because $\Pi MSM\Pi : \mathbb{X} \to \mathbb{X}$ is symmetric and positive definite, and the first term in (20) is symmetric and positive semi-definite.

The method actually implemented consists of solving first for $\boldsymbol{V}$ using (20), next solving $\vec{M}\vec{\boldsymbol{V}} = \vec{N}\boldsymbol{V}$ for $\vec{\boldsymbol{V}}$ and finally updating $\vec{\boldsymbol{X}}$ via $\vec{\boldsymbol{X}} + \tau\vec{\boldsymbol{V}}$.

5. Graph Formulation and Error Analysis

Let us now discuss the case of Γ being a graph. To this end, let $\Omega \subseteq I\!R^{d-1}$, $d \in \{2,3\}$, and $\Gamma(t) := \{(x, u(t,x)) \mid x \in \Omega\} \subseteq I\!R^d$, with $t \in [0,T]$ for some $T > 0$. If $Q(u) := \sqrt{1+|\nabla u|^2}$ denotes the elementary surface area, then we have the following formulas for the geometric quantities ν, κ and V:

$$\nu = \frac{1}{Q(u)}(-\nabla u, 1)^T, \quad \kappa = \nabla\cdot\Big(\frac{\nabla u}{Q(u)}\Big), \quad V = \frac{\partial_t u}{Q(u)}, \tag{21}$$

with $\nabla = (\partial_{x_1}, \dots, \partial_{x_{d-1}})^T$. According to (21), (1) can be written as the system

$$\frac{\partial_t u}{Q(u)} = -\Delta_S\kappa, \qquad \kappa = \nabla\cdot\Big(\frac{\nabla u}{Q}\Big). \tag{22}$$

This system has to be supplied with suitable initial and boundary conditions. We restrict to periodic boundary conditions for ease of presentation, but Neumann and Dirichlet boundary conditions can be handled as well, see [2]. In this vein, let Ω be a parallelogram and $\mathcal{X}$ be the subspace of $H^1(\Omega)$ with periodic boundary values. Then, (22) admits the following *variational formulation*: find u and κ with $u(t), \kappa(t) \in \mathcal{X}$ for a.e. t and $u(0) = u_0$ in a suitable sense fulfilling

$$\langle\partial_t u, \psi\rangle - \int_\Gamma \nabla_S\kappa\cdot\nabla_S\psi = 0 \qquad \forall\,\psi\in\mathcal{X}, \tag{23}$$

$$\langle\kappa, \varphi\rangle + \int_\Omega \frac{\nabla u\cdot\nabla\varphi}{Q(u)} = 0 \qquad \forall\,\varphi\in\mathcal{X}, \tag{24}$$

where $\langle\cdot,\cdot\rangle$ denotes the $L^2(\Omega)$ inner product; compare with §2. Note that, in contrast to the mean curvature flow of graphs [6, 7], (23) does not have the weight $Q(u)^{-1}$ because the equation is written on Γ.

If $\mathcal{X}_h$ is a finite element subspace of $\mathcal{X}$, then a *semi space discrete* scheme is obvious from (23) and (24). In particular, upon taking $\psi = 1 \in \mathcal{X}_h$ we deduce the exact volume conservation $\frac{d}{dt}\int_\Omega u_h = 0$. We can also prove stability as well as coercivity properties, which are crucial in deriving the following *error estimate* [2].

Theorem 5.1 (A Priori Error Estimate for the Semidiscrete Scheme). *Let u, κ be sufficiently smooth in $[0,T]$, and let $k \geq 1$ be the polynomial degree of $\mathcal{X}_h$. Then, there exists a constant $C > 0$, depending on the regularity of u, κ and T only, such that*

$$\begin{aligned} \sup_{t\in[0,T]} \Big(\|(u-u_h)(t)\|^2_{L^2(\Omega)} + \int_{\Gamma_h(t)} |\nabla_S u - \nabla_S u_h)|^2 \Big) &\leq C\, h^{2k}, \\ \int_0^T \Big(\|(\kappa-\kappa_h)(t)\|^2_{L^2(\Omega)} + \int_{\Gamma_h(t)} |\nabla_S \kappa - \nabla_S \kappa_h|^2 \Big)\, dt &\leq C\, h^{2k}. \end{aligned} \tag{25}$$

Since (25) involves the tangential gradients $\nabla_S u$ and $\nabla_S \kappa$, the order of convergence $O(h^k)$ is optimal. This is corroborated by the numerical experiments of [2], which are obtained via a *semi-implicit* time discretization similar to [7]. It consists in writing the equations for step $n+1$ on the current surface Γ_n, which linearizes the system and allows for a Schur complement solver similar to that in §4.

6. Implementation and Simulations

In this section we explain briefly the implementation of both the parametric and graph formulations within ALBERT [12, 13], and document their performance.

6.1. Mesh Regularization

FIGURE 1. Pathological *ear* formation in the evolution of a $4 \times 1 \times 1$ prism by surface diffusion. Surface at time $t = 0$ and $t = 0.07$.

The geometric flow by parametric surface diffusion is not as gentle as the corresponding mean curvature flow [8], and leads to severe mesh distortions. Even if the parametric formulation of §3 allows corners and edges, which are rather singular for surface diffusion, they give rise to fast node motion and mesh distortion. This is illustrated by the creation of *ears* during the evolution of a $4 \times 1 \times 1$ prism in Figs. 1 and 2 towards a ball. This is clearly a numerical artifact and cannot be cured by mesh refinement and/or coarsening. We use mesh regularization instead, which consists in replacing each node by the projection onto the surface of a weighted average of the nodes belonging to its finite element star, see also [11].

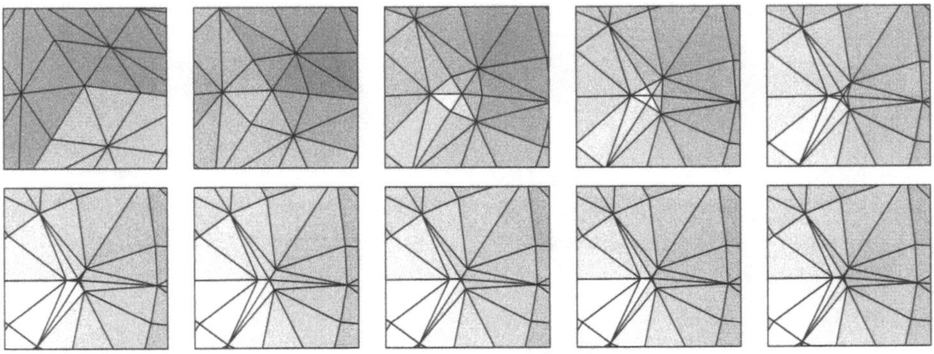

FIGURE 2. Steps towards the pathological formation of ears. Zoom into a vertex of the initial prism with surfaces at $t = k \times 10^{-3}$ for $0 \leq k \leq 9$. After 6 time steps some triangles collapse into segments, thereby making the mesh degenerate and producing numerical artifacts.

This regularization is equivalent to one time step of the discretized (tangential) heat equation. Its beneficial effect is is displayed by the following simulations of Figs. 3–6, all computed with forcing term $\epsilon = 0$.

6.2. Example 1: Smoothing Effect

We illustrate the evolution of a $4 \times 1 \times 1$ prism towards a ball with the same volume in Fig. 3, thereby showing the smoothing effect of surface diffusion.

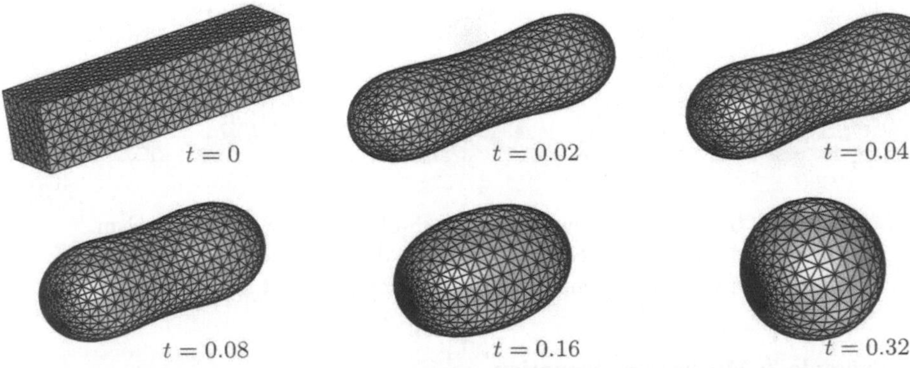

FIGURE 3. Smoothing effect of surface diffusion. Evolution of a $4 \times 1 \times 1$ prism towards a ball with equal volume at various time instants.

6.3. Example 2: Pinch-Off in Finite Time

Surface diffusion is not always a regularizing flow and may in fact create singularities in finite time depending on the initial configuration. This is shown in Figs. 4–6 which correspond to prisms with larger aspect ratios than before. Figs. 4–5 display

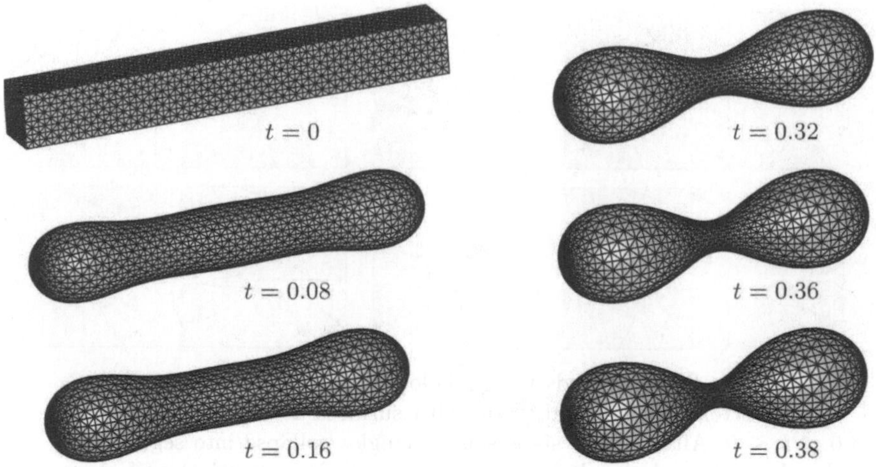

FIGURE 4. Pinch-off in finite time. Evolution of an $8 \times 1 \times 1$ prism at various time instants leading to a dumbbell and cusp formation.

the evolution of an $8 \times 1 \times 1$ prism towards a dumbbell and cusp formation in finite time. Increasing the aspect ratio of the prism seems to be an effective mechanism to produce several simultaneous cusps, as reported in Fig. 6.

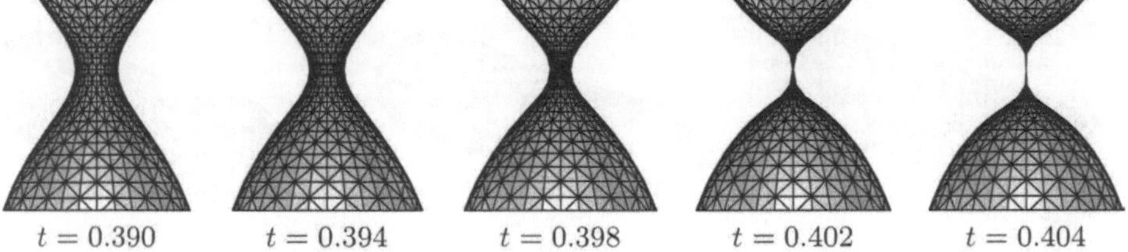

FIGURE 5. Detailed view of the pinch-off for the $8 \times 1 \times 1$ prism. Mesh regularization appears to cure mesh distortion until the moment of pinch-off when the elements are rather elongated but not degenerate.

6.4. Example 3: Mushroom Formation

We consider the graph formulation of §5 of a curve in 2d starting from the initial condition $u_0(x) = 1 + \delta(x)$, where δ is a steep perturbation with zero meanvalue (see Figure 7). Since the slope seems to become vertical around $t = 4.8 \times 10^{-5}$ in Fig. 7, the classical solution might cease to exist in finite time.

To investigate the formation of singularities in finite time, we use the parametric formulation with the same initial data, upon embedding the graph of u_0 into a closed curve (see Figure 8 top left). For the time scale of Fig. 7, the effect of

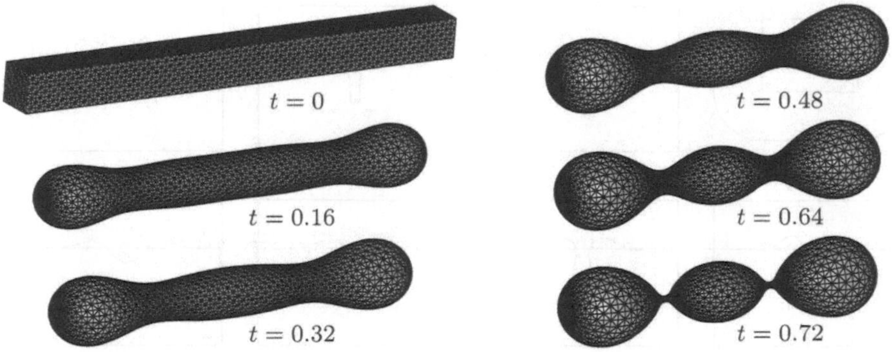

FIGURE 6. Evolution of a $12 \times 1 \times 1$ prism towards two simultaneous cusps revealing that the number of singularities depends on the aspect ratio of the initial prism.

this extension is negligible. Fig. 8 displays a sequence of solutions obtained for the same eight time instants of Fig. 7. It is worth noticing the striking similarity of the solutions obtained with both methods. Even though the parametric solution develops a *mushroom* shape at $t = 9.6 \times 10^{-5}$, and thus the solution to the graph formulation is questionable thereon, they still exhibit an excellent quantitative agreement for $t > 9.6 \times 10^{-5}$ (compare the last two plots of Figs. 8 and 7).

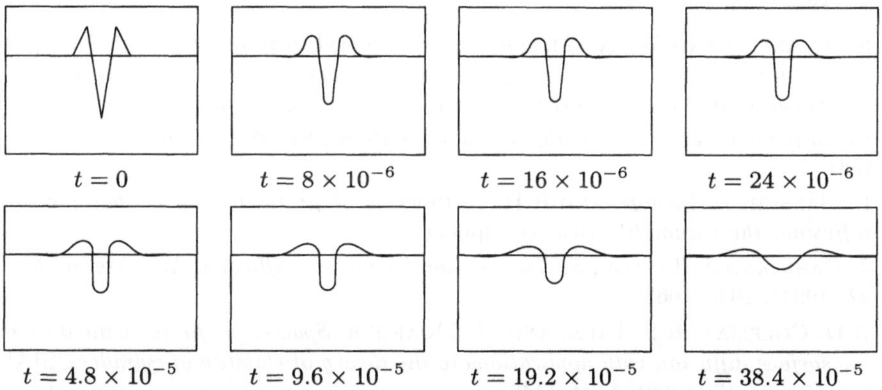

FIGURE 7. Evolution of a *graph* in 2d starting from $u_0(x) = 1 + \delta(x)$ with a steep perturbation $\delta(x)$. In all plots for various times t, the x-axis ranges from -1 to 1, and the y-axis ranges from 0 to 1.5.

Acknowledgments

We would like to thank K. G. Siebert for his assistance in incorporating new data structures to handle surfaces in $\mathbb{R}^3$ and curves in $\mathbb{R}^2$ within ALBERT [12, 13]. We

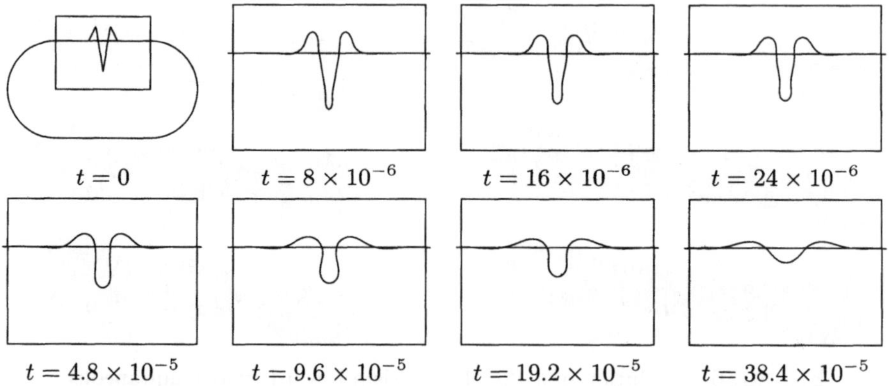

FIGURE 8. Mushroom formation. Parametric evolution of a curve in 2d with the same initial condition and the same times of Fig. 7. The rectangles in thin lines are $[-1, 1] \times [0, 1.5]$. The effect of embedding the graph of Fig. 7 into a close curved (see top-left picture) is negligible for the time scale of this evolution.

were all partially supported by the NSF-DAAD international cooperation grant INT-9910086.

References

[1] R. J. ASARO AND W. A. TILLER, *Surface morphology development during stress corrosion cracking: Part I: via surface diffusion*, Metall. Trans., **3** (1972), 1789–1796.

[2] E. BÄNSCH, P. MORIN, AND R.H. NOCHETTO, *Surface diffusion of graphs: variational formulation, error analysis and simulations*, SIAM J. Numer. Anal. (submitted).

[3] E. BÄNSCH, P. MORIN, AND R.H. NOCHETTO, *A finite element method for surface diffusion: the parametric case*, (to appear).

[4] J. CAHN AND J. TAYLOR, *Surface motion by surface diffusion*, Acta Metall. Mater., **42** (1994), 1045–1063.

[5] B.D. COLEMAN, R.S. FALK, AND M. MOAKHER, *Space-time finite element methods for surface diffusion with applications to the theory of stability of cylinders*, SIAM J. Sci. Comp., **17** (1996), 1434–1448.

[6] K. DECKELNICK AND G. DZIUK, *Convergence of a finite element method for the non-parametric mean curvature flow*, Numer. Math. **72** (1995), 197–222.

[7] K. DECKELNICK AND G. DZIUK, *Error estimates for a semi-implicit fully discrete finite element scheme for the mean curvature flow of graphs*, Interfaces and Free Boundaries **2**, (2000), 341–359.

[8] G. DZIUK, *An algorithm for evolutionary surfaces*, Numer. Math., **58** (1991), 603–611.

[9] K. DECKELNICK, G. DZIUK, AND C. ELLIOTT, *Semidiscrete finite elements for axially symmetric surface diffusion*, CMAIA Research Report No 2002/05 (2002).

[10] U.F. MAYER, *Numerical solutions for the surface diffusion flow in three space dimensions*, Comp. Appl. Math. (to appear).

[11] A. SCHMIDT, *Computation of three dimensional dendrites with finite elements*, J. Comput. Physics, **125** (1996), 293–312.

[12] A. SCHMIDT AND K.G. SIEBERT, *ALBERT: An Adaptive Hierarchical Finite Element Toolbox*, Documentation, Preprint 06/2000 Universität Freiburg, 244 p; available online from http://www.mathematik.uni-freiburg.de/IAM/ALBERT.

[13] A. SCHMIDT AND K.G. SIEBERT, *ALBERT-Software for scientific computations and applications*, Acta Math. Univ. Comenian. (N.S.), **70** (2001), 105–122.

[14] B.J. SPENCER, S.H. DAVIS, AND P.W. VOORHEES, *Morphological instability in epitaxially-strained dislocation-free solid films: nonlinear evolution*, Phys. Rev. B, **47** (1993), 9760–9777.

(E. Bänsch) Weierstrass Institute for Applied Analysis and Stochastics
Mohrenstrasse 39, 10117 Berlin,
and Freie Universität Berlin, GERMANY
E-mail address: `baensch@wias-berlin.de`

(P. Morin) Instituto de Matemática Aplicada del Litoral, Güemes 3450, 3000 Santa Fe,
and Departamento de Matemática, Facultad de Ingeniería Química,
Universidad Nacional del Litoral, Santa Fe, ARGENTINA
Partially supported by CONICET of Argentina and NSF Grant DMS-9971450.
E-mail address: `pmorin@math.unl.edu.ar`

(R. H. Nochetto) Department of Mathematics and Institute for Physical Science and Technology,
University of Maryland,
College Park, MD 20742, USA
Partially supported by NSF Grants DMS-9971450 and DMS-0204670.
E-mail address: `rhn@math.umd.edu`

International Series of Numerical Mathematics, Vol. 147, 65–74

Crystal Growth and Impingement in Polymer Melts

Martin Burger

1. Introduction

Crystallization of polymeric materials is a solidification process in strong interaction with heat conduction. Both the basic mechanisms involved in the solidification from a melt, namely the *nucleation* and *growth* of crystals, are strongly influenced by the temperature and its variation. Vice versa, the latent heat is rather large for polymeric materials, so that it causes a considerable change in the heat transfer process.

The interaction of solidification and heat conduction is a well-known phenomenon in models of phase-change, and involves important problems in the theory of free and moving boundary problems. In typical examples of moving boundary problems for the heat equation such as the well-studied *Stefan problem* (cf. [18]), the unknown boundary is assumed to be an isotherm, which leads in a mathematical model to a homogeneous Dirichlet condition for the (appropriately scaled) temperature. For polymers, the situation is different, since the phase change does not take place at a fixed temperature (or with kinetic undercooling close to this temperature), but in a rather large temperature range between the *thermal melting point* T_m and the *glass transition temperature* T_g.

The nucleation of new crystals can be modeled as a heterogeneous Poisson process in space and time, with rate $\alpha = \alpha(x,t)$. For many important materials such as isotactic Polypropylene (i-pp), a model for the nucleation rate in dependence of the temperature u is given by $\alpha(x,t) = \frac{\partial}{\partial t}\tilde{N}(u(x,t))$ (cf. [6]), with some material function $\tilde{N}$, which varies only in the interval (T_g, T_m), i.e., no crystals nucleate if the temperature is below the glass transition temperature or above the melting point. Usually it is assumed that new crystals nucleate as small spherical objects with initial radius R_0 determined from classical nucleation theory. The nucleation process has been well investigated with respect to its stochastic properties

2000 *Mathematics Subject Classification.* Primary: 35R35, Secondary: 80A22.

Key words and phrases. Crystal Growth, Moving Boundaries, Level Set Method.
Partial support is acknowledged to the Austrian National Science Foundation FWF under grant SFB F 013/08. The author thanks Prof. Vincenzo Capasso, Dr. Alessandra Micheletti and Dr. Claudia Salani (University Milano) for useful and stimulating discussions.

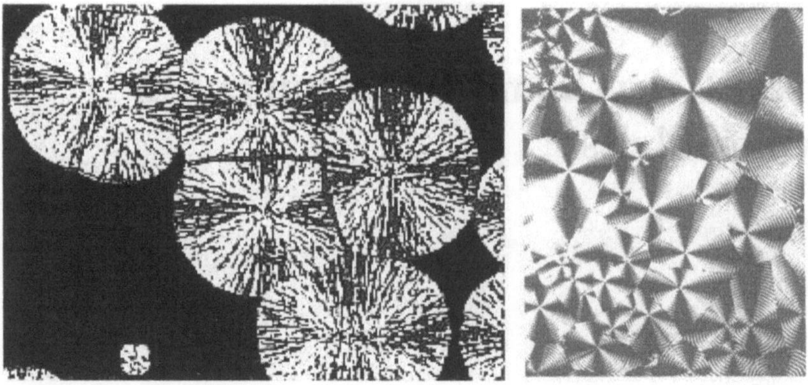

FIGURE 1. Impingement of polymer crystals (right) and a final tesselation from an experiment with isotactic Polypropylene.

(cf. [9, 20]), its averaging (cf. [5, 8]), its numerical simulation (cf. [19]), as well as the determination of the nucleation rate from indirect measurements (cf. [2, 7]).

The growth of the crystalline phase $\Theta = \bigcup_{i=1}^N \Theta_i$ ($\Theta_i(t)$ being the single crystals), is determined via a nonlinear *Gibbs–Thompson relation* for the velocity V_N and the temperature u,

$$V_N = G(u) = \frac{1}{L}\left[\frac{\partial u}{\partial n}\right] \qquad \text{on } \Gamma := \partial\Theta - \partial\Omega, \tag{1.1}$$

where G is a given material function, L is the latent heat, and $[\frac{\partial \cdot}{\partial n}]$ denotes the jump of the normal derivative on the boundary. For a certain temperature interval (e.g. for $u \geq 80°$ C in the case of isotactic Polypropylene) the function G is nonincreasing, so that the relation can be rewritten as $u = G^{-1}(V_N)$, which is a nonlinear version of the typical condition used for kinetic undercooling. Surprisingly, the function G usually reaches a maximum at some temperature (depending on the material) and is non-decreasing below it, which is confirmed by some experiments (cf. [21, 25]). The experimental background for this growth model in a nonisothermal situation was provided by Schulze and Naujeck [26], who considered the growth of a crystal in an environment with strong temperature gradients and found that the obtained shapes coincided with the ones predicted by a minimal-time principle. In [8], the growth model was reformulated in arbitrary spatial dimension as (1.1) (cf. [28] for an overview of different formulations of crystal growth models). Motivated by the physics of the problem we shall assume that G is a smooth function with compact support, which is nonnegative and globally bounded.

The heat transfer in the material is described by the heat equation in $\Omega - \partial\Theta$, in general with parameters (such as conductivity, density, and capacity) depending on the phase. For simplicity we restrict our attention to the case of all parameters being constant, which yields, together with the second identity in (1.1), the nonlinear parabolic equation

$$c\frac{\partial u}{\partial t} - k\Delta u = L\delta_\Gamma G(u) \qquad \text{in } \Omega \times (0,T), \tag{1.2}$$

where δ_Γ denotes the Dirac delta distribution on Γ. This equation is supplemented with appropriate initial conditions at time $t = 0$, i.e.,

$$u = u_0 \qquad \text{in } \Omega \times \{0\}, \tag{1.3}$$

and with boundary conditions on $\partial\Omega \times (0,T)$; for simplicity we restrict our attention to Dirichlet conditions of the form

$$u = u_D \qquad \text{on } \partial\Omega \times (0,T). \tag{1.4}$$

We finally note that (1.2) has an equivalent enthalpy formulation of the form

$$\frac{\partial e}{\partial t} - k\Delta u = 0, \qquad e = cu - L\chi, \tag{1.5}$$

with χ being the indicator function of Ω.

In the following we shall discuss the mathematical analysis and numerical simulation of the initial-boundary value problem (1.2), (1.3), (1.4), coupled to the growth law (1.1). We assume that the nucleation events $\{(X_j, T_j)\}_{j=1,\dots,N}$, representing the origin and time of nucleations, as well as the initial shape of a nucleated crystal (a small spherulite), are given a-priori, analysis and simulation of the stochastic aspects of nucleation shall be discussed elsewhere.

2. Mathematical Analysis

In the following we shall be concerned with the mathematical analysis of (1.1) coupled to the heat transfer problem (1.2), (1.3), (1.4). The analysis of this problem is challenging in the single crystal case already, where only local-in-time existence and uniqueness could be proven so far by Friedman and Velazquez [11] using fixed-point arguments and regularity (of the temperature and the free boundary) in a Hölder class $C^{2+\alpha}$, $\alpha > 0$.

In presence of multiple crystals, the mathematical analysis of the crystal growth via (1.1) leads to additional difficulties:

1. Because of the non-equilibrium condition at the interface, there is no weak formulation of the problem (such as for the two-phase Stefan problem, cf. [18]) that would circumvent the problem of dealing with the interface explicitly.
2. Even before impingement the growth of a fixed crystal is influenced by all the other crystals via changes in the temperature, i.e., we have to deal with a nonlocal front growth problem.
3. One cannot expect much regularity of the free boundary Γ even if all parameters and initial shapes are analytic. This can be seen from a simple case with $G \equiv 1$ and two spherulitic crystals. Before impingement, the shape $\Gamma(t)$ is analytic, but the impingement creates a non-Lipschitz

singularity. For each time after the impingement event, the shape $\Gamma(t)$ is only locally Lipschitz. Thus, the best type of regularity we can expect is "locally Lipschitz almost everywhere".

A detailed analysis can be carried out in the spatially one-dimensional case, where regularity of the free boundary plays no role since all crystals are compact intervals and $\Gamma(t)$ consists of a finite number of points. Global existence and uniqueness for this case has been obtained recently (cf. [3]); here we only state the main result:

Theorem 1. *Let $\Omega \subset \mathbf{R}^1$ be a bounded interval, let $T \in \mathbf{R}^+$ be arbitrary and let $(X_j, T_j) \in \Omega \times [0,T]$, $j = 1, \ldots, M$ be given nucleation events. Then the coupled problem* (1.1), (1.2), (1.3), (1.4) *has a unique weak solution*

$$u \in L^\infty(0,T; C^{0,1}(\Omega)) \quad \text{and} \quad \Gamma \in L^\infty([0,T]; \mathcal{K}_M),$$

where $\mathcal{K}_M$ is the set of all subsets of Ω with cardinality less or equal to $2M$ equipped with the Hausdorff metric. Moreover, for each time interval $[S_1, S_2]$ that does not contain a nucleation or impingement event, we have

$$u \in C(S_1, S_2; C^{0,1}(\Omega)) \quad \text{and} \quad \Gamma \in C^1([0,T]; \mathcal{K}_m).$$

In multi-dimensional situations, a promising approach is the representation of the growing crystalline phase via the level set approach, i.e.,

$$\Theta(t) = \{\, x \in \overline{\Omega} \mid \phi(x,t) \leq 0 \,\}, \tag{2.1}$$

where ϕ is a continuous functions. A weak formulation of the evolution of $\Omega(t)$ and $\Gamma(t) = \partial\Omega(t)$ via (1.1) is provided by computing a viscosity solution of the first-order Hamilton–Jacobi equation (cf. e.g. [10, 17] for details on viscosity solutions)

$$\frac{\partial \phi}{\partial t} + V|\nabla \phi| = 0, \qquad \text{in } \mathbf{R}^d \times (0,T), \tag{2.2}$$

where V is the extension of the velocity V_N to the whole space $\mathbf{R}^d$ and the initial value $\phi(.,0)$ is a uniformly continuous function satisfying (2.1) for $t = 0$. This approach, originally introduced for numerical purposes by Osher and Sethian [23], does not rely on any parametrization and allows to treat topological changes in an automatic way. We refer to the monographs by Osher and Fedkiw [22] and by Sethian [27] for an overview of the level set method and its applications.

Nucleation events can be included by restarting the level set evolution at each nucleation time with a new level set function including the nucleated object, and therefore we shall restrict our attention to a time interval $(0,T)$ without nucleation in the sequel, assuming that some initial shape $\Theta(0)$ is given. If we assume that a temperature $u \in C(0,T; C^{0,1}(\Omega))$ is given and if the domain Ω has a sufficiently smooth boundary, we can find some extension $\tilde{u}$ on $\mathbf{R}^N \times (0,T)$ such that the (natural) extension velocity defined by $V = G(\tilde{u})$ is bounded, nonnegative, and Lipschitz continuous. For such an evolution we can collect the following results:

- **Well-posedness:** For each uniformly continuous initial value, the level set equation (2.2) has a uniformly continuous unique viscosity solution (cf. [10]).
- **Stable dependence:** The solutions of the level set equation depend in a Lipschitz-continuous way (in the norm of $C(0,T;C^{0,1}(\Omega)))$ on the velocity V (cf. [12]).
- **Non-fattening:** If we start with a level set function such that $\Gamma(0) = \{ x \in \Omega \mid \phi(x,0) = 0 \}$ has zero Lebesgue measure, then, for all $t > 0$, the set $\Gamma(t)$ coincides with the zero level set of $\phi(.,t)$, which has zero Lebesgue measure (cf. [1]). A recent result by Ley [14] shows that even an estimate of the form $|\nabla\phi| \geq \eta > 0$ holds in the viscosity sense in a neighborhood of $\Gamma(t)$.
- **Partial regularity:** If we assume in addition that $V \geq \rho$ in $\Omega \times (0,T)$ for some $\rho \in \mathbf{R}^+$ (which is the case in the temperature range of interest), then for almost every $t \in (0,T)$, the set $\Gamma(t)$ is locally Lipschitz almost everywhere with respect to the $\mathcal{H}^{d-1}$-measure (cf. [4]).

Hence, by the level set approach we obtain a well-posed evolution, which yields exactly the maximal regularity we expected.

An interesting open problem for future research is the coupling of the level set evolution with the heat transfer model (1.2)–(1.4). A promising approach to global existence seems to be the use of a relaxation model as for the Stefan problem (cf. e.g. [29]) and (or) some smoothing of the form $\chi^\epsilon = f^\epsilon(\phi)$, with f^ϵ approximating

$$f(s) = \begin{cases} 1 & \text{if } s < 0 \\ [0,1] & \text{if } s = 0 \\ 0 & \text{if } s > 0 \end{cases} .$$

For the smoothed problem, the map $\phi \mapsto u$ is continuous and compact, which allows, together with the above stable dependence results for the solutions of the level set equation, to use a fixed point argument to deduce the existence of a solution. Moreover, analogous a-priori estimates as in [29] can be obtained for the temperature u in the norm of $L^1(0,T;W^{2,1}(\Omega))$, which might allow to deduce existence by similar weak-star convergence techniques as in [29].

3. Numerical Simulation

The numerical simulation of crystal growth consists of two main tasks, namely the numerical solution of the level set equation (2.2) and of the heat transfer problem (1.2)-(1.4). In addition, we have to implement the coupling of these problems via the velocity $V = G(u)$ and the source term in (1.2).

For solving first-order Hamilton–Jacobi equations like (2.2), several schemes have been proposed over the last two decades, most of them using finite differences. A frequently used class of solvers are *essentially nonoscillatory schemes* (cf. [24]), originally developed for conservation laws. This type of schemes allows high order approximation of solution, can be implemented efficiently, and is explicit in time (like the majority of schemes for conservation laws and Hamilton–Jacobi equations). As any other explicit scheme, there is a stability bound on the time step τ in dependence of the spatial grid size h, namely the so-called *CFL-condition*

$$\tau \sup_x |V(x,t)| < h. \tag{3.1}$$

Since the velocity $V = G(u)$ is usually rather small for polymeric materials, this stability bound is not restrictive and the actually chosen time step is usually far below the one allowed by (3.1).

Another consequence of an explicit time scheme for the level set equation is the possibility to use its result as an input for the time step in the heat equation, which should be implicit, since the stability bound for a parabolic equation ($\tau \sim h^2$) is quite restrictive. Thus, having computed ϕ at time $t+\tau$, we can apply the (horizontal) method of lines to the heat equation (1.2) or (1.5) and use the the values of ϕ to assemble the right-hand side, leading to an elliptic boundary value problem in each time step. For the spatial discretization of the arising elliptic problem, we have several choices. A first possibility is to choose a finite element grid matching the interface, which leads to high accuracy but also to an enormous numerical effort, since the grid has to be changed in each time step. More efficient are methods with non-matching grids such as the *immersed interface method* (cf. [15, 16]),which is a finite difference method on a rectangular grid with additional correction terms in those cells including parts of the interface, or a method approximating the interface terms by smoothed quantities, which is frequently used in the context of level set methods (cf. [22]). For the latter, the simplest possibility consists in using the enthalpy formulation (1.5) and approximating χ by $\chi^\epsilon = f^\epsilon(\phi)$, with f_ϵ as in the previous section, which yields an equation of the form

$$c\frac{\partial u}{\partial t} - k\Delta u = -L\delta^\epsilon(\phi)\frac{\partial \phi}{\partial t}, \tag{3.2}$$

with δ^ϵ approximating the Dirac delta distribution. After time discretization, we can employ a finite element method on a mesh fixed during the evolution or at least during several time steps. The mesh generation should be fine around the initial interface and consequently also around the interface after some time steps, provided the motion is not too fast. Only if the mesh around the interface becomes too coarse after some time steps, the domain has to be remeshed. In this way, the numerical effort can be reduced significantly. For computations with a large number of crystals (see [19] for such simulations), it seems of advantage to use a fine uniform grid (fixed during the evolution), and to apply fast Poisson solvers for the solution of the elliptic problem.

Finally, we show a numerical experiment with two crystals in a domain scaled to the cell $[-1,1]^2$. We use material parameters in a range typical for isotactic Polypropylene (cf. [5] for numerical values), and use a temperature scaling such that $L = 1$ (around 70° unscaled). The (unscaled) initial value is a uniform temperature $u_0 = 124°$ C. The heat transfer parameters, which are typically small for polymers, are chosen as $c = k = 10^{-2}$. All solvers were implemented in the software system MATLAB, for the heat equation we use a finite element discretization with $n = 19968$ triangles. The level set equation is solved on a uniform grid with 256×256 elements using a weighted ENO-scheme of fifth order in space and third order in time, with a Godunov flux, as proposed in [13].

In the first example, shown in Figure 2, we use a uniform Dirichlet value of $u = 88°C$ at the boundary. In this case, one can observe a strong reheating effect due to the release of latent heat at the crystal boundary and the free boundary is almost an isotherm of the temperature. Moreover, the absence of strong temperature gradients in the crystals causes the shape to stay close to a spherical shape. Consequently, the crystals heat each other at a single point and the contact interface propagates outward from this point.

The behavior in the second example (shown in Figure 3) is different due to different boundary values of the temperature, which are very low on the left and upper part of the rectangle (around 70° C) and rather high at the right and lower part (around $140°C$). This yields a temperature gradient mainly in diagonal direction, with some changes due to reheating. In particular, the temperature around the lower crystal is so high (and the value of $G(u)$ so low) that there is almost no growth, while the second crystal grows to a large, non-spherical shape. The impingement starts from two different contact points in this case and leaves a small liquid part inside the crystalline phase for some time.

References

[1] G. Barles, H.M. Soner, P.E. Souganidis, *Front propagation and phase-field theory,* SIAM J. Cont. Optim. **31** (1993), 439–469.

[2] M. Burger, *Iterative regularization of an identification problem arising in polymer crystallization*, SIAM J. Numer. Anal. **39** (2001), 1029–1055.

[3] M. Burger, *Growth of multiple crystals in polymer melts* (2001), submitted.

[4] M. Burger, *Growth fronts of first-order Hamilton–Jacobi equations,* SFB-Report 02-8 (University Linz, 2002), and submitted.

[5] M. Burger, V. Capasso, *Mathematical modelling and simulation of non-isothermal crystallization of polymers*, Math. Mod. Meth. Appl. Sci. **11** (2001), 1029–1054.

[6] M. Burger, V. Capasso, G. Eder, *Modelling crystallization of polymers in temperature fields*, ZAMM **82** (2002), 51–63.

[7] M. Burger, V. Capasso, H.W. Engl, *Inverse problems related to crystallization of polymers,* Inverse Problems **15** (1999), 155–173.

[8] M. Burger, V. Capasso, C. Salani, *Modelling multi-dimensional crystallization of polymers in interaction with heat transfer*, Nonlinear Anal. B, Real World Appl. **3** (2002), 139–160.

[9] V. Capasso, C. Salani, *Stochastic birth-and-growth processes modelling crystallization of polymers in a spatially heterogenous temperature field*, Nonlinear Anal. B, Real World Appl. **1** (2000), 485–498.

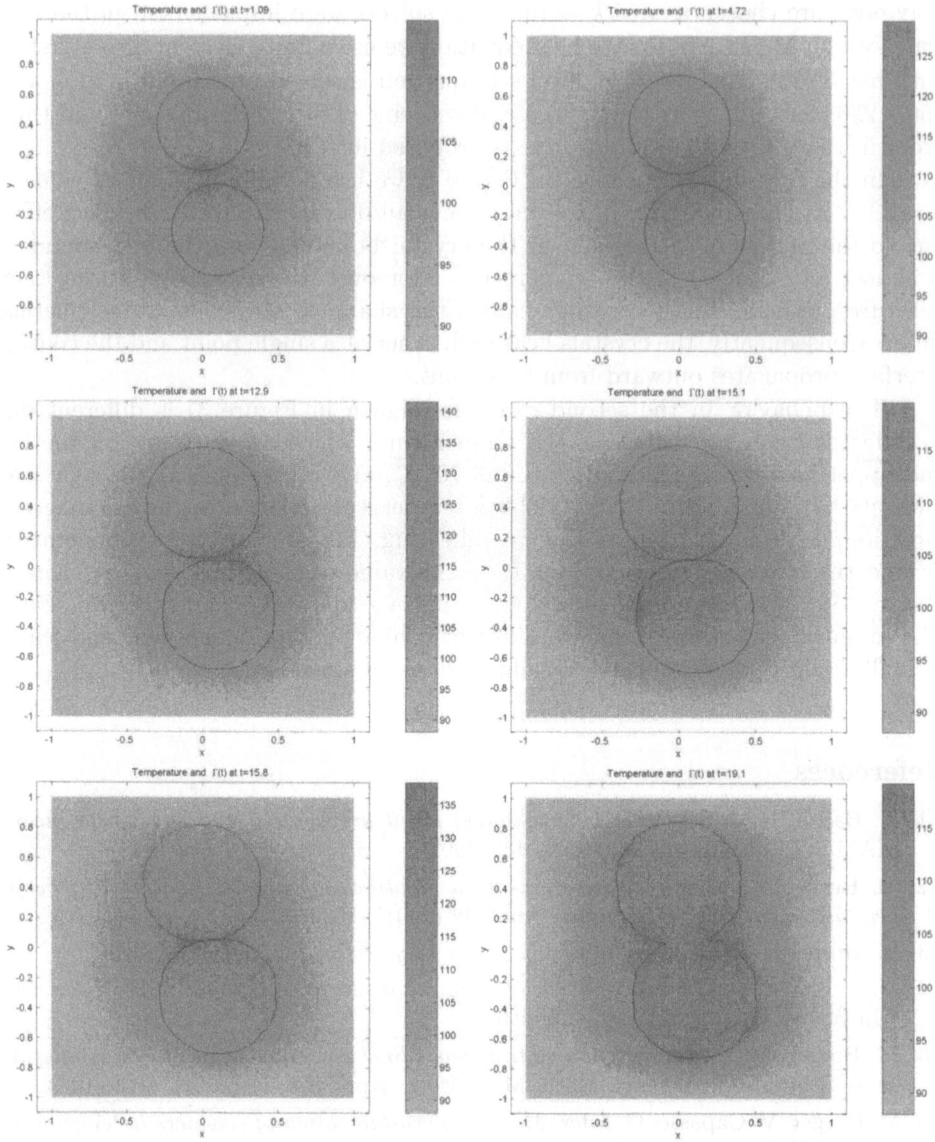

FIGURE 2. Color plot of the temperature and shape of the growing crystals in the first example: at the initial stage ($t = 1.09$, $t = 4.72$), short time before ($t = 12.9$), at ($t = 15.1$) and after impingement ($t = 15.8, t = 19.1$).

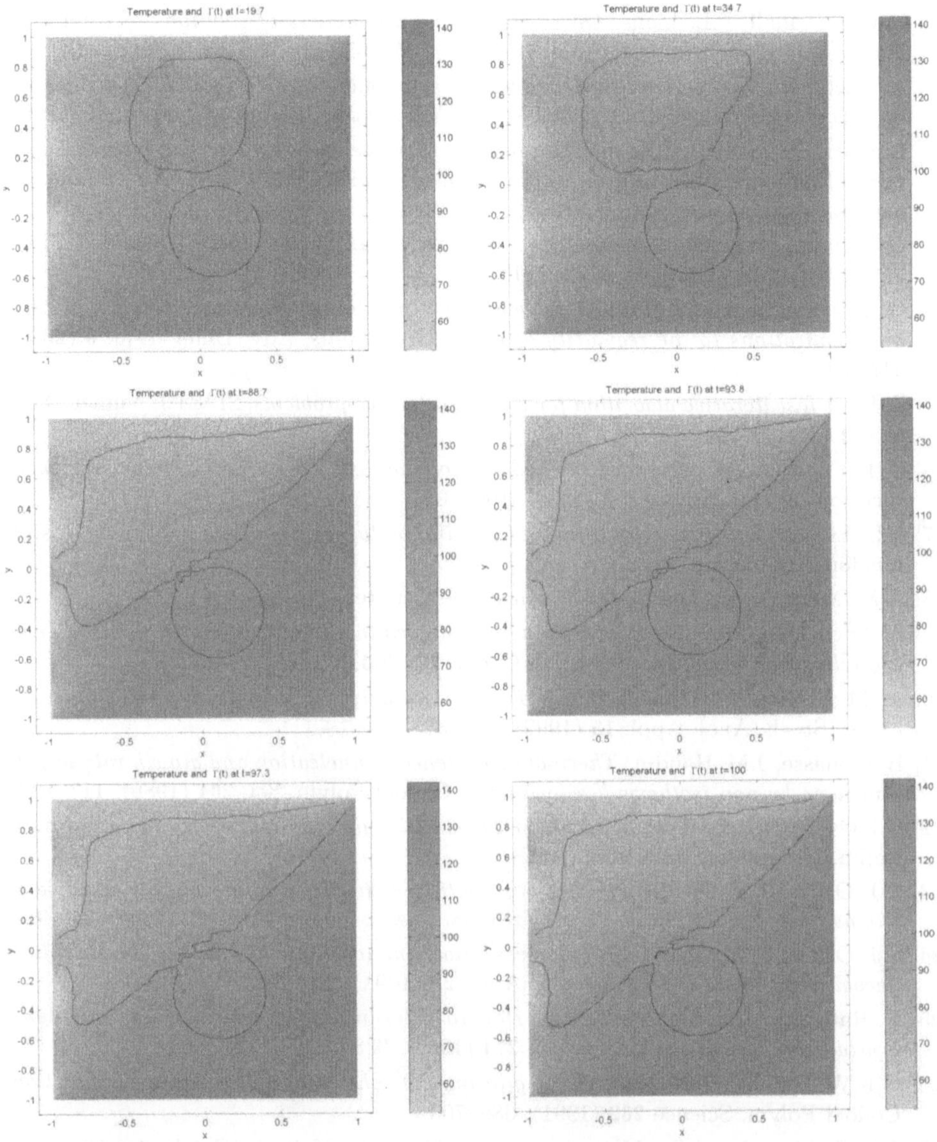

FIGURE 3. Color plot of the temperature and shape of the growing crystals in the second example: at the initial stage ($t = 19.7$, $t = 34.7$), short time before ($t = 88.7$), at ($t = 93.8$) and after impingement ($t = 97.3$, $t = 100$).

[10] M.G. Crandall, P.L. Lions, *On existence and uniqueness of solutions of Hamilton–Jacobi equations*, Nonlinear Anal. **10** (1986), 353–370.

[11] A. Friedman, J.L. Velazquez, *A free boundary problem associated with crystallization of polymers in a temperature field*, Indiana Univ. Math. Journal **50** (2001), 1609–1650.

[12] E.R. Jakobsen, K.H. Karlsen, *Continuous dependence estimates for viscosity solutions of fully nonlinear degenerate parabolic equations*, CAM Preprint 01-03 (Department of Mathematics, UCLA, 2001).

[13] G.S. Jiang, D. Peng, *Weighted ENO-schemes for Hamilton–Jacobi equations*, SIAM J. Sci. Comput. **21** (2000), 2126–2143.

[14] O. Ley, *Lower bound gradient estimates for first-order Hamilton–Jacobi equations and applications to the regularity of propagating fronts*, Adv. Differ. Equ. **6** (2001), 547–576.

[15] Z. Li, *A fast iterative algorithm for elliptic interface problems*, SIAM J. Numer. Anal. **35** (1998), 230–254.

[16] Z. Li, Y. Shen, *A numerical method for solving heat equations involving interfaces*, Electron. J. Diff. Eqns., **C 3** (1999), 100–108.

[17] P.L. Lions, *Generalized Solutions of Hamilton–Jacobi Equations* (Pitman, Boston, London, Melbourne, 1982).

[18] A.M. Meirmanov, *The Stefan Problem* (De Gruyter, Berlin, 1992).

[19] A. Micheletti, M. Burger, *Stochastic and deterministic simulation of nonisothermal crystallization of polymers*, J.Math.Chem. **30** (2001), 169–193.

[20] A. Micheletti, V. Capasso, *The stochastic geometry of polymer crystallization processes*, Stoch. Anal. Appl. **15** (1997), 355–373.

[21] B. Monasse, J.M. Haudin, *Thermal dependence of nucleation and growth rate in polypropylene by non-isothermal calorimetry*, Colloid Polym. Sci. **264** (1986), 117–122.

[22] S.J. Osher, R.P. Fedkiw, *Level Set Methods and Dynamic Implicit Surfaces* (Springer, Berlin, Heidelberg, New York, 2002).

[23] S.J. Osher, J.A. Sethian, *Fronts propagating with curvature-dependent speed: algorithms based on Hamilton–Jacobi formulations*, J. Comp. Phys. **79** (1988), 12–49.

[24] S.J. Osher, C.W. Shu, *High-order essentially nonoscillatory schemes for Hamilton–Jacobi equations*, SIAM J. Numer. Anal. **28** (1991), 907–922.

[25] E. Ratajski, H. Janeschitz-Kriegl, *How to determine high growth speeds in polymer crystallization*, Colloid Polym. Sci. **274** (1996), 938–951.

[26] G.E.W. Schulze, T.R. Naujeck, *A growing 2D spherulite and calculus of variations*, Colloid Polym. Science **269** (1991), 689–703.

[27] J.A. Sethian, *Level Set Methods and Fast Marching Methods* (Cambridge University Press, 2nd ed., Cambridge, 1999).

[28] J.E. Taylor, J.W. Cahn, C.A. Handwerker, *Geometric models of crystal growth*, Acta Metall. Mater. **40** (1992), 1443–1472.

[29] A. Visintin, *Models of phase-relaxation*, Diff. Int. Equations **14** (2001), 1469–1486.

Industrial Mathematics Institute, Johannes Kepler Universität Linz, Altenbergerstr. 69, 4040 Linz, Austria

E-mail address: `burger@indmath.uni-linz.ac.at`

International Series of Numerical Mathematics, Vol. 147, 75–86

Moving Bands and Moving Boundaries in an Hybrid Model for the Crystallization of Polymers

Vincenzo Capasso, Ramon Escobedo, and Claudia Salani

Abstract. Crystallization is a mechanism of phase change in polymeric materials; it consists of at least two processes, a nucleation (birth) process describing the time and the location of spots in the material where crystals start growing, and a growth process of the nucleated crystals.

Both processes are coupled to temperature, and viceversa the temperature field is coupled with the crystallization process due to the production of latent heat.

All the processes described above are of a random nature, since the birth process and the consequent geometry of the crystalline phase are random. However, under typical industrial conditions (many and small crystals), a multiple scale assumption can be made so that a deterministic approximation (homogeneization) for the spatial density of the crystalline phase at the (macro)scale of the temperature field is possible.

Thus, at the macroscale the whole crystallization process may be modelled as a reaction-diffusion system in which a deterministic ODE for the crystalline density is coupled with a deterministic PDE for the temperature field via the kinetic parameters of birth and growth on the one hand and the latent heat on the other hand.

Numerical simulations of this reaction-diffusion system show that the solution exhibits an advancing moving band of crystallization in the mass distribution, accompanied by a moving boundary in the temperature field, both advancing with a decreasing velocity.

1. Introduction

Crystallization is a mechanism of phase change in polymeric materials; as for various similar processes, it consists of at least two processes, a nucleation (birth) process describing the time and the location of spots in the material where crystals start growing, and a growth process of the nucleated crystals.

Both processes are coupled to temperature, since in general the kinetic parameters of birth and of growth depend significantly upon the local temperature field, and viceversa the temperature field is coupled with the crystallization process due to the production of latent heat, and the possible dependence of the parameters of the heat conduction upon the phase (liquid or solid) [1].

Growth processes may be very complicated, but in the sequel the restriction to the case of spherulitic growth is made, which is also a good assumption for the crystallization of relaxed polymer melts.

All the processes described above are of a random nature, since the birth process and the consequent geometry of the crystalline phase are random. However, under typical industrial conditions, a multiple scale assumption can be made, due to the fact that at the typical spatial scale of the temperature field (macroscale), many and small crystals are present, thus allowing a deterministic approximation (homogeneization) at the macroscale for the spatial density of the crystalline phase and consequently of the temperature field. This approach has been carried out in a rigorous way in [7] via suitable "laws of large numbers".

Thus, at the macroscale, the whole crystallization process may be modelled as a reaction-diffusion system in which a deterministic ODE for the crystalline density is coupled with a deterministic PDE for the temperature field via the kinetic parameters of birth and growth on the one hand and the latent heat on the other hand. In particular, we have introduced a critical temperature above which no nucleation and no growth can take place. Also, we have taken into account the possibility that the rate functions may present a jump at the critical temperature. Numerical simulations of this reaction-diffusion system show that the solution exhibits an advancing moving band of crystallization in the mass distribution, accompanied by a moving boundary in the temperature field, both advancing with a decreasing velocity. The fact that the speed is not constant, that is, that two different space segments of the same length have different times of crystallization, is a relevant aspect for the industrial process, and it is greatly interesting from the mathematical point of view.

On the other hand, if we wish to obtain geometric information about the crystallization process such as the evolution equations for the interface densities, we may couple the averaged deterministic temperature field with the stochastic birth-and-growth process (hybrid model) (see [14] and Sec.4).

2. A rigorous geometric approach

In this section we describe the crystallization process in a given deterministic field of temperature, neglecting at the moment the influence of the crystallization process on temperature, which we shall discuss later.

Crystallization is a stochastic process in time and space due to the combination of birth (nucleation) and growth of crystals. The first one is modelled as a marked point process, and the whole process is described by a dynamic Boolean model. The nucleation process is modelled as a stochastic spatially marked point process N on an underlying probability space $(\Omega, \mathcal{A}, P)$, with marks in the physical space E, a compact subset of $\mathbb{R}^d, d = 1, 2, 3$ (we shall denote by $\mathcal{B}_{\mathbb{R}_+}$, resp. $\mathcal{E}$, the σ-algebra of Borel sets on $\mathbb{R}_+$, resp. on E). The marked point process (MPP) N is a random measure given by $N = \sum_{n=1}^{\infty} \epsilon_{T_n, X_n}$ where T_n is an $\mathbb{R}_+$-valued random variable representing the time of birth of the $n-$th nucleus, X_n is an E-valued random variable representing the spatial location of the nucleus born at time T_n, and $\epsilon_{t,x}$ is the Dirac measure on $\mathcal{B}_{\mathbb{R}_+} \times \mathcal{E}$. The crystalline phase at time $t > 0$ is given by a random set $\Theta^t = \bigcup_{T_j \le t} \Theta^t_j$ (see [1], Sec.5.1), which is the union of all the crystals born at times T_j and locations X_j and freely grown up to time t. The integer valued random variable $N([0,t] \times B)$ counts the (random) number of nucleation events occurred in the time-space region $[0,t] \times B \in \mathcal{B}_{\mathbb{R}_+} \times \mathcal{E}$. We assume that the infinitesimal probability of birth of a new crystal during the time interval $[t, t+dt]$, in the free spatial interval $[x, x+dx] \subset E$, is given by (see also [14])

$$\nu_0(dt \times dx) = \alpha(x,t)\ dx\ dt.$$

The *(degree of) crystallinity* $\xi(x,t)$, at location $x \in E$ and time $t > 0$, is the probability that, at time t, x is covered (or "captured") by some crystal, i.e.

$$\xi(x,t) := P(x \in \Theta^t) = E[I_{\Theta^t}(x)].$$

As a complement to the degree of crystallinity (or probability of capture) $\xi(x,t)$, one can define the *survival function* or *porosity* of the point x at time t, which is the probability $p_x(t)$ that the point x is not yet crystallized at time t, i.e., $p_x(t) = P(x \notin \Theta^t) = 1 - \xi(x,t)$. Correspondingly, we may introduce the hazard function $h(x,t)$ as the rate of capture, i.e., the rate of the capture probability,

$$h(x,t) = \lim_{\Delta t \to 0} \frac{P(x \in (\Theta^{t+\Delta t} \setminus \Theta^t))}{\Delta t}.$$

We can see [1] that

$$\frac{\partial}{\partial t}\xi(x,t) = (1 - \xi(x,t))h(x,t).$$

Under the assumption of growth along the normal to the external surface due to a given field of growth rates $G(x,t)$, and assumptions of sufficient regularity of the field of birth rates $\alpha(x,t)$, it can be shown [5] that for almost any $x \in E$ and any $t \in \mathbb{R}_+$

$$h(x,t) = G(x,t) S_{ex}(x,t) \tag{1}$$

where S_{ex} is the mean value of the extended surface of the crystals, defined as follows in $\mathbb{R}^d$ (λ^d denotes the usual Lebesgue measure).

$$S_{ex}(x,t) = \lim_{r \downarrow 0} E\left[\sum_{T_j < t} \frac{\lambda^{d-1}(\partial\Theta^t_j \cap B_r(x))}{\lambda^d(B_r(x))}\right].$$

Clearly $S_{ex}(x,t)$ represents the expected value of the spatial density of the sum of the surfaces of the individual crystals which are born and develop independently of each other. From the above considerations, we have in this case

$$\frac{\partial}{\partial t}\xi(x,t) = G(x,t) S_{ex}(x,t)(1 - \xi(x,t)).$$

This formula may be seen as an extension of the Kolmogorov-Avrami-Evans formula to the spatially heterogeneous case.

In order to obtain an explicit evolution equation for $\xi(x,t)$, we need to couple the previous one with an evolution equation for $S_{ex}(x,t)$. This has been done in [4] as follows.

In $\mathbb{R}^d$ we introduce first the (stochastic) analogue $u(x,t)$ of $S_{ex}(x,t)$ such that for any $B \in \mathcal{B}_{\mathbb{R}^d}$, we have

$$\int_B u(x,t)\lambda^d(dx) = \sum_{T_j<t} \lambda^{d-1}(\partial\Theta_j^t \cap B).$$

In $\mathbb{R}^2$, with given fields $\alpha(x,y)$ and $G(x,y)$, we have the following system for u

$$\begin{cases} \dfrac{\partial u}{\partial t} = \mathrm{div}(Gv) + \sum_{T_j=t} S_j^2 & \text{in } E \times \mathbb{R}_+ \\ \dfrac{\partial v}{\partial t} = \nabla(Gu) & \text{in } E \times \mathbb{R}_+ \\ u = 0 & \text{in } E \times \{0\} \\ v = 0 & \text{in } E \times \{0\} \\ u + v \cdot \mathbf{n} = 0 & \text{on } \partial E \times \mathbb{R}_+, \end{cases} \tag{2}$$

where the only source of stochasticity is due to the nucleation $\sum_{T_j=t} S_j^2$; S_j^2 is the generalized function that describes the stochastic event of birth of the j-th crystal in $\mathbb{R}^2$ (system (2) has to be understood in a weak sense).

Since the system is linear and

$$S_{ex}(x,t) := E[u(x,t)],$$

we can obtain the deterministic equations

$$\begin{cases} \dfrac{\partial S_{ex}}{\partial t} = \mathrm{div}(GE[v]) + E\Big[\sum_{T_j=t} S_j^2\Big] \\ \dfrac{\partial E[v]}{\partial t} = \nabla(GS_{ex}), \end{cases} \tag{3}$$

subject to the corresponding boundary and initial conditions. In $\mathbb{R}^2$ (see [4])

$$E\Big[\sum_{T_j=t} S_j^2\Big] = 2\pi G(x,t) \int_0^t \alpha(x,t)ds.$$

Once available the hazard function and the growth rate, we may obtain further information about geometric measures regarding the crystallization process, such as the volume densities of interfaces of crystals at all Haussdorff dimensions (see [14] and [6]).

3. Coupling with temperature

In an experimental situation where spatial heterogeneities are caused only by the heat transfer in the material, we may assume that growth and nucleation rates are determined as material functions of the temperature [9].

$$G(x,t) = \widetilde{G}(T(x,t)), \qquad \alpha(x,t) = \widetilde{\alpha}(T(x,t)). \tag{4}$$

Viceversa, the increase of the crystalline phase influences the heat transfer process in the material because of the release of latent heat, i.e., the temperature T is determined by the heat transfer model

$$\frac{\partial}{\partial t}(\rho c T) = \text{div}\,(\kappa \nabla T) + \frac{\partial}{\partial t}(h \rho I_{\Theta^t}), \qquad \text{in } \mathbb{R}^+ \times E, \tag{5}$$

$$\frac{\partial T}{\partial n} = \beta(T - T_{out}), \qquad \text{on } \mathbb{R}^+ \times \partial E, \tag{6}$$

where $\frac{\partial}{\partial n}$ denotes the (outer) normal derivative on ∂E.

Here $\frac{\partial}{\partial t}(I_{\Theta^t})$ denotes the (distributional) time derivative of the indicator function I_{Θ^t} of the crystalline phase at time t, noting that Equation (5) has to be understood in the sense of distributional derivatives. The parameter ρ denotes the density, c the heat capacity, κ the heat conductivity, β the heat transfer coefficient and h the latent heat due to the phase change.

The parameters in the heat equation in general may depend upon the phase.

Thus the heat transfer model is now a random partial differential equation, since all parameters and the latent heat depend upon the random variable I_{Θ^t}. A direct consequence is the stochasticity of the temperature field, whose evolution influences (via (4)) the crystallization process itself (see also [3]).

Under these circumstances, System (2) is *doubly* stochastic, and as much nonlinear with respect to the operator E (expected value). In particular, we notice that

$$E[\text{div}(Gv)] \neq \text{div}(E[G]E[v])$$

since now $G(T(x,t))$ is a stochastic quantity itself.

To overcome this complication, we may take into account the fact that under typical industrial conditions (many and small crystals), we deal with a double scale process, so that one may approximate the stochastic model presented above in two ways. If we wish to obtain geometric information about the crystallization process such as the evolution equations for the interface densities, we approximate the stochastic temperature field with an averaged deterministic one and so go back to the approach described in Section 2 (Hybrid model). If we are just interested in the evolution of the crystallinity (volume density of the crystal phase), then we may approximate the stochastic system with a fully deterministic one, as done in [7] by using an interacting particle system coupled with the temperature field, and proceed via suitable "laws of large numbers".

4. The deterministic reaction-diffusion model

In the deterministic model the geometry is lost. We can make use of this fact assuming that a symmetry exists in the y-direction, so that the model can be studied in the one-dimensional case. For completeness, we reproduce here the model obtained in [1]. It consists of the following equations for the crystallinity $y(x,t)$ and the temperature $T(x,t)$, for $x \in [0, L]$ and $t > 0$,

$$\frac{\partial y}{\partial t} = \beta(y)\, b_G(T) + v_0\, \kappa(y)\, b_N(T) \tag{7}$$

$$\frac{\partial T}{\partial t} = \sigma \frac{\partial^2 T}{\partial x^2} + a_G\, \beta(y)\, b_G(T), \tag{8}$$

and of the initial and boundary conditions

$$y(x,0) = y_{\text{ini}}(x), \quad x \in [0, L], \tag{9}$$

$$T(x,0) = T_{\text{ini}}(x), \quad x \in [0, L], \tag{10}$$

$$T(0,t) = T_{\text{app}} \quad \forall t \geq 0, \tag{11}$$

$$\frac{\partial T(L,t)}{\partial x} = 0 \quad \forall t \geq 0, \tag{12}$$

where $\kappa(y)$ and $\beta(y)$ reproduce the features of crystallization phenomena, respectively, the start of the nucleation, and the aggregation and saturation of nuclei (see [7]):

$$\beta(y) = \begin{cases} y(1-y) & \text{in } [0,1] \\ 0 & \text{outside} \end{cases}, \quad \kappa(y) = \begin{cases} (1-y)^2 & \text{in } [0,1] \\ 0 & \text{outside} \end{cases}. \tag{13}$$

The constants σ, v_0 and a_G are the diffusion coefficient, the initial mass and the non-isothermal factor, respectively. The functions $b_N(T)$ and $b_G(T)$ are, respectively, the nucleation and growth rates for a temperature T; they are typically given by exponential functions with a negative exponent:

$$b_N(T) = N_0\, e^{-\beta_N (T - T_{\text{ref}})}, \qquad b_G(T) = G_0\, e^{-\beta_G (T - T_{\text{ref}})}, \tag{14}$$

where N and G are the nucleation and growth factors, β_N and β_G the nucleation and growth exponents, and T_{ref} is a temperature of reference (usually $T_{\text{ref}} = 0$). They have been obtained via statistical regression analysis in [7] and Ref. 6 therein.

It has been shown in [10] that a threshold temperature of freezing T_f must be introduced, above which no nucleation and no growth can take place: $b_N(T \geq T_f) = b_G(T \geq T_f) = 0$. The reason is that without this threshold, the full crystallization of the sample would be only a matter of time for every applied temperature,which may be not realistic. The introduction of T_f can be done simply by truncating the exponentials in (14), so that the rate functions will be discontinuous functions of T, or by modifying (14) to make the exponentials decay continuously to $(T_f, 0)$. To our understanding, there is no experimental basis to decide which option is the most realistic one. The second option can be preferred because of the continuity, but Nature is sometimes not continuous, and the discontinuous option is in fact the general case (the other one being a particular case of discontinuity of size 0), and it is mathematically more interesting, the casuistry being (not surprisingly) more rich.

For simplicity, we can assume that both rate functions can be reduced to a general rate function $\theta(T)$ such that $b_N(T) = N_0\,\theta(T)$ and $b_G(T) = G_0\,\theta(T)$, where

$$\theta(T) = \frac{1}{1 - a_0 e^{-T_f/10}} \begin{cases} e^{-\beta T} - a_0 e^{-\beta T_f/10} & \text{if } T < T_f \\ 0 & \text{if } T \geq T_f \end{cases}, \tag{15}$$

a_0 being a measure of the discontinuity of θ at $T = T_f$.

From the mathematical point of view, this model defines a Free Boundary Problem (FBP) for the temperature field. Let us illustrate this by means of a simple model of heat diffusion. Let $L = \pi/2$ and $u(x,t)$ be the solution of

$$u_t - u_{xx} = 0, \tag{16}$$

$$u(x,0) = \sin x, \tag{17}$$

$$u(0,t) = 0, \tag{18}$$

$$u_x(L,t) = 0. \tag{19}$$

In this problem it is not difficult to establish the existence, the uniqueness and the monotonicity of the solution. In fact, u is given by $u(x,t) = e^{-t}\sin x$. Then, for a given value $a \in (0,1)$ and for each $t \in [0, \ln\frac{1}{a}]$, there exists a unique abscissa $x_b(t) \in [0,L]$ such that $u(x_b(t),t) = a$:

$$x_b(t) = \arcsin(ae^t). \tag{20}$$

At a given instant of time, $x_b(t)$ determines a boundary in the interval $[0,L]$, in the sense that it separates $[0,L]$ into two intervals with some qualitative difference, namely, the sign of $u - a$. Eq. (20) gives the time evolution of this boundary value. In this case, x_b evolves with an always positive and increasing velocity, and it reaches the end of the interval at $t = \ln\frac{1}{a}$, see Fig. 1.

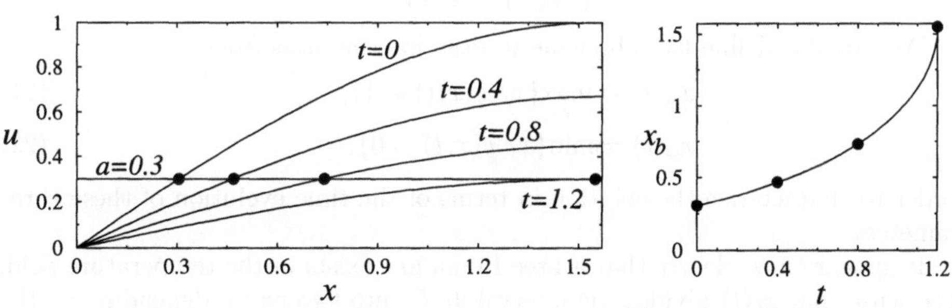

FIGURE 1. Left: Solution $u(x,t)$ of the heat equation at four instants of time. At the height $a = 0.3$ the boundary point $x_b(t)$ moves with an increasing velocity and arrives to $x = \frac{\pi}{2}$ at $t \approx 1.2$. Right: time evolution of $x_b(t)$.

In a similar way, a boundary value $x_b(t)$ is expected to be found in the polymer crystallization temperature field. Our ideal resolution of the polymer problem is to provide the same description as in the simple heat diffusion problem. Nevertheless, for such a problem it is more complicated to obtain analytical expressions or even estimates of the velocity and of the arrival time.

A first attempt in this direction has been recently presented in [10], where Eqs. (7) and (8) have been solved numerically with the conditions (9)–(12) and the parameter values shown in Table 1.

TABLE 1. Typical values and units of the parameters.

Symbol	Physical meaning	Typical value	Units
$T_{\rm app}$	applied temperature	0–100	°C
σ	diffusion coefficient	0.01	$m^2\ s^{-1}$
$T_{\rm ini}$	initial temperature	100	°C
$y_{\rm ini}$	initial crystallinity	0	–
T_f	nucleation temperature	40	°C
T_G	growth temperature	40	°C
$T_{\rm ref}$	reference temperature	0	°C
N_0	nucleation factor	1000	s^{-1}
G_0	growth factor	0.5	s^{-1}
β_N	nucleation exponent	0.1	$(^\circ C)^{-1}$
β_G	growth exponent	0.1	$(^\circ C)^{-1}$
v_0	initial mass	0.01	–
a_G	non-isothermal factor	2500	°C
L	length of the sample	10^{-3}	m

The result is that the temperature field exhibits an advancing boundary point at height T_f, approximately. This moving boundary is accompanied by an advancing front of crystallization in the mass distribution. Assuming that $T(x,t)$ is a monotone increasing function of x at every time t, as in the simple heat diffusion problem, $x_b(t)$ can be defined as the unique value such that

$$T(x_b(t),t) = T_f.$$

We can also define two abscissae to describe the mass front,

$$x_\alpha(t) = \max_t\{x/\ y(x,t)=1\} \tag{21}$$

$$x_\omega(t) = \min_t\{x/\ y(x,t)=0\}, \tag{22}$$

in order to characterize the solution in terms of the time evolution of these three parameters.

It appears now clearly that a Free Boundary exists in the temperature field, in the sense that $x_b(t)$ divides the interval $[0,L]$ into two parts, depending on the degree of crystallization: at time t, the part $x < x_b(t)$ is almost fully crystallized $(y=1)$, whereas the part $x > x_b(t)$ is almost empty of crystals $(y=0)$.

The following Free Boundary Problem can be formulated:

The free boundary $x_b(t)$ is a time-dependent parameter which divides $[0,L]$ into two intervals $I_1 = [0,x_b]$ and $I_2 = [x_b,L]$. Inside each interval, the temperature

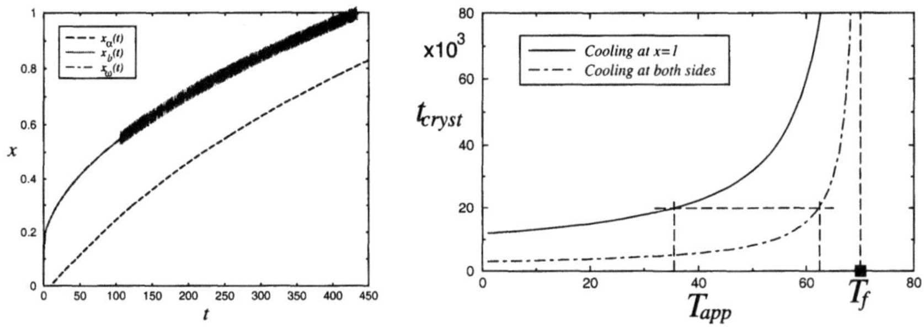

FIGURE 2. Left: Time evolution of the three coordinates which characterize the solution: the moving boundary $x_b(t)$ of the temperature field and the mass band $(x_\alpha(t), x_\beta(t))$. The velocity is slightly decreasing. Right: Crystallization time for two strategies of cooling in a slightly different case ($\sigma = 0.002$, $T_f = 70$, $N_0 = 20$ and $G_0 = 5$): cooling only at one side of the sample (solid line), and at both sides (dot-dashed line).

$T(x,t)$ solves the heat equation $T_t = \sigma T_{xx}$ with the initial condition (10) and the boundary conditions (11) and (12). The degree of crystallinity is such that $y = 1$ in I_1 and $y = 0$ in I_2, and $x_b(t)$ describes the crystallization process depicted in Fig. 2.

Still, we need an equation for the evolution of $x_b(t)$ and the boundary conditions for $T(x,t)$ at $x_b^+(t)$ and $x_b^-(t)$. The difficulty in completing the formulation lies in that the regions where $y = 1$ and $y = 0$ are in fact separated by a narrow band, and its width may not allow us to reduce the mass front to a point. It has been shown numerically that the mass front may be considered more a band than a layer [10], and an asymptotic argument for sufficiently large samples has recently established that what happens in the interior of the band can not be neglected [11].

Fig. 2(Left) shows the monitoring of the crystallization process by means of the three abscissae defined above. It is worth noticing that both the band of crystallization and the boundary in the temperature field advance with a decreasing velocity, which implies that two different space segments of the same length have different times of crystallization. We have also obtained numerically the arrival time defined in the simple heat diffusion problem; see Fig. 2(Right). The variation of the crystallization time as a function of the applied temperature is, as expected, monotonically increasing, but not linear. Thus, a better strategy of cooling can be suggested: namely, to cool the sample at both sides with a smaller (i.e. less expensive) temperature. The difference with the first strategy is as greater as the applied temperature is smaller, that is, as smaller is the time (i.e. the money!) spent in the process.

Analytical results are more difficult to obtain. For instance, we have provided a numerical temperature field that can be used in the hybrid stochastic model. We have used this field in a stochastic simulation of model (2), and we have displayed the result in Fig. 3, where we represent the degree of crystallization at different times.

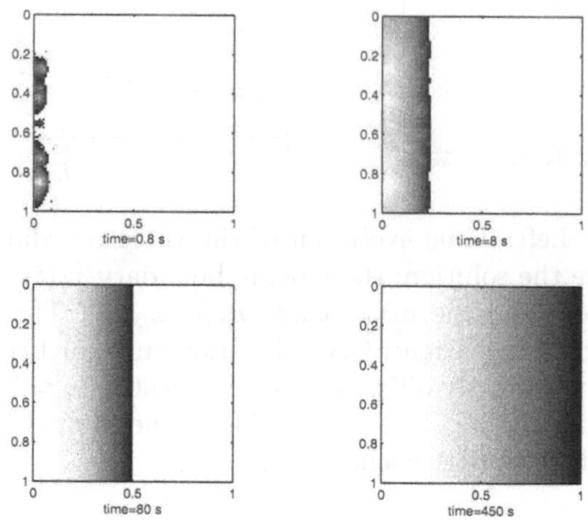

FIGURE 3. Evolution of the crystalline phase.

At the beginning of the process crystals develop randomly where the temperature is lower than T_f, but after a short time interval (less then 10 sec.), the crystallization front becomes deterministic and follows the evolution of the temperature field. In Figure 4, we can see the crystalline front compared with the moving boundary $x_b(t)$ of the temperature field.

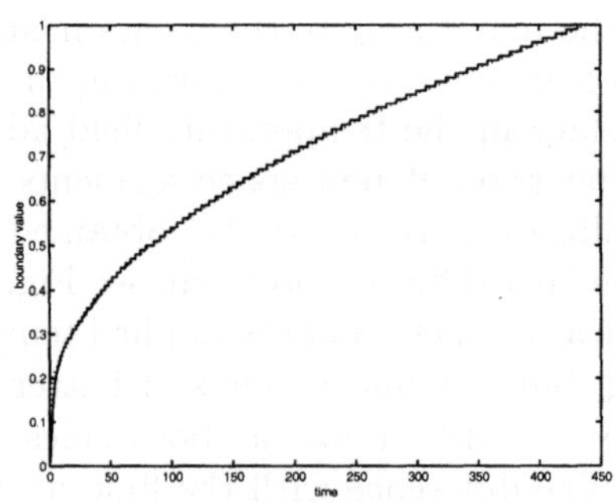

FIGURE 4. Crystalline front (solid line) and moving boundary $x_b(t)$ of the temperature field (dotted line).

Acknowledgments

Financial support is acknowledged to the EU under the TMR-Network *Differential Equations in Industry and Commerce*, grant N.FMRXCT97-0117, to the Italian MURST/cofin, program "Stochastic Processes with Spatial Structure", and to the Italian CNR contract n. 98.03635.ST74.

References

[1] V. Capasso, Ed., *Mathematical Modelling fo Polymer Processing. Polymerization, Crystallization, Manufacturing.* Mathematics in Industry, ECMI Subseries, Vol. 2, Springer-Verlag, Heidelberg, 2003.

[2] M. Avrami, *Kinetics of phase change. Part I*, J. Chem. Phys., **7** (1939), 1103–112.

[3] M. Burger,*Crystal Growth and Impingement in Polymer Melts*, this volume.

[4] M. Burger, V. Capasso, C. Salani, *Modeling multidimensional crystallization of polymers in interaction with heat transfer,* Nonlinear Analysis: Real World Applications, **3** (2002), 139–160.

[5] V. Capasso, A. Micheletti, *The hazard function of an inhomogeneous birth-and-growth process,* submitted.

[6] V. Capasso, A. Micheletti, M. Burger,*Densities of n-facets of incomplete Johnson-Mehl tessellation generated by inhomogeneous birth-and-growth processes,* submitted.

[7] V. Capasso, D. Morale and C. Salani, *Polymer Crystallization Processes via Many Particle Systems*, in [1], 243–259.

[8] V. Capasso, C. Salani, *Stochastic-birth-and-growth processes modeling crystallization of polymers with spatially heterogeneous parameters,* Nonlinear Analysis: Real World Applications, **1** (2000), 485–498.

[9] G. Eder, H. Janeschitz-Kriegl, *Structure development during processing: crystallization,* in Materials Science and Technology, Vol. 18 (H. Meijer, Ed.), Verlag-Chemie, Weinheim, 1997.

[10] R. Escobedo, V. Capasso, *Moving bands and moving boundaries with decreasing speed in polymer crystallization*, submitted.

[11] R. Escobedo, V. Capasso, *Reduction of a Mathematical Model for Polymer Crystallization*, in Progress in Industrial Mathematics at ECMI2002, Springer (to appear).

[12] V.R. Evans, *The laws of expanding circles and spheres in relation to the lateral growth rate of surface films and the grain-size of metals*, Trans. Faraday Soc., **41** (1945), 365–374.

[13] A.N. Kolmogorov, *On the statistical theory of the crystallization of metals*, Bull. Acad. Sci. USSR, Math. Ser. **1** (1937), 355–359.

[14] A. Micheletti, V. Capasso, *The stochastic geometry of the crystallization process of polymers*, this volume.

MIRIAM – Milan Research Centre
for Industrial and Applied Mathematics
and Dept.of Mathematics, University of Milan
via Saldini 50, 20133 Milan, Italy
E-mail address: Vincenzo.Capasso@mat.unimi.it,
escobedo@mat.unimi.it, salani@mat.unimi.it

International Series of Numerical Mathematics, Vol. 147, 87–101

Upscaling of Well Singularities in the Flow Transport through Heterogeneous Porous Media

Zhiming Chen and Xingye Yue

Abstract. We consider the steady flow transport through highly heterogeneous porous media driven by extraction wells. We develop a new upscaling technique lumping the small-scale details of the medium property into a few representative macroscopic parameters on a coarse scale, which preserve the large-scale behavior of the medium and are more appropriate for simulations. The method is based on the recently introduced over-sampling multiscale finite element method and on the introduction of new base functions that locally resolve the well singularities. Numerical experiments are carried out for flow transport in both periodic and randomly generated log-normal permeabilities to demonstrate the efficiency and accuracy of the proposed method.

1. Introduction

The central difficulty in the modelling of subsurface flow and transport is to account for the spatial variability in the parameters used to characterize the relevant physical properties of the natural porous media. In realistic situations, the precise spatial distribution of the parameters which are required to characterize the problem is never available due to the lack of enough data. Thus, sophisticated geological and geostatistical modelling tools are used in practice to generate highly detailed medium parameters, which are based on some site-specific measurements and on experience from other sites. There exists a vast literature on the upscaling or homogenization techniques that lump the small-scale details of the medium into a few representative macroscopic parameters on a coarse scale, which preserve the large-scale behavior of the medium and are more appropriate for reservoir simulations. We refer to the book of Christakos [3] for more information on the random field modelling of the natural porous medium parameters, and to the recent review paper [15] on the existent upscaling techniques in the engineering literature.

2000 *Mathematics Subject Classification.* Primary 65F10; Second 65F30.
The first author was supported in part by China NSF under the grant 10025102 and by China MOS under the grant G1999032802.

The study of steady flow through highly heterogeneous porous media driven by extraction wells is of great importance in hydrology and petroleum reservoir engineering. It is observed in the engineering literature (cf. e.g. [5], [12], and the references therein) that in the near-well region, many of the existing upscaling methods do not provide satisfactory results. The explanation may be that the standard upscaling methods generally assume that the pressure field is slowly varying, which is clearly not true in the proximity of the flowing wells [5]. This fact may also be accounted for mathematically on behalf of the homogenization theory for second order elliptic equations with periodic coefficients. In the homogenization theory, multiscale convergence is ensured under the assumption that the source should be at least in H^{-1}, so that the solution is bounded in the Sobolev space H^1. As we will see below, however, well singularities can be modelled as Dirac sources and thus, in the near-well region the solution behaves like a Green function which is not uniformly bounded in H^1.

The recently introduced multiscale finite element method [8, 9] for solving elliptic equations with oscillating coefficients provides an effective way to capture the large scale structures of the solutions on a coarse mesh without resolving all the fine scale structures. The central idea of the method is to incorporate the local small scale information of the leading order differential operator into the finite element bases. It is through these multiscale bases and the finite element formulation that the effect of small scales on the large scales is correctly captured. In section 2, we will describe one engineering upscaling technique and discuss its relation with the multiscale finite element method.

Let $\Omega \subset \mathbf{R}^2$ be a bounded domain with Lipschitz boundary Γ. Let us denote by $B(x_0, r)$ the disk centered at x_0 with radius $r > 0$. Let $B_j = B(\bar{x}_j, \delta)$, $1 \le j \le N$, be mutually disjoint subdomains inside Ω that are occupied by the wells. Denote by $\Omega_\delta := \Omega \backslash (\cup_{j=1}^N \bar{B}_j)$. We consider the following single phase pressure equation which is formed by combining Darcy's law with the conservation of mass

$$-\text{div}\,(K(x)\nabla u_\delta) = 0 \quad \text{in } \Omega_\delta, \tag{1}$$

where u_δ is pressure and K is the permeability, which is typically highly variable in space. We will impose a homogeneous Dirichlet boundary condition $u_\delta|_\Gamma = 0$ on the exterior boundary. The other kinds of boundary conditions can be dealt with similarly without any essential difficulties. On the well boundary $\Gamma_j = \partial B_j$, two quantities are of particular importance in practical applications: the well bore pressure $u_\delta|_{\Gamma_j}$ and the well flow rate $\int_{\Gamma_j} K \frac{\partial u_\delta}{\partial \nu} ds$, where ν is the unit outer normal to $\partial\Omega_\delta$. The boundary condition to be imposed on Γ_j is either the Dirichlet boundary condition which fixes the well bore pressure α_j (assumed to be constant)

$$u_\delta|_{\Gamma_j} = \alpha_j, \tag{2}$$

or the following mixed boundary condition which fixes the well flow rate q_j

$$u_\delta|_{\Gamma_j} = c_j = \text{const}, \qquad \int_{\Gamma_j} K \frac{\partial u_\delta}{\partial \nu} ds = \mathrm{q}_j. \tag{3}$$

The constant c_j in (3) is unknown.

The purpose of this paper is to develop a complete multiscale coarse grid algorithm, based on the multiscale finite element method, for solving steady flow problem involving well singularities in heterogeneous porous media. The additional well singularities of the problem are resolved locally by adding finite element base functions. The final coarse grid model is based on a variational formulation which is different from the heuristical techniques in [5, 12]. This rigorous variational coarse grid model allows us to establish a complete multiscale error analysis for locally periodic medium properties based on the homogenization theory. We emphasize that, as pointed out in [6], the spatial periodicity assumption does not a priori restrict the applicability of the results only to media exhibiting such a strict repetitive spatial ordering in the properties of interest. Indeed, the numerical experiments carried out for random log-normal relative permeabilities demonstrate clearly the applicability of our method beyond the periodic structures.

Since the size of the wells δ is negligible in practical situations, the first approximation we can make is to replace (1)–(3) by the following problem

$$-\operatorname{div}(K(x)\nabla u) = \sum_{j=1}^{N} \mathrm{q}_j \delta_{\bar{x}_j} \quad \text{in } \Omega \tag{4}$$

with the boundary condition $u|_\Gamma = 0$. Here $\delta_{\bar{x}_j}$ is the Dirac measure at $\bar{x}_j$, $1 \le j \le N$. Denote by $K_j = K(\bar{x}_j)$, $\phi_j = -\dfrac{1}{2\pi K_j} \ln|x - \bar{x}_j|$ and $U = u - \sum_{j=1}^{N} \mathrm{q}_j \phi_j$. Note that the flow rates q_i in (4), for $i \in I_D$, are unknown. They are determined through the following additional conditions which are obtained by requiring $u|_{\Gamma_i} \approx \alpha_i$ as the approximation of the boundary conditions (2):

$$-\frac{\mathrm{q}_i}{2\pi K_i} \ln\delta + \sum_{j \ne i} \mathrm{q}_j \phi_j(\bar{x}_i) + U(\bar{x}_i) = \alpha_i, \qquad i \in I_D. \tag{5}$$

We derive the following error estimate between the solution u_δ of the original problem (1)–(3) and the solution u of (4)–(5)

$$\max_{x \in \bar{\Omega}_\delta} |u - u_\delta| \le C\delta |\ln\delta| \sum_{j=1}^{N} |\mathrm{q}_j|.$$

To our knowledge, this error estimate is the first in the literature, although the convergence of the solution u_δ to u as $\delta \to 0$ is well-known in certain sense (cf. e.g. [10]). Instead of presenting the full proof of the estimate, which can be found in [2], we will present in section 3 the key ideas of the proof in the simplified situation of a single well and constant permeability. In this case, the proof is more transparent and the estimate is sharper (without the logarithmic factors).

To introduce our multiscale algorithm, we first introduce an equivalent variational formulation for (4), which is suitable for our multiscale approximation. Let $\Omega_j, 1 \le j \le N$, be mutually disjoint subdomains inside Ω such that $\bar{B}_j \subset\subset \Omega_j$.

Let G_j be the Green function associated with the domain Ω_j

$$-\mathrm{div}\,(K(x)\nabla G_j) = \delta_{\bar{x}_j} \quad \text{in } \Omega_j, \qquad G_j|_{\Sigma_j} = 0, \tag{6}$$

where $\Sigma_j = \partial\Omega_j$. Now, for any $v \in C_0^\infty(\Omega)$ we have

$$\begin{aligned} v(\bar{x}_j) &= \int_{\Omega_j} -\mathrm{div}\,(K(x)\nabla G_j) v dx \\ &= \int_{\Omega_j} K(x)\nabla G_j \nabla v dx - \int_{\Sigma_j} K\frac{\partial G_j}{\partial \nu} v ds. \end{aligned}$$

On the other hand, from (4) we know that

$$\sum_{j=1}^{N} \mathrm{q}_j v(\bar{x}_j) = \int_\Omega K(x)\nabla u \nabla v dx.$$

Let $G_j = 0$ for $x \in \Omega\backslash\bar{\Omega}_j$ and $\zeta = u - \sum_{j=1}^N \mathrm{q}_j G_j$, then we know that $\zeta \in H_0^1(\Omega)$ satisfies the following variational form

$$\int_\Omega K(x)\nabla\zeta\nabla v dx = -\sum_{j=1}^{N} \mathrm{q}_j \int_{\Sigma_j} K\frac{\partial G_j}{\partial \nu} v dx \qquad \forall v \in C_0^\infty(\Omega). \tag{7}$$

Denote by $w_j = G_j + \dfrac{1}{2\pi K_j}\ln|x - \bar{x}_j|$ in Ω_j, then the complementary condition (5) to determine q_i for $i \in I_D$ becomes

$$-\frac{\mathrm{q}_i}{2\pi K_i}\ln\delta + \mathrm{q}_i w_i(\bar{x}_i) + \zeta(\bar{x}_i) = \alpha_i, \quad i \in I_D. \tag{8}$$

We note that for ζ the singularities of the original solution u are removed and we will use the over-sampling multiscale finite element method to discretize it on a coarse grid.

In section 5, we will discuss the implementation of our algorithm and show some numerical results for randomly generated coefficients to demonstrate the accuracy of the method.

2. Upscaling of the permeability

The purpose of this section is to show that one of the well-known engineering upscaling techniques (see e.g. [11]) is equivalent to the multiscale finite element method proposed in [7, 8]. We remark that the multiscale finite element method is shown to be convergent under the condition that the permeability is locally periodic $K_\varepsilon(x) = K(x, x/\varepsilon)$, where $K(x,\cdot)$ is periodic with respect to the second variable. As a consequence, the convergence of the engineering approach described in this section is guaranteed.

Let $\mathcal{M}_H$ be a finite element mesh of Ω with the mesh size H much larger than the ε, the characteristic length representing the small scale variability of the media. Usually, ε is equal to the correlation length in the statistical random field modelling of the media. Let W_H be the standard conforming linear finite element space over $\mathcal{M}_H$ and $W_H^0 = W_H \cap H_0^1(\Omega)$. In the engineering literature, the problem

$$-\operatorname{div}(K_\varepsilon(x)\nabla u_\varepsilon) = f \quad \text{in } \Omega, \quad u_\varepsilon = 0 \quad \text{on } \Gamma \tag{9}$$

is approximated by the homogenized or upscaled problem: find $u_H^* \in W_H^0$ such that

$$\int_\Omega K^*(x)\nabla u_H^* \nabla v_H dx = \int_\Omega f v_H dx \quad \forall v_H \in W_H^0, \tag{10}$$

K^* being piecewise constant on the coarse mesh $\mathcal{M}_H$. The so-called effective permeability matrix K^* on each $T \in \mathcal{M}_H$ is defined as follows. For any $G \in \mathbf{R}^2$, let θ_ε be the solution of the problem

$$-\operatorname{div}(K_\varepsilon \nabla \theta_\varepsilon) = 0 \text{ in } T, \quad \theta_\varepsilon|_{\partial T} = G \cdot x.$$

A simple integration by parts implies that

$$\overline{\nabla\theta_\varepsilon} = \frac{1}{|T|}\int_T \nabla\theta_\varepsilon dx = \frac{1}{|T|}\int_{\partial T} Gx \cdot \nu ds = \frac{1}{|T|}\int_T \nabla(Gx)dx = G. \tag{11}$$

On the other hand, since $Q = \overline{K_\varepsilon \nabla\theta_\varepsilon} = \frac{1}{|T|}\int_T K_\varepsilon \nabla\theta_\varepsilon dx$ is linear in G, there exists a matrix K^* such that

$$Q = K^* \cdot G \iff -\overline{K_\varepsilon \nabla\theta_\varepsilon} = -K^*\overline{\nabla\theta_\varepsilon}. \tag{12}$$

The multiscale finite element method introduced in [7, 8] is based on multiscale finite element base functions. For any $T \in \mathcal{M}_H$, let ψ_i, $i = 1, 2, 3$, be the linear nodal bases. Define ϕ_i^T, $i = 1, 2, 3$, as the solution of the local problem

$$-\operatorname{div}(K_\varepsilon \nabla \phi_i^T) = 0 \quad \text{in } T, \quad \phi_i^T|_{\partial T} = \psi_i.$$

Denote by $V(T) = \operatorname{span}\{\phi_i^T, i = 1, 2, 3\}$ and by V_H the finite element space

$$V_H = \{v_H \in H^1(\Omega) : v_H|_T \in V(T)\}.$$

Then the multiscale finite element approximation to (9) is: find $u_H \in V_H^0 = V_H \cap H_0^1(\Omega)$ such that

$$\int_\Omega K_\varepsilon \nabla u_H \nabla v_H dx = \int_\Omega f v_H dx \quad \forall v_H \in V_H^0. \tag{13}$$

We have the following proposition concerning the relationship of these two upscaling methods.

Proposition 2.1. *Let A^* and A be the stiffness matrices corresponding to the discrete problems* (10) *and* (13), *respectively. Then we have $A^* = A$.*

Proof. It is clear that we only need to prove that the element stiffness matrices coincide, that is, for any $T \in \mathcal{M}_T$ and $i, j = 1, 2, 3$,

$$\int_T K^* \nabla\psi_i \nabla\psi_j dx = \int_T K_\varepsilon \nabla\phi_i^T \nabla\phi_j^T dx.$$

To do that, we note that by (11) we have $\overline{\nabla\phi_i^T} = \nabla\psi_i$. Since $\nabla\psi_i, \nabla\psi_j$ are constant over T, we get by (12) that

$$\int_T K^*\nabla\psi_i\nabla\psi_j dx = |T|K^*\overline{\nabla\phi_i^T}\nabla\psi_j = \int_T K_\varepsilon\nabla\phi_i^T\nabla\psi_j dx,$$

which, after integration by parts and using the definition of the multiscale base functions, yields

$$\begin{aligned}\int_T K^*\nabla\psi_i\nabla\psi_j dx = \int_{\partial T}(K_\varepsilon\nabla\phi_i^T\cdot\nu)\psi_j ds &= \int_{\partial T}(K_\varepsilon\nabla\phi_i^T\cdot\nu)\phi_j^T ds\\ &= \int_T K_\varepsilon\nabla\phi_i^T\nabla\phi_j^T dx.\end{aligned}$$

This completes the proof. □

3. The mathematical modelling

Let Ω be a convex polygon in $\mathbf{R}^2$ with boundary Γ. Let $B(x_0,r) = \{x \in \mathbf{R}^2 : |x-x_0| < r\}$. Let $B_j = B(\bar{x}_j,\delta)$, $1 \le j \le N$, be mutually disjoint subdomains inside Ω which are occupied by the wells. Set $\Gamma_j = \partial B_j$ and $\Omega_\delta = \Omega\backslash(\cup_{j=1}^N \bar{B}_j)$. We consider the following problem

$$-\mathrm{div}\,(K(x)\nabla u_\delta) = 0 \quad \text{in } \Omega_\delta, \tag{14}$$

$$u_\delta|_{\Gamma_j} = \text{const}\,, \qquad \int_{\Gamma_j} K\frac{\partial u_\delta}{\partial\nu}ds = \mathrm{q}_j \quad \text{for } j \in I_M, \tag{15}$$

$$u_\delta|_{\Gamma_j} = \alpha_j \quad \text{for } j \in I_D, \qquad u_\delta|_\Gamma = 0. \tag{16}$$

Here I_D, I_M are index sets such that $I_D \cup I_M = \{1,2,\cdots,N\}$, while α_j for $j \in I_D$ and q_i for $i \in I_M$ are known constants. We imposed the following conditions on the coefficient $K(x)$:

(H1) $K \in C^{0,1}(\bar{\Omega})$, and there exist constants $\lambda, \Lambda > 0$ such that $\lambda \le K(x) \le \lambda^{-1}$ and $|K(x)-K(y)| \le \Lambda|x-y|$ for any $x,y \in \bar{\Omega}$. As we are interested in highly variable coefficients, we can assume $\Lambda \ge 1$ without loss of generality.

As already mentioned in §1, we will consider the following approximation of the problem (14)–(16): find u and the constants $\{\mathrm{q}_i\}_{i\in I_D}$, such that

$$-\mathrm{div}\,(K(x)\nabla u) = \sum_{j=1}^N \mathrm{q}_j\delta_{\bar{x}_j} \quad \text{in } \Omega, \tag{17}$$

$$-\frac{\mathrm{q}_i}{2\pi K_i}\ln\delta + \sum_{j\ne i}\mathrm{q}_j\phi_j(\bar{x}_i) + U(\bar{x}_i) = \alpha_i \quad \text{for } i \in I_D, \tag{18}$$

$$u|_\Gamma = 0. \tag{19}$$

Here $\delta_{\bar{x}_j}$ is the Dirac measure at $\bar{x}_j$, $K_j = K(\bar{x}_j)$, $\phi_j(x) = -\dfrac{1}{2\pi K_j} \ln|x - \bar{x}_j|$, and $U = u - \sum_{j=1}^{N} \mathrm{q}_j \phi_j$. The relation (18) can be viewed as an approximation of the boundary condition $u|_{\Gamma_i} \approx \alpha_i$ for $i \in I_D$.

The following error estimate is derived in [2].

Theorem 3.1. *Let the assumption* (H1) *be satisfied. Let u_δ and u be the solutions of* (14)–(16) *and* (17)–(19), *respectively. Then there exists a constant $\gamma_0 \in (0,1)$ such that for $\delta\Lambda \leq \gamma_0$ the problem* (17)–(19) *has a unique solution. Moreover, there exists a constant C independent of δ, q_j and Λ such that*

$$\max_{x \in \bar{\Omega}_\delta} |u - u_\delta| \leq C\Lambda(1 + \ln\Lambda)^{1/2} \Big(\sum_{j=1}^{N} |\mathrm{q}_j| \Big) \delta |\ln\delta|,$$

and, for every $i \in I_D$,

$$|\mathrm{q}_i - \mathrm{q}_i^\delta| \leq C\Lambda(1 + \ln\Lambda)^{1/2} \Big(\sum_{j=1}^{N} |\mathrm{q}_j| \Big) \delta |\ln\delta|,$$

where

$$\mathrm{q}_i^\delta = \int_{\Gamma_i} K_\varepsilon \frac{\partial u_\delta}{\partial \nu} ds \quad \textit{for } i \in I_D.$$

The proof of this theorem is rather complicated. It depends on a sharp estimate between the well flow rate and the well bore pressure and an explicit dependence on p of the constant in the classical $W^{1,p}$ interior estimate for second order elliptic equations. Instead of presenting the full proof of the estimate which can be found in [2], we will discuss in the sequel the simplified situation of a single well flow in media with constant permeability.

The starting point of the proof is the following basic lemma.

Lemma 3.2. *Let $\Omega_\delta = \Omega \backslash \bar{B}(x_0, \delta)$, $\Gamma_\delta = \partial B(x_0, \delta)$ and $\Gamma = \partial\Omega$. Consider the problem:*

$$-\Delta v = 0 \ \text{ in } \Omega_\delta, \quad v = 1 \ \text{ on } \Gamma_\delta, \quad v = 0 \ \text{ on } \Gamma.$$

Then

$$\frac{2\pi}{\ln D_0/\delta} \leq \int_{\Gamma_\delta} \frac{\partial v}{\partial \nu} ds \leq \frac{2\pi}{\ln d_0/\delta},$$

where $d_0 = \min_{x\in\Gamma} |x - x_0|$ and $D_0 = \max_{x\in\Gamma} |x - x_0|$.

Proof. Let $r = |x - x_0|$. Consider the following barrier functions

$$v_\delta^+ = 1 - \frac{\ln(r/\delta)}{\ln(D_0/\delta)}, \ v_\delta^- = 1 - \frac{\ln(r/\delta)}{\ln(d_0/\delta)}.$$

Let $w_\delta^\pm = v_\delta^\pm - v$. Then it is clear that $-\Delta w_\delta^\pm = 0$ in Ω_δ, $w_\delta^\pm = 0$ on Γ_δ, and $w_\delta^+ \geq 0, w_\delta^- \leq 0$ on Γ. Thus, by the maximum principle we get $w_\delta^+ \geq 0$, $w_\delta^- \leq 0$

in Ω_δ. This yields $v_\delta^- \le v \le v_\delta^+$ in Ω_δ. By the definition of $v_\delta^\pm$, we obtain

$$-\frac{\ln r - \ln\delta}{r-\delta}\cdot\frac{1}{\ln(d_0/\delta)} \le \frac{v-1}{r-\delta} \le -\frac{\ln r - \ln\delta}{r-\delta}\cdot\frac{1}{\ln(D_0/\delta)} \quad \text{in } \Omega_\delta.$$

By letting $r \to \delta^+$, we get the desired estimate. □

A direct consequence of the basic lemma is the following corollary.

Corollary 3.3. *Consider the problem:*

$$-\Delta\zeta = 0 \ \ in\ \Omega_\delta, \quad \zeta|_\Gamma = 0, \quad \zeta|_{\Gamma_\delta} = \beta_\delta, \quad \int_{\Gamma_\delta}\frac{\partial\zeta}{\partial\nu}ds = \mathrm{q} > 0.$$

Then we have

$$\frac{\mathrm{q}}{2\pi}\ln\left(\frac{d_0}{\delta}\right) \le \beta_\delta \le \frac{\mathrm{q}}{2\pi}\ln\left(\frac{D_0}{\delta}\right).$$

Now we consider the problem with a single well and constant permeability

$$-\Delta u_\delta = 0 \ \text{ in } \Omega_\delta, \quad u_\delta|_\Gamma = 0, \quad u_\delta|_{\Gamma_\delta} = \alpha_\delta, \quad \int_{\Gamma_\delta}\frac{\partial u_\delta}{\partial\nu}ds = \mathrm{q}, \tag{20}$$

and its limit problem

$$-\Delta u = \mathrm{q}\delta_{x_0} \ \text{ in } \Omega_\delta, \quad u|_\Gamma = 0. \tag{21}$$

Let $u = -\frac{\mathrm{q}}{2\pi}\ln|x-x_0| + w$ and $v_\delta = u_\delta - u$. Then we have

$$-\Delta v_\delta = 0 \ \text{ in } \Omega_\delta, \quad v_\delta|_\Gamma = 0$$

$$v_\delta|_{\Gamma_\delta} = \alpha_\delta + \frac{\mathrm{q}}{2\pi}\ln\delta - w|_{\Gamma_\delta}, \quad \int_{\Gamma_\delta}\frac{\partial v_\delta}{\partial\nu}ds = 0.$$

The problem is now reduced to estimating $\| v_\delta \|_{0,\infty,\Omega_\delta}$. In order to do this, we first split v_δ into two parts: $v_\delta = v_1 + v_2$, where v_1 and v_2 satisfy

$$-\Delta v_1 = 0 \ \text{ in } \Omega_\delta, \quad v_1|_\Gamma = 0, \quad v_1 = w(x_0) - w \ \text{ on } \Gamma_\delta,$$

and

$$-\Delta v_2 = 0 \ \text{ in } \Omega_\delta, \quad v_2|_\Gamma = 0,$$

$$v_2|_{\Gamma_\delta} = \alpha_\delta + \frac{\mathrm{q}}{2\pi}\ln\delta - w(x_0) =: \alpha_\delta', \quad \int_{\Gamma_\delta}\frac{\partial v_2}{\partial\nu}ds = -\int_{\Gamma_\delta}\frac{\partial v_1}{\partial\nu}ds.$$

Since w is smooth around x_0, it is easy to see that $|w - w(x_0)| \le C_0\delta$ on Γ_δ, which, in turn, implies that $|v_1| \le C_0\delta$ in Ω_δ.

Let $\theta = C_0\delta v \pm v_1$, then $-\Delta\theta = 0$ in Ω_δ, $\theta = 0$ on Γ and $\theta \ge 0$ on Γ_δ. Thus $\theta \ge 0$ in Ω_δ and $\frac{\partial\theta}{\partial\nu} \le 0$ on Γ. Therefore

$$\begin{aligned}\left|\int_{\Gamma_\delta}\frac{\partial v_1}{\partial\nu}ds\right| &= \left|\int_{\Gamma}\frac{\partial v_1}{\partial\nu}ds\right| \le C_0\delta\left|\int_{\Gamma}\frac{\partial v}{\partial\nu}ds\right| \\ &= C_0\delta\left|\int_{\Gamma_\delta}\frac{\partial v}{\partial\nu}ds\right| \le \frac{2\pi C_0\delta}{\ln(d_0/\delta)}.\end{aligned}$$

Now by the third equality in the definition of v_2, we know that

$$\left| \int_{\Gamma_\delta} \frac{\partial v_2}{\partial \nu} ds \right| \leq \frac{2\pi C_0 \delta}{\ln(d_0/\delta)},$$

which implies by Corollary 3.3 that

$$|\alpha'_\delta| \leq \frac{1}{2\pi} \frac{2\pi C_0 \delta}{\ln(d_0/\delta)} \ln\left(\frac{D_0}{\delta}\right) \leq C\delta$$

A final application of the maximum principle gives $|v_2| \leq C\delta$. This proves the following theorem.

Theorem 3.4. *Let u_δ and u be the solutions of* (20) *and* (21), *respectively. Then there exists a constant C independent of δ such that the following estimate holds*

$$\max_{x \in \bar{\Omega}_\delta} |u_\delta - u| \leq C\delta.$$

4. The multiscale method

In this section, we are going to introduce our multiscale method to solve the problem (17)–(19), which is a good approximation of the original problem (14)–(16) for small δ according to the analysis in last section. We make the following local periodicity assumption for the coefficient:

(H2) $K(x) = K_\varepsilon(x) = K(x, x/\varepsilon)$ for some small parameter $\varepsilon > 0$, and we assume $K \in C^1(\bar{\Omega}; C^1_p(\mathbf{R}^2))$, where $C^1_p(\mathbf{R}^2)$ stands for the collection of all the $C^1(\mathbf{R}^2)$, periodic functions with respect to the unit square Y.

Our method is based on the over-sampling multiscale finite element method in [7], which reduces the so-called resonance error in the previous multiscale finite element method in [8, 9]. We first briefly recall the basic facts of the over-sampling multiscale finite element method. Let $\mathcal{M}_H$ be a regular and quasi-uniform triangulation of Ω and V_H the standard conforming linear finite element space over $\mathcal{M}_H$. For any $T \in \mathcal{M}_H$ with nodes $\{x_i^T\}_{i=1}^3$, let H_T denote the size of T, $P_1(T)$ the set of linear polynomials defined in T, and $\{\varphi_i^T\}_{i=1}^3$the basis of $P_1(T)$ satisfying $\varphi_i^T(x_j^T) = \delta_{ij}, i,j = 1,2,3$. For any $T \in \mathcal{M}_H$, we denote by $S = S(T)$ a macro-element which contains T and satisfies the following condition:

(H3) $H_S \leq C_1 H_T$ and $\text{dist}\,(\partial T, \partial S) \geq \delta_0 H_T$ for some positive constants C_1, δ_0 independent of H. The minimal angle of $S(T)$ is bounded from below by some positive constant θ_0 independent of H.

Let $MS(S)$ be the multiscale finite element space spanned by $\psi_i^S, i = 1,2,3$, $\psi_i^S \in H^1(S)$ being the solution of the problem

$$-\text{div}\,(K_\varepsilon(x)\nabla\psi_i^S) = 0 \quad \text{in } S, \qquad \psi_i^S|_{\partial S} = \varphi_i^S.$$

Here $\{\varphi_i^S\}_{i=1}^3$ is the nodal basis of $P_1(S)$ such that $\varphi_i^S(x_j^S) = \delta_{ij}$, $i,j = 1,2,3$. The over-sampling multiscale finite element base functions over T is defined by

$$\bar{\psi}_i^T = c_{ij}^T \psi_j^S|_T \quad \text{in } T,$$

with the constants c_{ij}^T chosen in such a way that

$$\varphi_i^T = c_{ij}^T \varphi_j^S|_T \quad \text{in } T.$$

The existence of the constants c_{ij}^T is ensured from the fact that $\{\varphi_j^S\}_{j=1}^3$ also forms the basis of $P_1(T)$. In practical applications, the base functions ψ_i^S of $MS(S)$ will be further approximated by solving (22) on a mesh of S, with a mesh size resolving ε and using linear finite element methods [7]. These base functions can be preliminarily computed to generate the coarse grid operator, which turns out to be only a small overhead of the overall computations. This is also a common practice in many upscaling methods [15].

Let $OMS(T) = \text{span}\{\bar{\psi}_i^T\}_{i=1}^3$ and $\Pi_T : OMS(T) \to P_1(T)$ the projection

$$\Pi_T \psi = c_i \varphi_i^T \quad \text{if } \psi = c_i \bar{\psi}_i^T \in OMS(T).$$

Let $\bar{X}_H$ be the finite element space

$$\bar{X}_H = \{\psi_H : \psi_H|_T \in OMS(T),\ \forall T \in \mathcal{M}_H\}$$

and define $\Pi_H : \bar{X}_H \to \Pi_{T\in\mathcal{M}_H} P_1(T)$ through the relation $\Pi_H \psi_H|_T = \Pi_T \psi_H$ for any $T \in \mathcal{M}_H$, $\psi_H \in \bar{X}_H$. The over-sampling multiscale finite element space is then defined as

$$X_H = \{\psi_H \in \bar{X}_H : \Pi_H \psi_H \in V_H \subset H^1(\Omega)\}.$$

In general, $X_H \not\subset H^1(\Omega)$ and the requirement $\Pi_H \psi_H \in V_H$ serves to impose some continuity of the functions $\psi_H \in X_H$ across the interelement boundaries.

For any $j = 1, 2, \cdots, N$, let Ω_j be a convex polygon inside Ω, which consists of a union of elements in $\mathcal{M}_H$ such that $\bar{B}_j \subset\subset \Omega_j$. We assume $\bar{\Omega}_i \cap \bar{\Omega}_j = \emptyset$ for any $i \neq j$. Let G_j^ε be the Green function

$$-\text{div}\,(K_\varepsilon(x)\nabla G_j^\varepsilon) = \delta_{\bar{x}_j} \quad \text{in } \Omega_j, \quad G_j^\varepsilon|_{\Sigma_j} = 0,$$

where $\Sigma_j = \partial\Omega_j$. Let u_ε and $\{\mathrm{q}_i\}_{i\in I_D}$ be the solution of the problem (17)–(19). Then the argument of §1 yields that $\zeta_\varepsilon = u_\varepsilon - \sum_{j=1}^N \mathrm{q}_j G_j^\varepsilon \in H_0^1(\Omega)$ and the constants $\{\mathrm{q}_i\}_{i\in I_D}$ satisfy

$$\int_\Omega K_\varepsilon(x)\nabla\zeta_\varepsilon \nabla\chi dx = -\sum_{j=1}^N \mathrm{q}_j \int_{\Sigma_j} K_\varepsilon \frac{\partial G_j^\varepsilon}{\partial\nu}\chi ds \quad \forall \chi \in H_0^1(\Omega),$$

$$-\frac{\mathrm{q}_i}{2\pi K_i}\ln\delta + \mathrm{q}_i w_i(\bar{x}_i) + \zeta_\varepsilon(\bar{x}_i) = \alpha_i \quad \forall i \in I_D.$$

Here and in the sequel G_j^ε is extended to be zero outside Ω_j, and $w_j = G_j^\varepsilon - \phi_j$, with $\phi_j = -\dfrac{1}{2\pi K_j}\ln|x - \bar{x}_j|$.

Let $X_H^0 = \{\psi_H \in X_H : \Pi_H \psi_H = 0 \text{ on } \Gamma\}$. Then we introduce the following discrete problem: find $\zeta_H \in X_H^0$ and $\{\mathrm{q}_i^H\}_{i \in I_D}$ such that

$$\sum_{T \in \mathcal{M}_H} \int_T K_\varepsilon(x) \nabla \zeta_H \nabla \chi_H dx = -\sum_{j=1}^{N} \mathrm{q}_j^H \int_{\Sigma_j} K_\varepsilon \frac{\partial G_j^\varepsilon}{\partial \nu} \hat{\chi}_H ds \quad \forall \chi_H \in X_H^0,$$
$$-\frac{\mathrm{q}_i^H}{2\pi K_i} \ln \delta + \mathrm{q}_i^H w_i(\bar{x}_i) + \zeta_H(\bar{x}_i) = \alpha_i \quad \forall i \in I_D,$$

where $\hat{\chi}_H = \Pi_H \chi_H \in V_H^0 = V_H \cap H_0^1(\Omega)$, and we set $\mathrm{q}_j^H = \mathrm{q}_j$ for $j \in I_M$ to simplify the notation.

We make the following assumption on the local domains Ω_j.

(H4) Let $H_j = \mathrm{dist}\,(\bar{x}_j, \Sigma_j)$: there exist constants $C_2, C_3 > 0$ such that $C_2 H_i \le H_j \le C_3 H_i$ for any $i, j = 1, 2, \cdots, N$.

In practical applications, the most interesting quantities are the well bore pressure for the wells prescribing the flow rate and the flow rate for the wells prescribing the well bore pressure. The following theorem establishes the error estimate for these quantities.

Theorem 4.1. *Let the assumptions* (H1)–(H4) *be fulfilled. Then there exists a constant C independent of ε, δ, H and H_j such that, for $j \in I_M$,*

$$\max_{x \in \Gamma_j} |\zeta_\varepsilon - \zeta_H(\bar{x}_j)| \le C\Big\{H + \frac{H^2}{H_j^2}\Big|\ln \frac{H}{H_j}\Big|^2 + \frac{\varepsilon}{H} + \sqrt{\frac{\varepsilon}{H_j}}\Big\}|\ln H| \sum_{i=1}^{N} |\mathrm{q}_i|,$$

and, for $j \in I_D$,

$$|\mathrm{q}_j - \mathrm{q}_j^H| \le C\Big\{H + \frac{H^2}{H_j^2}\Big|\ln \frac{H}{H_j}\Big|^2 + \frac{\varepsilon}{H} + \sqrt{\frac{\varepsilon}{H_j}}\Big\}|\ln H| \sum_{i=1}^{N} |\mathrm{q}_i|.$$

The proof of this theorem can be found in [2], in which, some sharp bounds on the homogenization of the Green functions are derived.

5. Numerical results

We start this section by summarizing the algorithm proposed in this paper for solving the problem (17)–(19), which is a good approximation of the original problem (14)–(16) when the size of the wells is negligible.

Algorithm 1. Given the well bore pressure α_i for $i \in I_D$ and the well flow rate q_j for $j \in I_M$, the following procedure finds the approximate well bore pressure α_j^H for $j \in I_M$, the approximate well flow rate q_i^H for $i \in I_D$, and the approximate pressure $u_H = \zeta_H + \sum_{j=1}^{N} \mathrm{q}_j^H G_j^h$, where $\mathrm{q}_j^H = \mathrm{q}_j$ for $j \in I_M$.

- For $j = 1, \cdots, N$, compute the discrete Green function G_j^h on each subdomain Ω_j, i.e. $G_j^h \in V_h^0(\Omega_j)$ such that

$$\int_{\Omega_j} K(x)\nabla G_j^h \nabla v_h dx = v_h(\bar{x}_j) \qquad \forall v_h \in V_h^0(\Omega_j). \tag{22}$$

Here $V_h^0(\Omega_j) = V_h(\Omega_j) \cap H_0^1(\Omega_j)$, and $V_h(\Omega_j)$ is the standard conforming linear finite element space over the mesh $\mathcal{M}_h(\Omega_j)$ of Ω_j, with the mesh size h resolving the scale of the permeability field $K(x)$. Compute the associate effective radius $\bar{r}_j$ from G_j^h according to the following relation

$$\frac{1}{2\pi K_j}\ln \bar{r}_j = \int_{\Omega_j}(K - K_j)\nabla\phi\nabla G_j^h dx - \int_{\Omega_j} K\nabla G_j^h \nabla\psi_h, \tag{23}$$

where $\phi = -\frac{1}{2\pi K_j}\ln|x - x_j|$ and $\psi_h \in V_h(\Omega_j)$, with nodal values defined by

$$\psi_h(x_k) = \begin{cases} \phi(x_k) & \text{if } x_k \in \partial\Omega_j, \\ 0 & \text{otherwise.} \end{cases}$$

Thus the approximate value of the Green function G_j on Γ_j is

$$\alpha_j^h = G_j^h(\bar{x}_j) - \frac{1}{2\pi K_j}\ln\frac{\delta}{\bar{r}_j}, \qquad K_j = K(\bar{x}_j).$$

- Use the over-sampling multiscale finite element method to find $\zeta_H \in X_H^0$ and $\{\mathrm{q}_i^H\}_{i\in I_D}$ such that

$$\sum_{T\in\mathcal{M}_H}\int_T K(x)\nabla\zeta_H\nabla\chi_H dx = -\sum_{j=1}^N \mathrm{q}_j^H \int_{\Sigma_j} K\frac{\partial G_j^h}{\partial\nu}\hat{\chi}_H ds \quad \forall\chi_H \in X_H^0,$$

$$\mathrm{q}_i^H\alpha_i^h + \zeta_H(\bar{x}_i) = \alpha_i \quad \text{for } i \in I_D.$$

- Compute the approximate well bore pressure α_j^H for $j \in I_M$ through the relation

$$\alpha_j^H = \zeta_H(\bar{x}_j) + \mathrm{q}_j\alpha_j^h.$$

The formula (23) defining the effective radius extends the well-known Peaceman method [13, 14] in the engineering literature, for five point finite difference discretization and constant permeability, to the general case and it is convergent for any finite element meshes and any heterogeneous porous media, as proved in [2].

Now we report one numerical example carried out for random log-normal permeability to show the accuracy and efficiency of our method. More extensive numerical experiments can be found in [2]. The computations are performed on the unit square $\Omega = (0,1)\times(0,1)$. We consider two wells $B_j = B(\bar{x}_j, \delta)$ with $x_1 = (\frac{1}{4}, \frac{3}{4})$, $x_2 = (\frac{3}{4}, \frac{1}{4})$. We impose the mixed boundary condition (3), which fixes the well flow rate q_j for both wells. Thus we have $I_D = \emptyset$ and the interested quantities are the well bore pressure (WBP) on each well. In the computations,

Well No.	Radius	'Exact' WBP	WBP	Relative error
1	1.E-05	-1.7278769	-1.7013823	0.1533E-01
1	1.E-04	-1.3540093	-1.3275152	0.1957E-01
1	1.E-03	-0.9801418	-0.9536476	0.2703E-01
2	1.E-05	3.7894742	3.8199386	0.7773E-02
2	1.E-04	2.8264341	2.8558985	0.1042E-01
2	1.E-03	1.8623940	1.8918584	0.1582E-01

TABLE 1. Results of the Algorithm 1 in the case of random log-normal field with $\sigma^2 = 1.69$ and $l_x = l_y = 0.01$.

we take $q_1 = -1$ and $q_2 = 1$, which corresponds to the situation in which the well B_1 is an extraction well and B_2 is an injection well.

The domain Ω_j is taken to be the square centered at $\bar{x}_j$ with the length of the sides being $2H_j$ so that $H_j = \text{dist}(\bar{x}_j, \partial\Omega_j)$. The discrete Green function G_j^h in (22) is computed on a uniform finite element mesh of size $h = H_j/512$.

We generate the random field $K(x)$ on a uniform 1024×1024 mesh by using the moving ellipse average technique [4] with the variance of the logarithm of the permeability $\sigma = 1.3$, and the correlation lengths $l_x = l_y = 0.01$ in the two space directions. Figure 1 shows a realization of the random permeability field. The "exact" well bore pressures computed by on the fine 1024×1024 mesh are $\alpha_1 = -1.7278769, \alpha_2 = 3.7894742$.

Table 1 shows the results computed by using the Algorithm 1 on the coarse 64×64 mesh with the size of the local domains $H_1 = H_2 = H$. The mesh size for solving the over-sampling base functions is $1/1024$. It clearly shows that the WBP by our method on the coarse 64×64 mesh is a very good approximation of the "exact" WBP on the fine 1024×1024 mesh.

References

[1] Z. Chen and T.Y. Hou, *A mixed multiscale finite element method for elliptic problem with oscillating coefficients,* Math. Comp. (to appear).

[2] Z. Chen and X. Yue, *Numerical homogenization of well singularities in the flow transport through heterogeneous porous media,* Multiscale Modeling and Simulation, submitted.

[3] G. Christakos, *Random Field Models in Earth Sciences,* Academic Press, San Diego, 1992.

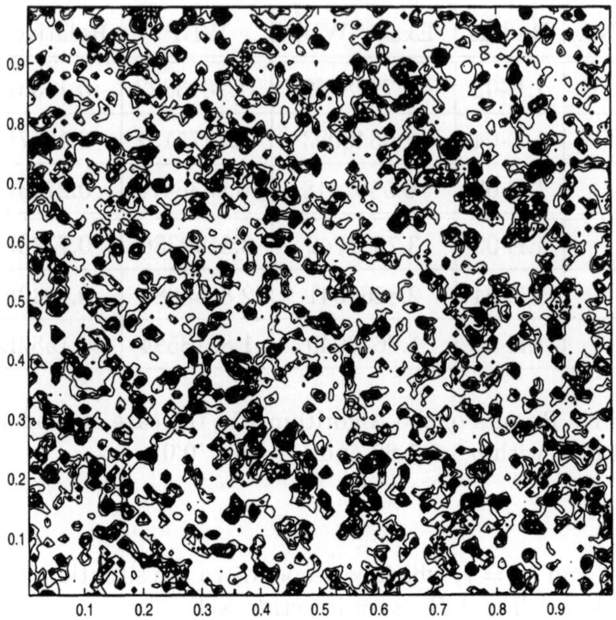

FIGURE 1. The random log-normal permeability field $K(x)$, The ratio of maximum to minimum is 5.0590E+04.

[4] L.J. Durlofsky, *Numerical calculation of equivalent grid block permeability tensors for heterogeneous porous media,* Water Resources Research **27** (1991), 699–708.

[5] L.J. Durlofsky, W.J. Milliken and A. Bernath, *Scale up in the Near-Well Region,* SPE 51940, Proceedings of the 15th SPE Symposium on Reservoir Simulation, Houston, February 14-17, 1999.

[6] B.B. Dykaar and P.K. Kitanidis, *Transmissivity of a heterogeneous formation,* Water Resource Research **29** (1993), 985–1001.

[7] Y.R. Efendiev, T.Y. Hou, and X.H. Wu, *The convergence of non-conforming multiscale finite element methods,* SIAM J. Numer. Anal. **37** (2000), 888–910.

[8] T.Y. Hou and X.H. Wu, *A multiscale finite element method for elliptic problems in composite materials and porous media,* J. Comput. Phys. **134** (1997), 169–189.

[9] T.Y. Hou, X.H. Wu, and Z. Cai, *Convergence of a multiscale finite element method for elliptic problems with rapidly oscillating coefficients,* Math. Comp. **68** (1999), 913–943.

[10] D.Q. Li and S.X. Chen, *On asymptotic behaviors of the equi-valued surface boundary value problem for the second order self-adjoint elliptic equations,* Fudan Journal (Natural Sciences), 6-14, December 1978. (in Chinese)

[11] K. Markov and L. Preziosi, eds., *Heterogeneous Media: Micromechanics Modeling, Methods and Simulations,* Birkhäuser, Boston, 2000.

[12] O. Mascarenhas and L.J. Durlofsky, *Scale up in the vicinity of horizontal wells,* Proceedings of the 20th Annual International Energy Agency Workshop and Symposium, Paris, September 22-24, 1999.

[13] D.W. Peaceman, *Interpretation of well-block pressures in numerical reservoir simulations,* SPEJ, 183–194, June 1978.

[14] D.W. Peaceman, *Interpretation of well-block pressures in numerical reservoir simulations with non-square grid blocks and anisotropic permeability,* SPEJ, 531–543, June 1983.

[15] Ph. Renard and G. de Marsily, *Calculating equivalent permeability: A review,* Advances in Water Resources, **20** (1997), 253–278.

(Z. Chen) LSEC, Institute of Computational Mathematics,
Chinese Academy of Sciences,
Beijing 100080, China.
E-mail address: `zmchen@lsec.cc.ac.cn`

(X. Yue) LSEC, Institute of Computational Mathematics, Chinese Academy of Sciences, Beijing 100080, China.
Current Address: Department of Mathematics, Suzhou University, Suzhou 215006, China.
E-mail address: `xyyue@lsec.cc.ac.cn`

International Series of Numerical Mathematics, Vol. 147, 103–112

On Plasma Expansion in Vacuum

Pierre Degond, Céline Parzani, and Marie-Hélène Vignal

Abstract. In this paper, we propose a model describing the expansion of a plasma in vacuum. Our starting point consists in a two-fluids Euler system coupled with the Poisson equation. Since numerical simulations of this model are very expensive, we investigate a quasi-neutral limit of it. We show that electron emission occurs at the plasma-vacuum interface. This emission is well modelled by a Child–Langmuir law. The difficulty consists in accounting for the motion of the plasma-vacuum interface. In this paper, we formally and numerically justify why electron emission produces a reaction pressure which slows the plasma expansion down.

1. Introduction

This paper is devoted to the study of a quasi-neutral plasma expansion in vacuum, and more particularly to the description of the motion of the plasma-vacuum interface. This configuration occurs in some diodes like the ones mentioned in [10]. So, let us consider an external electric field applied between two electrodes and assume that a quasi-neutral plasma constituted of ions and electrons is emitted from the cathode. This plasma undergoes a thermal expansion towards the anode while the electrons are emitted from the plasma-vacuum interface according to a Child–Langmuir law [3, 7]. The gap between the interface and the anode is decreasing with time and the extracted electron current is simultaneously increasing in accordance with the Child–Langmuir law.

Our starting point is a one-dimensional Euler model for each species (ions and electrons), coupled with the Poisson equation. However, due to the large plasma densities, numerical simulations of this model are very expensive in practice. So, we rather investigate a one-fluid Euler model obtained as a formal quasi-neutral limit of the initial two-fluids model. Moreover, during the thermal expansion of the plasma, electrons are emitted from the plasma-vacuum interface to the anode. The quasi-neutral model is unable to describe their motion between the plasma-vacuum interface and the anode but by means of a rescaling of the system, we show that this emission is well modelled by a Child–Langmuir law. Then, the boundary conditions at the interface $X(t)$, which connect the two models, are determined through two additional assumptions validated by numerical simulations. Namely, these assumptions are:

i) The electron emission is governed by the Child–Langmuir law with a maximal extracted electron current.

ii) Electron acceleration in vacuum produces a reaction pressure which acts on the plasma-vacuum interface and slows its motion down.

A similar asymptotic analysis is performed in [4] for the isentropic Euler system. Here we extend it to the full Euler system. We detail assumptions i) and ii) and numerically justify them by comparing simulations obtained on the one hand with the two-fluids Euler system and on the other hand with the quasi-neutral model.

2. Two-fluids Euler–Poisson model

Let us consider a pair of electrodes: the cathode is located at $\bar{x} = 0$ and the anode at $\bar{x} = L$. A plasma constituted of ions and electrons is emitted from the cathode. Let i and e be the index for ions and electrons. Then, $m_{i,e}$ are the masses of the particles, $N_{i,e}$ their densities, $U_{i,e}$ their velocities and $T_{i,e}$ their temperatures. The total energies $W_{i,e}$ are defined by

$$W_e = \frac{1}{\gamma - 1}\, P_e + \frac{m_e\, N_e\, U_e^2}{2}, \qquad W_i = \frac{1}{\gamma - 1}\, P_i + \frac{m_i\, N_i\, U_i^2}{2},$$

where $\gamma = 5/3$ and the pressure laws are given by $P_{i,e} = N_{i,e}\, k_B\, T_{i,e}$, k_B being the Boltzmann constant. We start with the Euler system for each species, coupled with the Poisson equation. Thus, the system for ions is given by:

$$\begin{cases} N_{i_{\bar{t}}} + (N_i\, U_i)_{\bar{x}} = 0, \\ m_i\, \left((N_i\, U_i)_{\bar{t}} + (N_i\, U_i^2)_{\bar{x}}\right) + P_{i_{\bar{x}}} = -q\, N_i\, \Phi_{\bar{x}}, \\ W_{i_{\bar{t}}} + (W_i\, U_i)_{\bar{x}} + (P_i\, U_i)_{\bar{x}} = -q\, N_i\, U_i\, \Phi_{\bar{x}}, \end{cases}$$

while electrons are described by:

$$\begin{cases} N_{e_{\bar{t}}} + (N_e\, U_e)_{\bar{x}} = 0, \\ m_e\, \left((N_e\, U_e)_{\bar{t}} + (N_e\, U_e^2)_{\bar{x}}\right) + P_{e_{\bar{x}}} = q\, N_e\, \Phi_{\bar{x}}, \\ W_{e_{\bar{t}}} + (W_e\, U_e)_{\bar{x}} + (P_e\, U_e)_{\bar{x}} = q\, N_e\, U_e\, \Phi_{\bar{x}}, \end{cases}$$

where $q > 0$ is the elementary charge.

The Poisson equation is used to take into account the evolution of the electric field, so that:

$$-\Phi_{\bar{x}\bar{x}} = \frac{q}{\varepsilon_0}\, (N_i - N_e),$$

where Φ is the electric potential and ε_0 is the vacuum permittivity.

We assume that a quasi-neutral plasma is emitted at $\bar{x} = 0$, with the same velocities and temperatures for the two species. So, the boundary conditions for the plasma are given by: $N_i(\bar{x} = 0) = N_e(0) = N_0$, $U_i(0) = U_e(0) = U_0$, $T_i(0) = T_e(0) = T_0$, and the ones for the electric potential are: $\Phi(0) = 0$, $\Phi(L) = \Phi_L$.

3. Quasi-neutral limit in the plasma region

Now, in order to get a quasi-neutral model, we scale these equations with the following characteristic sizes: the gap between the two electrodes L, the thermal ionic velocity $U_{th} = \sqrt{T_0/m_i}$, the time $\tau = L/U_{th}$, the density of the emitted plasma N_0, the temperature at the cathode T_0, the anode potential Φ_L and the internal energy at the cathode $W_0 = N_0 T_0$. This leads us to introduce some parameters:

- the mass ratio: $\varepsilon = \dfrac{m_e}{m_i}$,
- the ratio of the internal energy at the cathode to the applied potential energy: $\eta = \dfrac{T_0}{q\Phi_L} << 1$,
- the ratio of the applied potential energy to the coulombian interaction energy: $\lambda = \dfrac{\varepsilon_0 \Phi_L}{q N_0 L^2}$,
- the ratio of the emission velocity at the cathode to the thermal ionic velocity: $\alpha = \dfrac{U_0^2}{U_{th}^2}$.

Next, writing the Euler–Poisson system with the scaled variables: x, t, ϕ, $n_{i,e}$, $u_{i,e}$, $w_{i,e}$ and $p_{i,e}$ where $\bar{x} = Lx$, $\bar{t} = \tau\, t$, $\Phi = \Phi_L \phi$, $N_{i,e} = N_0\, n_{i,e}$, $U_{i,e} = U_0\, u_{i,e}$, $W_{i,e} = W_0\, w_{i,e}$ and $P_{i,e} = N_0\, T_0\, p_{i,e}$, we get:

$$\begin{cases} {n_i}_t + (n_i\, u_i)_x = 0, \\ (n_i\, u_i)_t + (n_i\, u_i^2)_x + {p_i}_x = -\dfrac{n_i\, \phi_x}{\eta}, \\ {w_i}_t + (w_i\, u_i + p_i\, u_i)_x = -\dfrac{n_i\, u_i\, \phi_x}{\eta}, \end{cases} \tag{1}$$

$$\begin{cases} {n_e}_t + (n_e\, u_e)_x = 0, \\ \varepsilon\left((n_e\, u_e)_t + (n_e\, u_e^2)_x\right) + {p_e}_x = \dfrac{n_e\, \phi_x}{\eta}, \\ {w_e}_t + (w_e\, u_e + p_e\, u_e)_x = \dfrac{n_e\, u_e\, \phi_x}{\eta}, \end{cases} \tag{2}$$

$$-\lambda\, \phi_{xx} = n_i - n_e, \tag{3}$$

with

$$w_i = \frac{1}{\gamma - 1}\, p_i + \frac{1}{2}\, n_i\, u_i^2, \quad w_e = \frac{1}{\gamma - 1}\, p_e + \frac{\varepsilon}{2}\, n_e\, u_e^2, \tag{4}$$

and the boundary conditions are given by:

$$\begin{cases} n_i(x = 0) = n_e(0) = 1, & u_i(0) = u_e(0) = \sqrt{\alpha}, \\ w_i(0) = \left(\dfrac{1}{\gamma - 1} + \dfrac{\alpha}{2}\right), & w_e(0) = \left(\dfrac{1}{\gamma - 1} + \dfrac{\varepsilon\, \alpha}{2}\right), \\ p_i(0) = p_e(0) = 1, & \phi(0) = 0, \quad \phi(1) = 1. \end{cases} \tag{5}$$

We are now interested in the limit $\eta \to 0$ with $\lambda = O(1)$ and $\alpha = O(1)$. These values agree with the ones computed in some high current diodes [10]. Since the forthcoming analysis is independent of whether ε is assumed to be small or not, we assume $\varepsilon = O(1)$.

So, let $X(t)$ be the plasma-vacuum interface position at time t. Then, when η goes to 0, we formally obtain that $n_e \to n$, $n_i \to n$, $u_e \to u$, $u_i \to u$ and $(w_i + w_e) \to w$. Furthermore, n, u and w, the density, the velocity and the total energy of the quasi-neutral fluid, satisfy:

$$\begin{cases} n_t + (n\,u)_x = 0, \\ (1+\varepsilon)\,\left((n\,u)_t + (n\,u^2)_x\right) + p_x = 0, \\ w_t + (w\,u + p\,u)_x = 0, \\ \phi(x) = 0, \end{cases} \tag{6}$$

for all $x \in [0, X(t)]$, where

$$w = \frac{1}{\gamma - 1}\,p + (1+\varepsilon)\,\frac{n\,u^2}{2} \quad \text{and} \quad p = p_i + p_e,$$

with the following boundary conditions:

$$\begin{cases} n(x=0) = 1, \quad u(0) = \sqrt{\alpha}, \\ w(0) = \dfrac{2}{\gamma - 1} + \dfrac{\alpha\,(1+\varepsilon)}{2}. \end{cases}$$

Let us notice that only the total pressure $p = p_i + p_e$ is given. To compute each pressure term, p_e for example, we need to solve the additional non-conservative equation (see [1] for more details):

$$p_{e\,t} + u_e\, p_{e\,x} + \gamma\, p_e\, u_{e\,x} = 0.$$

Moreover, the quasi-neutral model (6) deduced from (1, 5) in the limit $\eta \to 0$, is valid only when the density n is strictly positive. A rigorous proof of this limit was given in [2] assuming the density strictly greater than a positive constant. The same condition appears for the quasi-neutral limit of the drift-diffusion model in [8]. So, the quasi-neutral model (6) is no more valid close to the interface $X(t)$ where ions and electrons are moving apart. Therefore, a different asymptotic regime must be investigated in order to describe the behavior of the electrons in the region $[X(t), 1]$ (the "beam").

4. Child–Langmuir limit in the beam region

Under the action of the anode potential, the electrons are accelerated while the ions are slowed down. Beyond the interface $X(t)$, there are no ions. The electrons are described by equations (2, 3) with $n_i = 0$. Due to the external electric field, the electron velocity is increasing to reach a value of the order of $(\varepsilon\,\eta)^{-1/2}$. In order to investigate the behavior of the electrons in the beam, we must rescale the electron velocity. So for all $x \in [X(t), 1]$, we set $u_e = \bar{u}_e/\sqrt{\varepsilon\,\eta}$. This scaling transforms the system (2, 3) into

$$\begin{cases} \bar{n}_{e\,t} + \dfrac{1}{\sqrt{\varepsilon\,\eta}}\,(\bar{n}_e\,\bar{u}_e)_x = 0, \\ \varepsilon\left(\dfrac{1}{\sqrt{\varepsilon\,\eta}}\,(\bar{n}_e\,\bar{u}_e)_t + \dfrac{1}{\varepsilon\,\eta}\,(\bar{n}_e\,\bar{u}_e^2)_x\right) + p_{e\,x} = \dfrac{\bar{n}_e\,\bar{\phi}_x}{\eta}, \\ \dfrac{\eta\sqrt{\varepsilon\,\eta}}{\gamma-1}\,p_{e\,t} + \dfrac{\sqrt{\varepsilon\,\eta}}{2}\,\left(\bar{n}_e\,\bar{u}_e^2\right)_t + \dfrac{\gamma\,\eta}{\gamma-1}\,(p_e\,\bar{u}_e)_x + \dfrac{1}{2}\,(\bar{n}_e\,\bar{u}_e^3)_x = \bar{n}_e\,\bar{u}_e\,\bar{\phi}_x, \\ -\lambda\,\bar{\phi}_{xx} = -\bar{n}_e, \end{cases}$$

with the boundary conditions $\phi(X(t)) = 0$, $\phi(1) = 1$ and $u_e(X(t)) = 0$.

When η goes to 0, we formally get a stationary model (which only depends on time through the interface position: $X(t)$) such that for all $x \in [X(t), 1]$:

$$\begin{cases} (\bar{n}_e\,\bar{u}_e)_x = 0, \\ (\bar{n}_e\,\bar{u}_e^2)_x = \bar{n}_e\,\bar{\phi}_x, \\ \dfrac{1}{2}\,(\bar{n}_e\,\bar{u}_e^3)_x = \bar{n}_e\,\bar{u}_e\,\bar{\phi}_x, \\ -\lambda\,\bar{\phi}_{xx} = -\bar{n}_e. \end{cases} \tag{7}$$

Note that the third equation can be easily deduced from the other ones. This system is known as the Child–Langmuir model (see [3, 7]). Its solution depends on a free parameter, the current $\bar{j} = \bar{n}_e\,\bar{u}_e$, where $\bar{j}$ ranges in the interval $[0, \bar{j}_{CL}]$. The maximal value of $\bar{j}$, called the Child–Langmuir current, is given by the relation

$$\bar{j}_{CL} = \frac{4\sqrt{2}\,\lambda}{9\,(1-X)^2},$$

and is associated with the additional boundary condition $\bar{\phi}_x(X(t)) = 0$. We shall assume that the emitted current coincides with the Child–Langmuir current, and shall verify numerically that this hypothesis seems fulfilled.

5. A model for the plasma-beam interface

Now, we have to connect the quasi-neutral model (6) and the Child–Langmuir model (7). The boundary conditions at $X(t)$ result from the dynamics of the particles confined about the interface. A formal asymptotic analysis will be developed in a future work (see [5]). In the present work, we shall rely on some hypotheses. We set that the total momentum is preserved while the electrons are crossing the interface. The acceleration of the electrons towards the anode generates a reaction pressure which slows down the motion of the plasma-vacuum interface.

So, let us determine this pressure term. Let $\sigma = dX/dt$ be the interface velocity. We denote by $[f]$ the jump of the function f across the interface, by f_- the limit of f on the left of $X(t)$ and by f_+ the one on the right. The Rankine–Hugoniot relations applied to the conservation of the ion density across the interface leads to:

$$[n_i\,(u_i - \sigma)] = 0. \tag{8}$$

Similarly, using the total momentum equation, we have:

$$\left[\left(n_i\,u_i^2 + \varepsilon\,n_e\,u_e^2 + p_e + p_i - \frac{\lambda}{2\,\eta}\,|\phi_x|^2\right) - \sigma\,(n_i\,u_i + \varepsilon\,n_e\,u_e)\right] = 0. \tag{9}$$

Now, from equations (1, 3), the conservation of total energy can be written as:

$$(w_i + w_e)_t + (w_i\,u_i + p_i\,u_i + w_e\,u_e + p_e\,u_e)_x = -\frac{\lambda}{\eta}\,(\phi_x)_t\,\phi_x = -\frac{\lambda}{2\eta}(\phi_x^2)_t,$$

and applying the Rankine–Hugoniot conditions, we get:

$$\left[(w_i\,u_i + p_i\,u_i + w_e\,u_e + p_e\,u_e) - \sigma\,\left(w_i + w_e + \frac{\lambda}{2\eta}|\phi_x|^2\right)\right] = 0. \tag{10}$$

Moreover, we suppose that the quasi-neutral limit is still valid up to the interface $X(t)$, such that:

$$n_{e-} = n_-, \quad n_{i-} = n_-, \quad u_{i-} = u_-, \quad u_{e-} = u_-, \tag{11}$$

where n_- and u_- denote the limit of the quasi-neutral quantities on the left of the interface.
There are no ions beyond the interface, hence

$$n_{i+} = 0, \quad \text{and} \quad w_{i+} = 0. \tag{12}$$

With the condition of maximal current given by the Child–Langmuir law and the quasi-neutrality of the plasma, we get: $\phi_{x-} = \phi_{x+} = 0$. Next, assuming that the electron density has no jump across the interface, we obtain: $n_{e+} = n_-$. We have seen that the electron flow in the gap between $X(t)$ and the anode is given by the Child–Langmuir current such that: $j_{CL} = \bar{j}_{CL}/\sqrt{\varepsilon\,\eta} = n_{e+}\,(u_{e+} - \sigma)$. From (8) we deduce the interface velocity:

$$\frac{dX}{dt} = \sigma = u_-. \tag{13}$$

and (9) reduces to:

$$\varepsilon\,[n_e\,u_e\,(u_e - \sigma)] + [p_e + p_i] = 0.$$

Therefore, since $u_{e-} = u_- = \sigma$, $n_{e+}\,(u_{e+} - \sigma) = j_{CL}$, $p_{i+} = 0$ and $n_{e+} = n_-$, we have:

$$p_- = p_{e-} + p_{i-} = \varepsilon j_{CL} u_{e+} + p_{e+} = \varepsilon\,j_{CL}\left(\frac{j_{CL}}{n_-} + \sigma\right) + p_{e+}. \tag{14}$$

Using $w_{i+} = 0$, $u_{i-} = \sigma$ and $\phi_{x-} = \phi_{x+} = 0$, (10) becomes

$$[p_i\,u_i + w_e\,u_e + p_e\,u_e - \sigma\,w_e] = 0.$$

Now from (11) and (12), we deduce:

$$p_- u_- = p_{i-} u_{i-} + p_{e-} u_{e-} = w_{e+}\,(u_{e+} - \sigma) + p_{e+} u_{e+}\,,$$

then thanks to the definition of w_{e+} and to the relation $n_{e+}(u_{e+}-\sigma) = j_{CL}$, we get

$$p_- u_- = p_{e+}\left(\frac{\gamma}{\gamma-1}\frac{j_{CL}}{n_-}+\sigma\right)+\frac{\varepsilon}{2}j_{CL}\left(\frac{j_{CL}}{n_-}+\sigma\right)^2. \tag{15}$$

Solving the system (14) and (15), we determine p_{e+} and p_-. The pressure term p_{e+} is given by:

$$p_{e+} = \frac{\varepsilon}{2}\left(\frac{\gamma-1}{\gamma}\right)n_-\left(u_-^2-\left(\frac{j_{CL}}{n_-}\right)^2\right).$$

Therefore, substituting this expression into (14) , we can deduce the momentum flux p_- in the plasma resulting from the electronic emission at the interface and finally the energy flux $p_- u_-$.

6. Numerical validation of the asymptotic model

This section is devoted to the numerical study of the quasi-neutral model for the plasma (6) coupled with the Child–Langmuir model (7) for the "beam" by the momentum and energy fluxes p_- and $p_- u_-$. So, let us consider the quasi-neutral system (6) written in the conservative form (see [9],[11]):

$$\frac{\partial V}{\partial t}+\frac{\partial F(V)}{\partial x}=0, \quad V(0,x)=0, \quad V(t,0)=V_0 \tag{16}$$

where $V = (n,\ (1+\varepsilon)nu,\ w)^T$ and $F(V) = \left(nu,\ (1+\varepsilon)nu^2+p,\ wu+pu\right)^T$. Let us consider a uniform grid constituted by N cells $(m_k)_{k=1,\cdots,N}$ of size Δx. We denote by Δt the time step and we set $t^n = n\Delta t$ for all $n \geq 0$. Then for $n \geq 0$ and $k \in \{1,\cdots,N\}$, V_k^n is an approximation of V at time t^n on the cell m_k and X^n is the approximation of the plasma-vacuum interface position X at time t^n. Let us denote by m_{k_0} ($k_0 \in \{1,\cdots,N\}$) the cell such that $X^n \in m_{k_0}$. Using the evolution equation of the interface velocity (13), we get: $X^{n+1} = X^n + \Delta t u_{k_0}^n$. The system (16) is discretized with a classical finite volume method (see [11]). So, we describe the scheme only in the last cell occupied by the plasma, i.e. m_{k_0}.

Using the Green formula and taking into account the interface motion, we have

$$\left(X^{n+1}-X_{k_0-1/2}\right)\tilde{V}-\left(X^n-X_{k_0-1/2}\right)V_{k_0}^n = \Delta t\left(g_{k_0-1/2}^n - g_+^n\right)$$

where $g_{k_0-1/2}^n$ is the numerical flux through the interface between the cells m_{k_0-1} and m_{k_0} at time t^n. It is computed using classical schemes like the HLLE scheme or the "polynomial" upwind scheme (see [6],[11]). Moreover, g_+^n is the numerical flux through the interface, we set $g_+^n = \left(0,(p_-)^n,(p_-u_-)^n\right)^T$. Finally if $X^{n+1} \in m_{k_0}$, that is if the plasma-vacuum interface has not left the cell m_{k_0} during the time step, we set $V_{k_0}^{n+1} = \tilde{V}$, else if $X^{n+1} \in m_{k_0+1}$ we set $V_{k_0}^{n+1} = V_{k_0+1}^{n+1} = \tilde{V}$.

We compare the simulations given by the two-fluids and the asymptotic models. We use $\Delta x = 2.10^{-4}$, $\gamma = 5/3$, $\varepsilon = 5.10^{-1}$, $\eta = 10^{-4}$, $\lambda = 10^{-3}$ and $\alpha = 1$.

Figures 1 and 2 show the analysis of the plasma between the cathode and the interface $X(t)$. Contrary to a classical fluid expansion in vacuum, in our case the ionic fluid is slowing down. The interface velocity given by the asymptotic model seems to be correct in a qualitative way. The values numerically used for the momentum and energy fluxes p_- and p_-u_- are here: βp_- and βp_-u_-. The parameter β has to be adjusted according to the two-fluids simulations. Here, $\beta = 0.5$; note that without this correction the interface velocity given by the quasi-neutral model is a little lower than the one of the two-fluids model, but its order of magnitude is correct.

Figure 3 shows the electron dynamics in the beam. Plotting the electron velocity, we can see that the acceleration of electrons from the interface to the anode is well approximated by the Child–Langmuir law.

Figure 4 gives the electric potential computed from the Poisson equation on the one hand and by the Child–Langmuir model on the other hand. We can observe that the potential is equal to zero between the cathode and the interface, which justifies the quasi-neutral approach. We also show (right picture) that considering a constant Child–Langmuir current in the gap between $X(t)$ and the anode is a good approximation.

Acknowledgments: Support by the Commissariat l'Energie Atomique (DIF) and by the European network HYKE (EC contract HPRN-CT-2002-00282), is acknowledged. The authors wish to thank F. Assous, J. Segré and K. C. Le Thanh for stimulating discussions and encouragements.

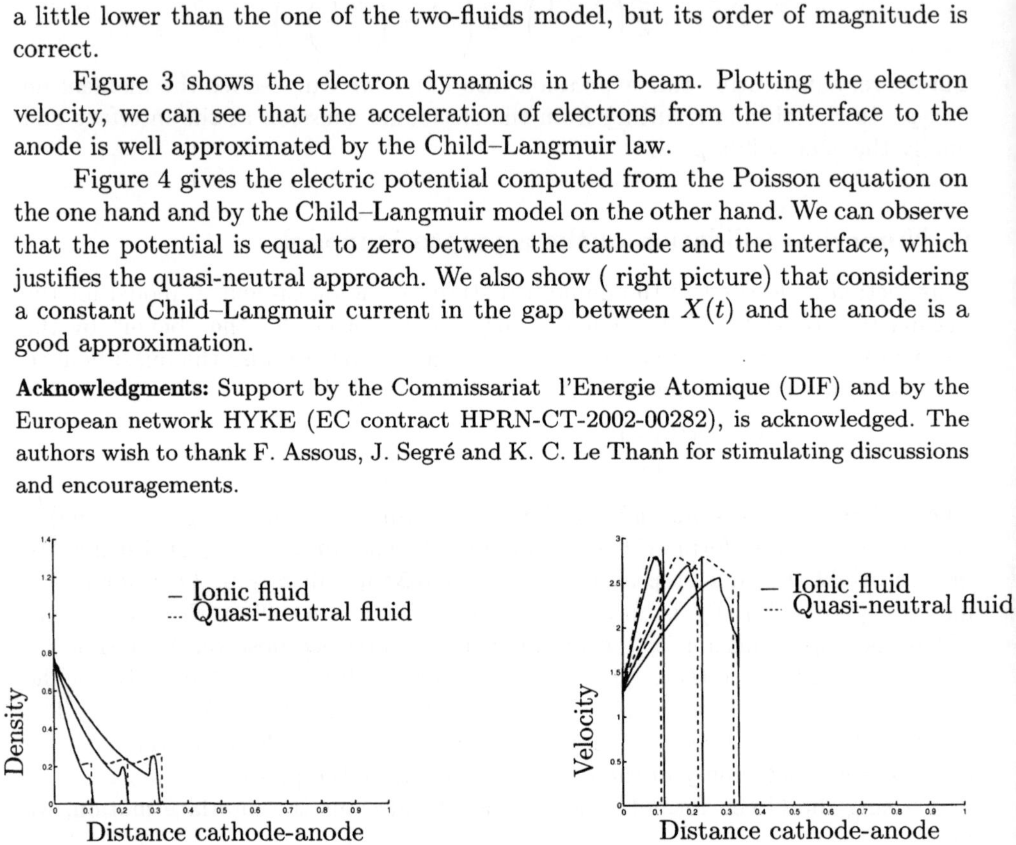

FIGURE 1. Densities and speeds of the ionic fluid given by the 2-fluids model compared to those given by the quasi-neutral model: values observed between the cathode $x = 0$ and the interface at times $t = 0.04\tau$, $t = 0.08\tau$, $t = 0.12\tau$.

References

[1] S. Cordier, *Modélisation mathématique et simulation numérique du plasma magnétosphérique,* thèse de doctorat de l'ENS Cachan.

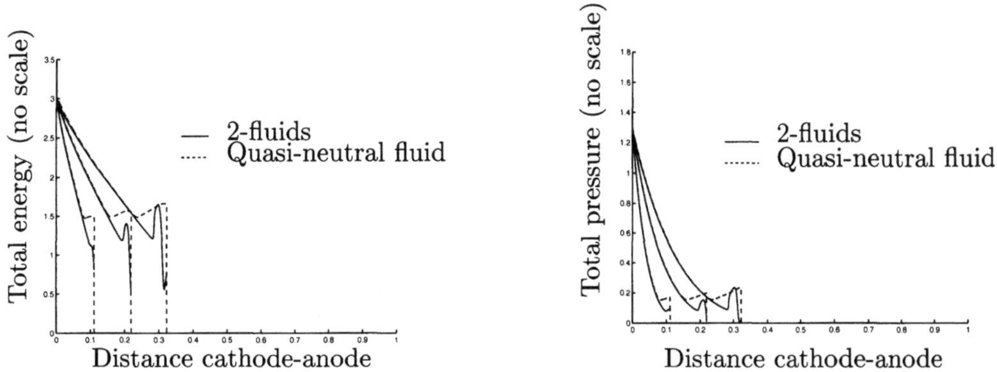

FIGURE 2. Total energy $(w_i + w_e)$ and pressure $(p_i + p_e)$ given by the 2-fluids model compared to those given by the quasi-neutral model: values observed between the cathode $x = 0$ and the interface at times $t = 0.04\tau$, $t = 0.08\tau$, $t = 0.12\tau$.

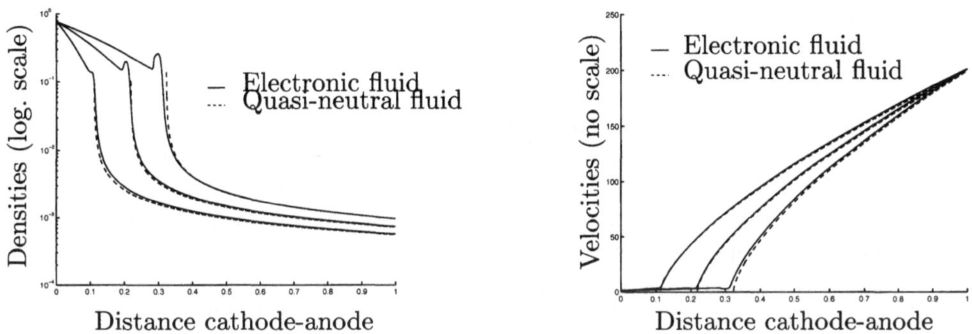

FIGURE 3. Densities and speeds of the electronic fluid given by the 2-fluids model compared to those given by the Child-Langmuir model: values observed between the interface and the anode $x = 1$ at times $t = 0.04\tau$, $t = 0.08\tau$, $t = 0.12\tau$.

[2] S. Cordier, E. Grenier, *Quasineutral limit of Euler–Poisson system arising from plasma physics,* Comm. Partial Differential Equations, **25** no 5-6 (2000), 1099–1113.

[3] P. Degond, *The Child–Langmuir Law in the Kinetic Theory of Charged-Particles. Part 1, Electron Flows in vacuum,* Advances in Kinetic Theor (B.Perthame, ed), World Scientific, Singapore, (1994), 3–44.

[4] P. Degond, C. Parzani, M.H. Vignal, *Un modèle d'expansion de plasma dans le vide,* C. R. Acad. Sci. Paris, **335** (2002), 399–404.

[5] P. Degond, C. Parzani, M.H. Vignal, *One dimensional modeling of plasma expansion in vacuum,* in preparation.

[6] P. Degond, F. Peyrard, G. Russo, Ph. Villedieu, *Polynomial upwind schemes for hyperbolic systems,* C. R. Acad. Sci. Paris, **328** (1999), 479–483.

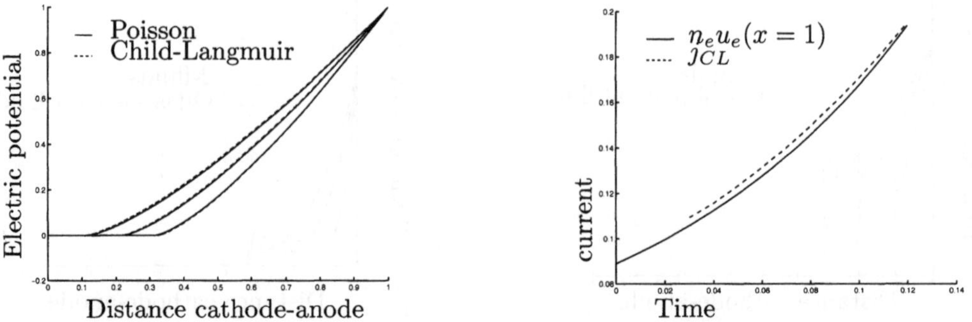

FIGURE 4. Electric potential computed on the one hand by the Poisson equation, on the other hand by the Child-Langmuir model: values observed between the interface and the anode $x = 1$ at times $t = 0.04\tau$, $t = 0.08\tau$, $t = 0.12\tau$ (left picture). Electronic current at the anode given by the Child-Langmuir model and by the 2-fluids model (right picture)

[7] P. Degond, P.A. Raviart, *An Asymptotic Analysis of the One-Dimensional Vlasov–Poisson System: the Child–Langmuir Law*, Asymptotic Analysis, **4** (1991), 187–214.

[8] I. Gasser, L. Hsiao, P.A. Markowich, S. Wang, *Quasineutral limit of a nonlinear drift diffusion model for semiconductors*, J. Math. Anal. Appl., **268** (2002), 184–199.

[9] E. Godlewski, P.A. Raviart, *Numerical Approximation of Hyperbolic Systems of Conservation Laws*, Springer, 1996.

[10] H. Sze, J. Benford, W. Woo, B. Harteneck, *Dynamics of a virtual cathode oscillator driven by a pinched diode*, Phys. Fluids, **29** no 11 (1986), 3873–3880.

[11] E.F. Toro, *Riemann Solvers and Numerical Methods for Fluid Dynamics*, Springer, 1999.

MIP,UMR 5640
University Paul Sabatier Toulouse 3,
118, route de Narbonne
31062 TOULOUSE cedex, FRANCE
E-mail address: {degond,parzani,mhvignal}@mip.ups-tlse.fr

International Series of Numerical Mathematics, Vol. 147, 113–130

Towards the Thermodynamic Modeling of Nucleation and Growth of Liquid Droplets in Single Crystals

Wolfgang Dreyer and Frank Duderstadt

Abstract. Stress assisted diffusion in single crystal Gallium Arsenide (GaAs) leads to the formation and growth of unwanted liquid arsenic droplets in a solid matrix. This process happens during the heat treatment of single crystal GaAs, which is needed for its application in opto-electronic devices, and it is of crucial importance to pose and answer the question if the appearance of droplets can be avoided. To this end we start a thermodynamic simulation of this process. Special emphasis is given to the influence of mechanical effects on chemistry, diffusion and interface motion in GaAs.

1. Introduction of the complete problem

Stress assisted diffusion in single crystal Gallium Arsenide (GaAs) leads to the formation and growth of unwanted liquid droplets in a solid surrounding, [14], [15], [4], [5]. This is an elaborate thermodynamic process involving chemical reactions, diffusion, motion of phase boundaries, and all these phenomena are strongly influenced by local mechanical stress fields. The resulting model consists of a coupled system of algebraic equations, partial differential equations and conditions at the appearing liquid/solid interfaces. The variables are the mole fractions of various constituents, which appear in semi-conducting GaAs, the mechanical strain, and a phase density, indicating the size distribution of droplets.
Single crystal GaAs contains the major elements gallium and arsen, and additionally various trace elements, which are of most importance in order to fabricate semi-conducting or semi-insolating GaAs, respectively. All constituents occupy three sublattices of face-centered cubic (fcc) symmetry. The sublattices are indicated by three greek letters: α, β, γ. The lattice α is dominantly occupied by gallium, while the arsenic is the major substance on the lattice sites of sublattice β. The sublattice γ indicates an interstitial lattice, where the dominant elements are vacancies. For a proper application of single crystal semi-conducting or semi-insolating GaAs as a wafer material, it is crucial that the trace elements are distributed homogeneously on the lattice sites of the three sublattices. In order to remove dislocations, which appear during the process of crystal growth, a special heat treatment of the wafer

becomes necessary. However, the heat treatment might start a diffusion process, where the trace elements move preferably towards the proximity of the dislocations. There the trace elements find thermodynamic conditions so that the formation of liquid droplets, dominantly formed by the arsenic, sets in. However, the arsenic precipitates appear also in regions free from dislocations.

The first Figure shows two graphs, which represent the morphology of single crystal Gallium Arsenide on the *mm*-scale.

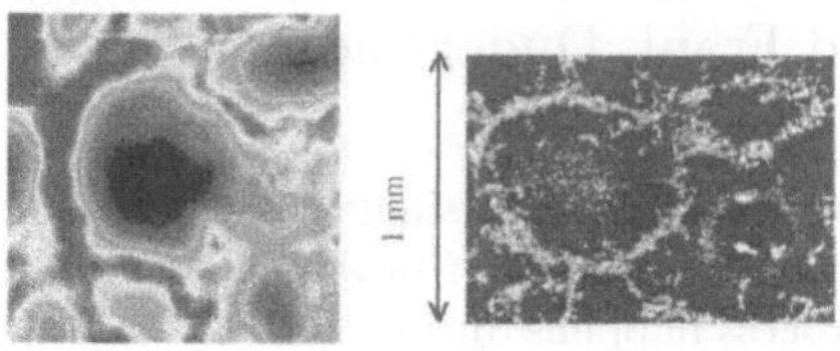

The graph on the left hand side is a photo-luminescence mapping, and shows the distribution of dislocations. The cellular structure is typical for Gallium Arsenide: the dislocations are dominantly arranged in rings, with an inner region, which is free of dislocations. The graph on the right hand side of the Figure shows the distribution of droplets. A careful study of this graph reveals that large droplets appear in the cell wall, i.e. in the vicinity of the dislocations. Adjacent to this ring of droplets, there is a region without droplets, and the center of the dislocation ring is occupied by small droplets. The number densities of small and large droplets are of the same order.

We are thus confronted with the problem to describe the thermodynamic nucleation and growth conditions of liquid droplets within a solid matrix. In particular, we will study the influence of local mechanical stresses on chemistry, diffusion and interface motion. Stress fields arise due to dislocations, eigenstrains due to point defects and due to misfit strain. The latter describes the phenomenon that the arsenic liquid phase, which appears as inclusions, has a larger specific volume than the solid phase.

2. Some physical and chemical properties of semi-insulating/semi-conducting GaAs

2.1. The phase diagram

The phase diagram of GaAs, which is shown in the following Figure, indicates the various phases and their boundaries. Below the liquidus line, we find, left and right of the stoichiometric line, i.e. (arsenic) composition $y = 0.5$, two regions where the solid GaAs crystal may coexist with the liquid. Right and left from $y = 0.5$, the crystal is As-rich and Ga-rich, respectively. These regions are bounded from below by the two eutectic lines at $29.5°C$ and $810°C$, respectively. The two regions below the eutectic lines indicate mixtures consisting of the solid GaAs crystal as matrix material, and containing solid Ga-rich and As-rich precipitates, respectively. In this study, we are exclusively concerned with thermodynamic processes in the two-phase region right from the stoichiometric line and above the 810°C eutectic.

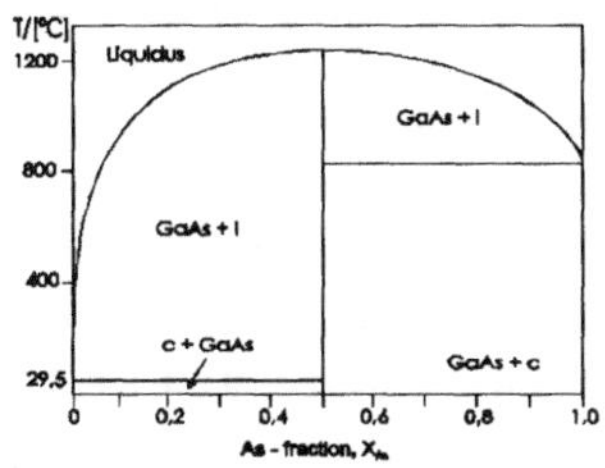

The GaAs crystal at the stoichiometric composition consists of two fcc sublattices that are displaced along the diagonal of the elementary cube, i.e. the crystal has the symmetry of Zinkblende. However, the ideal crystal at $y = 0.5$ is neither a semi-conductor nor a semi-insulator. The latter one may form if the crystal composition is slightly on the As-rich side. Moreover, there are further constituents appearing in GaAs. These are Oxygen (O), Silicon (Si), Bor (B) and Carbon (C) in very small quantities, but nevertheless these trace elements induce very important phenomena. Above the 810°C eutectic there is a region, which is called region of homogeneity, where the excess amount of As and the aforementioned trace elements behave like point defects that are homogenously distributed over the two already mentioned sublattices and on a further sublattice, which is formed by interstitial lattice sites.

The next Figure shows the region of homogeneity (H) as it was calculated on demand for Freiberg Compound Materials (FCM) by H. Wenzl and K.Mika -for details of the underlying model, see [15]. During a heat treatment of the crystal, the boundary of the region H may be crossed, as it is indicated in the Figure. If this happens, unwanted liquid As-rich precipitates appear. Thus the knowledge of the boundary of the region H is of crucial importance for the proper application of GaAs as a wafer material.

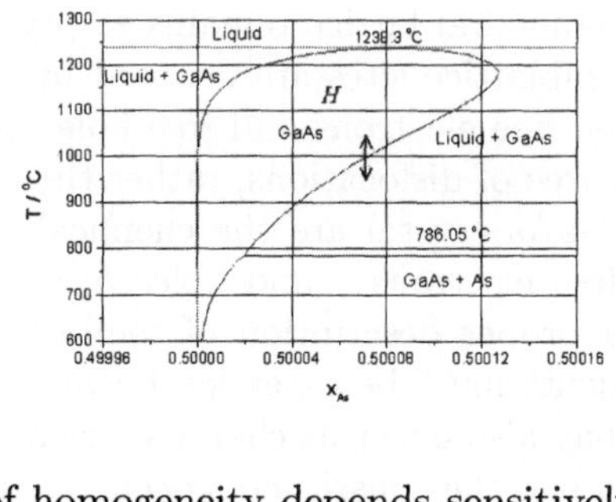

The calculated region of homogeneity depends sensitively on the assumed chemistry, i.e. on the chemical reactions which are taken into account.

One of the objectives of this study is the determination of the mechanical influence on the region of homogeneity. The mechanical influence arises due eigenstrains, which are induced by point defects and dislocations, and due to misfit strain, which is induced by a volume change during the solid/liquid phase transition. Regarding the non-mechanical parts of the thermodynamic theory, we rely on the chemical model given by Wenzl et. al.,[15], in collaboration with the wafer manufacturer FCM, and which will be discussed in detail in the following Section.

2.2. The chemical constitution of semi-insulating As-rich GaAs

The chemical model of Wenzl et. al. [15] is suited to describe the behaviour of semi-insulating As-rich GaAs that is fabricated by Freiberg Compound Materials (FCM).

The GaAs crystal is assumed to consist of three sublattices, indexed by $\{\alpha, \beta, \gamma\}$. Among them there are the already introduced two *fcc* sublattices, which are mainly occupied by Ga- and As-atoms, respectively, and the fcc interstitial sublattice, whose lattice sites are mostly occupied by vacancies.

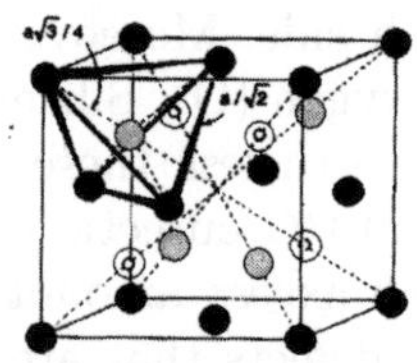

The above Figure gives a sketch of the elementary cell of the crystal. The lattice sites at the corners and faces of the cube form the sublattice α, which is mainly occupied by Ga atoms. In the interior of the elementary cell of the sublattice α there are further lattice sites, viz. four octahedral and eight tetrahedral lattice sites. Four lattice sites of the tetrahedral sites form the elementary cell of the β sublattice, indicated in the Figure by light balls. The remaining four tetrahedral sites form the sublattice γ. It is important to know that the available space of a lattice site of the β sublattice is equal to the corresponding space in the γ sublattice. The available space for octahedral sites is too small for an occupation by atoms, and thus the octahedral lattice remains empty.

The number densities of sublattice sites are denoted by n_α, n_β, n_γ. There holds $n_\alpha = n_\beta = n_\gamma = n_{SL}$. The free electrons and free holes do not need lattice sites, also in regions, which are free of dislocations, rather they form the electronic gas. The constituents of the GaAs crystal are the chemical elements *Ga*, *As*, *O*, *Si*, *B*, *C*, and vacancies *V*, free electrons *e* and holes *h*, which are also treated like chemical elements. For a proper description of semi-conducting/semi-insulating GaAs, the list of constituents must be extended because the elements *As*, *O*, *Si*, *C* and the vacancies *V* may also occur as charge carriers, which is indicated by a superscript. In addition, when the constituents occur on different sublattices they are considered as different constituents, and this is indicated by a subscript. There are thus 17 constituents: free electrons and holes, which do not occupy lattice sites, and 15 constituents on the three sublattices according to the following table.

Sublattice α	Ga_α	$As_\alpha^{(0)}$	$As_\alpha^{(+)}$	$V_\alpha^{(0)}$	$V_\alpha^{(3-)}$	$B_\alpha^{(0)}$	$Si_\alpha^{(+)}$
Sublattice β	As_β	$O_\beta^{(0)}$	$O_\beta^{(-)}$	$O_\beta^{(+)}$	$C_\beta^{(-)}$	$V_\beta^{(0)}$	
Sublattice γ	$As_\gamma^{(0)}$	$V_\gamma^{(0)}$					

The constituents Ga_α and As_β form the major portions on their privileged sublattices which determine the symmetry of the crystal. The other constituents are called (point) defects, which thus consist of antisite atoms, interstitial atoms, trace elements and vacancies.

The special choice of the constituents, and in particular the assumption that Ga is exclusively without charge on the sublattice α, has been suggested by FCM and is called Freiberger model. The underlying reasoning of the Freiberger model relies on experimental hints, and in particular on the objective to calculate a quite realistic homogeneity region that fits to the properties of FCM made semi-insulating GaAs. For different choices and resulting consequences we refer the reader to [15].

3. A list of simplified problems with increasing complexity

The last section has revealed that we have a quite complex thermochemical process under consideration. For this reason, it is appropriate to consider at first only very simple initial and boundary value problems that serve especially to design the mathematical model and to calibrate its parameters.

In particular, we consider the two situations sketched in the following Figure. The two pressure vessels, which are closed by a movable piston, set two thermodynamic systems with prescribed outer pressure p_0 and temperature T_0. The inner temperature is assumed to be homogeneous and equal to the outer temperature.

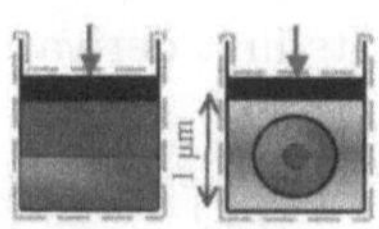

The example on the left hand side considers a liquid and a solid phase separated by a planar interface. This example is designed to study the influence of mechanical stresses on the chemical and interfacial equilibrium condition and on the diffusion process in the solid.

In particular, the example on the right hand side considers a solid ball on μm scale in contact with a gas consisting of the four arsenic constituents As, As_2, As_3, As_4. The solid ball contains at its center a liquid droplet, made exclusively from As and Ga, that can grow or shrink. The outer interface of the ball is in equilibrium with the gas, and we consider the case that only As_2 may cross the interface and become an interstitial As in the solid.

4. Thermodynamics of mixtures, Part 1: Variables

We consider a body B whose volume $V(t)$ may depend on time t. A space point, which is at some time t occupied by a *material point* P of B, is indicated by its cartesian coordinates $x = (x^i) = (x^1, x^2, x^3)$. The notion of *material point* means the smallest units of B that cannot be further resolved on the chosen space scale. The body B may consist of coexisting fluid and solid phases, which are both mixtures of various chemically different constituents.
The thermodynamic state of the fluid phase is described by the variables

$$\begin{aligned} &n_a = n_a(t,x), \quad a \in \{1,2,...,a_L\}, && \text{number densities of the constituents,}\\ &v^i = v^i(t,x), && \text{barycentric velocity of the mixture,}\\ &T = T(t), && \text{(absolute) temperature.} \end{aligned} \tag{1}$$

The total mass density of the mixture, ρ, is given by the sum over the number densities $\times$ molecular weights, M_a, and for some purposes it is useful to calculate the mixture velocity, v, from the velocities of the constituents, v^i_a. Thus we write

$$\rho = \sum_{a=1}^{a_L} M_a n_a \quad \text{and} \quad v^i = \sum_{a=1}^{a_L} \frac{M_a n_a}{\rho} v^i_a. \tag{2}$$

In order to describe the deformation of the solid phase we must at first introduce a reference configuration. To this end, we indicate at a fixed reference time t_0 any material point P of the solid phase by its cartesian coordinates $X = (X^1, X^2, X^3)$. The motion of P is described by the field $\chi^i(t,X) = (\chi^1(t,X), \chi^2(t,X), \chi^3(t,X))$, which gives the coordinates of P at time $t > t_0$:

$$x^i = \chi^i(t,X). \tag{3}$$

The field χ^i is called *motion*, and its first derivatives

$$\hat{v}^i = \frac{\partial \chi^i}{\partial t} \quad \text{and} \quad F^{ij} = \frac{\partial \chi^i}{\partial X^j} \quad \text{with} \quad J = \det(F^{ij}) \neq 0 \tag{4}$$

are the *velocity* and the *deformation gradient*, respectively. The condition on the Jacobian J guarantees that (3) is invertible.
It is well known that the Jacobian is related to the total mass density by

$$J = \frac{\rho_R}{\rho}, \tag{5}$$

where ρ_R is the mass density in the reference configuration.
The solid phase may consist of a mixture with a_S constituents, and thus the thermodynamic state of the solid phase is described by the variables

$$x^i = \chi^i(t,X), \quad n_a = n_a(t,x), \quad T = T(t). \tag{6}$$

5. Thermodynamics of mixtures, Part 2: General Constitutive Model

We consider a GaAs crystal consisting of two coexisting phases, which are formed by a nonviscous mixture of liquids and a thermoelastic solid mixture, respectively. There are eigenstrains due to the presence of point defects, misfit strain and dislocations. However, dislocations are not considered in the current paper.
Beyond the variables, further important quantities are needed to describe the properties of mixtures. These are: the specific internal energy, u, the specific entropy, s, the specific free energy, $\psi = u - Ts$, and the chemical potentials of the constituents, μ_a, [10], [11].
Furthermore, there are three measures of stress, viz. the Cauchy stress, σ^{ik}, and the first and second Piola Kirchhoff stress, respectively, $\hat{\sigma}^{ik}$ and t^{ik}:

$$\sigma^{ik} = \sigma^{ki}, \quad \hat{\sigma}^{ik} = J\overset{-1}{F}{}^{ij}\sigma^{jk}, \quad t^{ik} = J\overset{-1}{F}{}^{ij}\overset{-1}{F}{}^{kl}\sigma^{jl}, \tag{7}$$

where $\overset{-1}{F}{}^{ij}$ are the components of the tensor $(\mathbf{F}^{-1})^T$. The Cauchy stress appears naturally in a thermodynamic description where the reference configuration of the solid phase does not explicitly appear. The first Piola Kirchhoff stress is a simply measurable quantity, and appears naturally in the description that relies on the reference configuration. Finally, the second Piola Kirchhoff stress is needed to formulate constitutive laws relating the stress to the variables, see [13].
Let us decompose the Cauchy stress into its trace and the traceless part, the latter one being indicated by angle brackets around the indices; for example we write

$$\sigma^{ik} = -p\delta^{ik} + \sigma^{<ik>} \quad \text{with} \quad p = -\frac{1}{3}\sigma^{nn}. \tag{8}$$

The quantity p is called pressure.
There are also various measures of the deformation of a body. The most important one is the right Cauchy Green tensor, C^{ik}, and its unimodular restriction, which we denote by c^{ik}:

$$C^{ik} = F^{mi}F^{mk} \quad \text{and} \quad c^{ik} = J^{-2/3}C^{ik} \quad \text{so that} \quad \det(c^{ik}) = 1. \tag{9}$$

All these quantities are related to each other by the Gibbs equation, which reads in fluids as well as in thermoelastic solids

$$d\rho\psi = -\rho s dT + \frac{1}{2}J^{-1/3}t^{ik}dc^{ik} + \sum_a \mu_a dn_a. \tag{10}$$

The Gibbs equation implies

$$\rho s = -\frac{\partial \rho\psi}{\partial T}, \quad t^{ik} = 2J^{1/3}\frac{\partial \rho\psi}{\partial c^{ik}}, \quad \text{and} \quad \mu_a = \frac{\partial \rho\psi}{\partial n_a}. \tag{11}$$

Furthermore, there holds the Gibbs/Duhem equation

$$\rho\psi + p = \sum_{a=1} \mu_a n_a. \tag{12}$$

For the solid phase the knowledge of the free energy density $\rho_s\psi_s(T, n_1,...,n_{a_s}, c^{ik})$ is sufficient to relate the quantities we have introduced to the variables.
In a mixture of isotropic fluids we have $c^{ik} = \delta^{ik}$, and the second Piola/Kirchhoff stress drops out of the Gibbs equation. Thus for the liquid phase the knowledge of the free energy density $\rho_L\psi_L(T, n_1,...,n_{a_L})$ is sufficient to relate the aforementioned quantities to the variables.
The proof of the Gibbs equation for fluid mixtures and for the solids that only consist of a single substance can be found in [10]. The generalization to a solid mixture, viz. eqn. (10), has not been published elsewhere, but it can be derived following the strategies that are outlined by I. Müller in [10].

6. Thermodynamics of mixtures, Part 3: Explicit representations of free energies, chemical potentials, pressure and stresses

In general, there are three competing additive contributions to the free energy of a mixture: (i) entropy, (ii) interaction or mixing energy, and (iii) mechanical energy. If only the entropic contribution is considered, the mixture is called an *ideal mixture*. In the sequel, we will combine the two first contributions in the so called chemical part of the free energy, and we write

$$\rho\psi = \rho\psi^{chem} + \rho\psi^{mech} \qquad \text{and accordingly} \qquad \mu_a = \mu_a^{chem} + \mu_a^{mech}. \tag{13}$$

6.1. Chemical potentials and pressure of the liquid phase

For the sake of simplicity, we assume that the liquid phase only consists of Ga and As. We give at once the explicit constitutive laws for the chemical potentials and the pressure. Then, the free energy density can be calculated by means of the Gibbs/Duhem equation (12).

$$\begin{aligned} \mu_{Ga}^{chem} &= \mu_{Ga}^0(T) + RT\,\mathrm{Log}(1-y) + (L_0 + L_1(3-4y))y^2, \\ \mu_{As}^{chem} &= \mu_{As}^0(T) + RT\,\mathrm{Log}(y) + (L_0 + L_1(1-4y))(1-y)^2. \end{aligned} \tag{14}$$

The newly introduced quantity $y = n_{As}/n$ is the arsenic mole fraction. In both the above formulae, the first terms give the standard chemical potentials. The second terms are the entropic contributions, and the third terms give the mixing energy according to the Redlich/Kister model, L_0, L_1 measuring the strength of the mixing energy. R is the universal gas constant. For details, see [12].
The mechanical parts of the chemical potentials are nonclassical contributions. They describe the *interfacial misfit*, which arises because the transition from the solid phase to the liquid phase is accompanied by a change of the specific volume. In fact, arsenic rich liquid needs more space than solid GaAs. In order to describe

this phenomenon, let us at first introduce the function

$$\frac{1}{\rho^{therm}(T,y)} = v_{As}(T)\frac{M_{As}y}{M_{As}y+M_{Ga}(1-y)} + v_{Ga}(T)\Big(1-\frac{M_{As}y}{M_{As}y+M_{Ga}(1-y)}\Big), \quad (15)$$

where $v_{As}(T)$ and $v_{Ga}(T)$ are the specific volumes of pure As and pure Ga, respectively. Next we choose as a reference density of the liquid $\rho_R = \rho^{therm}(T_R, 1)$, and we set $\rho_*(T) = \rho_s^{therm}(T)$, where ρ_s^{therm} is the density of stoichiometric solid GaAs which can be read off from a table. Finally, we measure the interfacial misfit by

$$\frac{\rho_*(T)-\rho_R}{\rho_R}. \quad (16)$$

There result the mechanical chemical potentials

$$\mu_{Ga}^{mech} = \frac{M_{Ga}}{\rho_R} K_L Log\left(\frac{\rho}{\rho_*(T)}\right) \quad \text{and} \quad \mu_{As}^{mech} = \frac{M_{As}}{\rho_R} K_L\, Log\left(\frac{\rho}{\rho_*(T)}\right), \quad (17)$$

which rely on the simple constitutive law for the liquid pressure:

$$p = p_R + K_L\left(\frac{\rho-\rho_*(T)}{\rho_R}\right). \quad (18)$$

Recall that the liquid density is denoted by ρ, while K_L is the bulk modulus of the liquid.

6.2. The free energy of the solid phase

The solid phase of GaAs consists of three fcc sublattices α, β, γ which are occupied by seventeen different constituents according the description of Section 2.2. However, there are only fifteen unknown number densities, because the trace elements Si, C and B are prescribed.

The chemical part of the free energy density of the solid phase is given by

$$\rho\psi^{chem} = \sum_{a=1}^{17} n_a\mu_a^0(T) + RT\sum_{a=1}^{17} n_a \log(\frac{n_a}{n_{SL}}) + \frac{q}{2}\sum_{a,b=1}^{17} I_{ab}(T) n_a \frac{n_b}{n_{SL}}. \quad (19)$$

The free energy density $\rho\psi^{chem}$ consists of three contributions, which are the standard chemical potential $\mu_a^0(T)$, consisting of the energy and entropy constants (first term), entropy (second term) and mixing energy (third term). The constant q, (= 4 in the current study), is the coordination number and the temperature dependent quantities I_{ab} denote the interaction energies. We take only nearest neighbour interactions into account. Furthermore, we assume that electrons, holes and the other charged constituents have only entropic free energies, which is a fairly good approximation. Examples for numerical values for the nonzero interaction energies can be found in [14] and [15].

The free energy density $\rho\psi^{mech}$ describes elastic energy including eigenstrains and it is given by

$$\rho\psi^{mech} = \frac{1}{4J}(J^{2/3}c^{ij} - C_*^{ij})K^{ijkl}(J^{2/3}c^{kl} - C_*^{kl}) - p_R\sqrt{\det(c^{ij})}_{|\det(c^{ij})=1}. \quad (20)$$

K^{ijkl} is the stiffness matrix and C_*^{ij} is a measure for the eigenstrains. The form of $\rho\psi^{mech}$ has been chosen in such a way that the Gibbs equation implies Hookes law for the second Piola/Kirchhoff stress, viz.

$$t^{ij} = \frac{1}{2}K^{ijkl}(J^{2/3}c^{kl} - C_*^{kl}) - p_R J^{1/3}\overset{-1}{c^{ki}} \quad (21)$$

The matrix C_*^{kl} describes the various eigenstrains due to point defects, see the next Section for details.
Next we calculate from (19) and (20) the chemical and mechanical parts of the chemical potentials. There results

$$\mu_a^{chem} = \mu_a^0(T) + RT\log\left(\frac{n_a}{n_{SL}}\right) + q\sum_{b=1}^{17} I_{ab}(T)\frac{n_b}{n_{SL}} - \frac{q}{2}\sum_{a,b=1}^{17} I_{ab}(T)\frac{n_a n_b}{n_{SL}^2}, \quad (22)$$

for the chemical part of the chemical potential, while its mechanical part reads

$$\mu_a^{mech} = -\frac{M_a}{3\rho_S J}\left(\frac{1}{4}(J^{2/3}c^{ij} - C_*^{ij}) - C_*^{ij}\right)K^{ijkl}(J^{2/3}c^{kl} - C_*^{kl}). \quad (23)$$

7. Thermodynamics of mixtures, Part 4: Eigenstrains due to point defects

There are three different mechanisms that give rise to eigenstrains, which may appear due to (i) dislocations, (ii) point defects, and (iii) interfacial misfit.
The role of eigenstrains due to dislocations have been discussed in the introduction and will not considered anymore in the current study.
The eigenstrains due to interfacial misfit can be attributed either to the liquid phase or to the solid phase, see for example [7]. In the present study we have chosen the liquid phase in order to incorporate the interfacial misfit, see Section 6.1.
There remain to discuss eigenstrains due to the various point defects on lattice sites of the two sublattices α and β, viz. $As_\alpha^{(0)}$, $As_\alpha^{(+)}$, $V_\alpha^{(0)}$, $V_\alpha^{(3-)}$, $B_\alpha^{(0)}$, $Si_\alpha^{(+)}$ on sublattice α, $O_\beta^{(0)}$, $O_\beta^{(-)}$, $O_\beta^{(+)}$, $C_\beta^{(-)}$ on sublattice β. Note that the Freiberger model does not lead to eigenstrains on sublattice γ, because the defect $As_\gamma^{(0)}$ needs the same space as an As atom on the sublattice β. Uncharged vacancies on interstitials have no effect at all. The defects on sublattices α and β induce eigenstrains, which are different for the neutral and the charged defects. The neutral defects except the vacancies can be fairly well approximated as elastic balls, whereas the charged defects and the vacancies behave as they were rigid balls.

Let us decompose the function C_*^{ij} as a product of thermal eigenstrains and eigenstrains due to defects:

$$C_*^{ij} = C_{therm}^{ik} C_{defects}^{kj}. \tag{24}$$

The thermal part is obviously given by

$$C_{therm}^{ik} = (1 + \alpha(T - T_R))^2 \delta^{ik}, \tag{25}$$

where α denotes the coefficient of thermal expansion.
A simple calculation relying on elastic and rigid balls yields explicit expressions for the corresponding eigenstrains, which can be written as a product of elastic and rigid eigenstrains:

$$C_{defects}^{kj} = \prod_a \Big(1 + \frac{n_a}{3n_{SL}}\Big)^{Z_a} \prod_b \Big(1 + \frac{n_b}{3n_{SL}}\Big)^{Z_b} \delta^{kj}. \tag{26}$$

for $a \in \{As_\alpha^{(0)}, B_\alpha^{(0)}, O_\beta^{(0)}\}$ and $b \in \{As_\alpha^{(+)}, V_\alpha^{(0)}, V_\alpha^{(3-)}, Si_\alpha^{(+)}, O_\beta^{(-)}, O_\beta^{(+)}, C_\beta^{(-)}, V_\beta^{(0)}\}$.
The constants Z_a and Z_b are given by

$$Z_a = \frac{\delta_a}{1 + \delta_a} \qquad \text{and} \qquad Z_b = \frac{4\mu}{3K}\delta_b. \tag{27}$$

μ and K are the shear and bulk modulus of the solid phase, and

$$\delta_a = \{-0.055, -0.119, -0.317\} \tag{28}$$

$$\delta_b = \{-0.119, -0.119, -0.119, -0.119, 0.479, -0.317, -0.317, -0.317\}. \tag{29}$$

The difference of the two representations for eigenstrains arises from the fact that elastic eigenstrains are characterized by continuous normal stresses at the interface between an elastic ball and the solid phase, while rigid eigenstrains are due to a prescribed normal displacement at the interface between a rigid ball and the solid phase.

8. Thermodynamics of mixtures, Part 5: Field equations

The variables of the thermodynamic model are the five components of the strain field and the fourteen number densities of the constituents in the solid phase, while in the liquid phase the only variables are the two number densities of Ga and As, because for simplicity we do not allow the other constituents of the solid to cross the interface.
The field equations for the variables rely on several assumptions:
(i) Mechanical equilibrium is reached immediately, so that the strain field is determined by the quasistatic momentum balance for the given number densities.

$$\frac{\partial \sigma^{ik}}{\partial x^k} = 0. \tag{30}$$

(ii) Local chemical equilibrium in the solid phase is reached much faster than diffusional equilibrium, so that for the given total number densities of Ga and As, the other number densities are determined by the 14 nonlinear algebraic equations representing the laws of mass actions. These equations will be derived in Section 9 and are given by (40), (41), and (44)-(49). They have the generic form

$$F_A(n_1, n_2, ...n_{14}; T, n^S_{Ga}, n^S_{As}) = 0, \qquad A \in \{1, ..14\}. \tag{31}$$

We write the solution of these equations as

$$n_a = \tilde{n}_a(T, n^S_{Ga}, n^S_{As}). \tag{32}$$

(iii) Thus only a single diffusion equation in the solid is needed. Furthermore, this equation simplifies due to the fact that only the mobility of the interstitial arsenic is important. The diffusion equation reads

$$\frac{\partial n^S_{As}}{\partial t} = M\Delta(\tilde{\mu}_{As^{(o)}_\gamma} - \tilde{\mu}_{Ga_\alpha}), \tag{33}$$

where M denotes the mobility of the interstitial $As^{(o)}_\gamma$ and n^S_{As} denotes the total number density of the As content of the solid. $\tilde{\mu}_{As^{(o)}_\gamma}$ and $\tilde{\mu}_{Ga_\alpha}$ are the chemical potentials of the indicated substances. However, the equations (32) has been used to eliminate those constituents that are in thermal equilibrium with n^S_{Ga}, n^S_{As}, and this procedure is indicated by the tilde. The exploitation of the diffusion equation is carried out in [4].

(iv) The liquid is assumed to be in equilibrium and its number densities can be read off from the liquidus line of the phase diagram.

(v) The form of the interfacial jump conditions depends on the interfacial mobilities. If the interfacial mobilities are much larger than the bulk mobility of $As^{(o)}_\gamma$, local equilibrium at the interface can be assumed, and the jump conditions are the ones that will be derived in the next two sections, see equations $(43)_2$, (51)and (52). However, currently the magnitudes of the interfacial mobilities are not known, and it may happen that they have the same order as the bulk mobility. In this case the interfacial jump conditions result from the Becker/Döring theory of nucleation, see [5] for details.

9. Thermodynamics of mixtures, Part 6: The Global Laws of Energy and Entropy

In this Section we derive the equilibrium conditions for two coexisting phases. We consider the systems from Section 3 consisting of a container, which is closed by a piston. At time t, the volume $V(t)$ includes either the liquid/solid or the liquid-droplet/solid/gas. We assume that the outer pressure p_0 is given. Furthermore we prescribe the outer temperature T_0 and assume that T_0 is constant throughout $V(t)$. The dashed closed line indicates a control volume which we introduce in order to apply the global balance laws of energy E and entropy S to the described system:

$$\frac{dE}{dt} + p_0 \frac{dV}{dt} = \dot{Q}, \quad \text{and} \quad \frac{dS}{dt} \geq \frac{\dot{Q}}{T_0}. \tag{34}$$

The quantity $\dot{Q}$ denotes the heat power, which may enter, $\dot{Q} > 0$, or leave, $\dot{Q} < 0$, the system, so that a constant temperature T_0 is guaranteed. The equality sign holds in equilibrium, whereas in nonequilibrium, the variation of the entropy is greater than the ratio of supplied heat and temperature. This statement expresses CLAUSIUS' version of the second law of thermodynamics, [2].
Elimination of the heat power leads to the main inequality

$$\frac{dA}{dt} \leq 0 \quad \text{with the definition} \quad A = E - T_0 S + p_0 V. \tag{35}$$

The newly defined quantity, A, is called the available free energy or availability. We conclude that in dynamic processes that run at constant outer pressure, constant temperature and constant total mass, the availability must decrease and assumes its minimum at equilibrium.
The availability contains the combination $\Psi = E - T_0 S$ denoting the Helmholtz free energy, whereas $A = \Psi + p_0 V$ give the Gibbs free energy G only if there is an overall constant pressure $p = p_0$ in the interior of the volume V. However, we will not meet this situation in the two considered examples.
The evaluation of the main inequality requires an explicit knowledge of the functional dependence of energy, entropy and volume on the variables that may change in the process under consideration. In the next sections we will introduce and discuss the functions we need.
The two systems considered in Section 3 consist of solid, S, liquid, L, and gas, G, phases, which occupy different regions, so that the total volume at time t is given by

$$V(t) = V_S(t) + V_L(t), \qquad V(t) = V_S(t) + V_L(t) + V_G(t), \tag{36}$$

for the left and right example, respectively. The total free energies of the two examples can be additively decomposed into the free energies of the phases and the surface free energy of the interface boundaries

$$\Psi = \Psi_S(t) + \Psi_L(t) + \Psi_G(t) + \sigma O_L. \tag{37}$$

The surface energy, which is also called surface tension, is denoted by σ and the liquid/solid interface area is O_L.
The free energies of the phases can be represented by the volume integrals

$$\Psi_S(t) = \int\limits_{V_S} \rho\psi dV, \qquad \Psi_L(t) = \int\limits_{V_L} \rho\psi dV, \qquad \Psi_G(t) = \int\limits_{V_G} \rho\psi dV, \tag{38}$$

where ψ is the specific free energy and ρ is the mass density of the mixture.

10. The laws of mass action and interface conditions

We consider the first problem from Section 3, where a liquid phase is connected to a solid phase by a plane interface. We assume that both phases are in equilibrium and we pose the problem to determine the corresponding equilibrium conditions by exploiting the minimum condition for the availability.
Thus we determine the minimum of the function

$$A = \rho_s \psi_s V_s + \rho_L \psi_L V_L + p_0 (V_s + V_L) \tag{39}$$

under several side conditions.
Conservation of atoms

$$\begin{aligned} N_{Ga} &= n_{Ga_\alpha} V_S + n^L_{Ga} V_L, \\ N_{As} &= (n_{As_\beta} + n_{As^{(0)}_\alpha} + n_{As^{(+)}_\alpha} + n_{As^{(0)}_\gamma}) V_S + n^L_{As} V_L, \\ N_O &= (n_{O^{(0)}_\beta} + n_{O^{(-)}_\beta} + n_{O^{(+)}_\beta}) V_S, \qquad N_{Si} = n_{Si^{(+)}_\alpha} V_S, \\ N_B &= n_{B^{(0)}_\alpha} V_S, \qquad N_C = n_{C^{(-)}_\beta} V_S. \end{aligned} \tag{40}$$

The three sublattices have the same number of lattice sites:

$$\begin{aligned} & n_{As_\beta} + n_{O^{(0)}_\beta} + n_{O^{(-)}_\beta} + n_{O^{(+)}_\beta} + n_{C^{(-)}_\beta} - \\ & (n_{Ga_\alpha} + n_{As^{(0)}_\alpha} + n_{As^{(+)}_\alpha} + n_{V^{(0)}_\alpha} + 3 n_{V^{(3-)}_\alpha} + n_{B^{(0)}_\alpha} + n_{Si^{(+)}_\alpha}) = 0 \\ & n_{As_\beta} + n_{V^{(0)}_\beta} + n_{O^{(0)}_\beta} + n_{O^{(-)}_\beta} + n_{O^{(+)}_\beta} + n_{C^{(-)}_\beta} - (n_{V^{(0)}_\gamma} + n_{As^{(0)}_\gamma}) = 0. \end{aligned} \tag{41}$$

Conservation of charges

$$n_{O^{(-)}_\beta} + n_{C^{(-)}_\beta} + 3 N_{V^{(3-)}_\alpha} + n_e - (n_{As^{(+)}_\alpha} + n_{Si^{(+)}_\alpha} + n_{O^{(+)}_\beta} + n_h) = 0. \tag{42}$$

We can get rid of these nine constraints by introducing Lagrange multipliers. The variables are the volumes V_S and V_L of the solid and liquid phase, respectively, the number densities n^L_{Ga}, n^L_{As} of the liquid, and the fourteen number densities, $n_{Ga_\beta}, n_{As^{(0)}_\alpha}, n_{As_\beta}, n_{As^{(0)}_\gamma}, n_{As^{(+)}_\alpha}, n_{O^{(0)}_\beta}, n_{O^{(-)}_\beta}, n_{O^{(+)}_\beta}, n_{V^{(0)}_\alpha}, n_{V^{(3-)}_\alpha}, n_{V^{(0)}_\beta}, n_{V^{(0)}_\gamma}$, n_e, n_h, of the solid.
The minimization with respect to V_S and V_L yields the conditions for mechanical equilibrium, which obviously read

$$p_L = p_0, \qquad \sigma^{33} = -p_L. \tag{43}$$

Furthermore there result two groups of algebraic equations. The first group contains the so called laws of mass action that determine the number densities of the solid phase for given $(n^S_{Ga}, n^S_{As}, n_O, n_C, n_{\mathrm{Si}}, n_B)$. The laws of mass action read

$$\mu_{As_\beta} - \mu_{As^{(0)}_\gamma} - \mu_{V^{(0)}_\beta} + \mu_{V^{(0)}_\gamma} = 0, \tag{44}$$

$$\mu_{As^{(0)}_\alpha} - \mu_{As^{(0)}_\gamma} - \mu_{V^{(0)}_\alpha} + \mu_{V^{(0)}_\gamma} = 0, \tag{45}$$

$$\mu_{V^{(0)}_\alpha} + \mu_{V^{(0)}_\beta} + \mu_{V^{(0)}_\gamma} = 0, \tag{46}$$

$$\mu_{O^{(+)}_\beta} - \mu_{O^{(0)}_\beta} + \mu_e - \frac{M_e}{\rho_S}\sigma^{<33>} = 0, \qquad \mu_{O^{(-)}_\beta} - \mu_{O^{(0)}_\beta} - \mu_e + \frac{M_e}{\rho_S}\sigma^{<33>} = 0, \tag{47}$$

$$\mu_{V^{(3-)}_\beta} - \mu_{V^{(0)}_\alpha} - 3\mu_e + 3\frac{M_e}{\rho_S}\sigma^{<33>} = 0,$$

$$\mu_{As^{(+)}_\alpha} - \mu_{As^{(0)}_\alpha} + \mu_e - \frac{M_e}{\rho_S}\sigma^{<33>} = 0, \tag{48}$$

$$\mu_e + \mu_h - 2\frac{M_e}{\rho_S}\sigma^{<33>} = 0. \tag{49}$$

The equations of the second group are jump conditions at the interface that relate the solid densities to the number densities of the liquid. The jump conditions read

$$\begin{aligned} &\mu_{Ga_\alpha} - \frac{M_{Ga}}{\rho_S}\sigma^{<33>} - \mu_{V_\alpha} - \mu^L_{Ga} = 0, \\ &\frac{1}{3}(\mu_{As^{(0)}_\alpha} + \mu_{As_\beta} + \mu_{As^{(0)}_\gamma}) - \frac{M_{As}}{\rho_S}\sigma^{<33>} - \mu^L_{As} = 0. \end{aligned} \tag{50}$$

11. Interfacial conditions at curved interfaces

In this section we generalize the interfacial jump condition to the case where the liquid and the solid phase are separated by a curved interface. The following results can be obtained for spherical droplets by minimizing (37) or for nonspherical droplets by methods that are introduced in [3]. The mechanical equilibrium condition (43) now reads

$$-(\sigma^{ik}\nu^i\nu^k + p_L) = 2\sigma K_M, \tag{51}$$

where σ denotes the surface tension, and K_M is the mean curvature, with $K_M = -1/r$ for the spherical liquid droplet discussed in Section 3. The jump conditions regarding the chemical potential at a curved interface read

$$\begin{aligned} &\mu_{Ga_\alpha} - \frac{M_{Ga}}{\rho_S}\sigma^{<ik>}\nu^i\nu^k - \mu_{V_\alpha} - \mu^L_{Ga} = 0, \\ &\frac{1}{3}(\mu_{As^{(0)}_\alpha} + \mu_{As_\beta} + \mu_{As^{(0)}_\gamma}) - \frac{M_{As}}{\rho_S}\sigma^{<ik>}\nu^i\nu^k - \mu^L_{As} = 0. \end{aligned} \tag{52}$$

12. On the influence of misfit strain on the phase diagram

The equilibrium conditions that were derived in the last section can now be used to calculate the liquidus line and the homogeneity region of the phase diagram. In particular, we will determine the influence of mechanical effects on these parts of the phase diagram. To this end, we need to know values for the standard chemical potentials, the interaction coefficients and the Redlich/Kister coefficients. Furthermore we need the elastic constants and thermal expansion coefficients. The parameters that are needed for the liquidus line are taken from [12]. The parameters that are needed for the determination of chemical equilibrium of the solid phase are read off from different sources: the data for (49) are from [14]. The data for (44), (45), (46) and (48) are from [15], but they have been slightly modified. Data regarding the oxygen were reported from [9]. The elastic and thermal expansion parameters can be calculated from the data given in [1]. The temperature dependent data for liquid arsenic are read off from [8]. The misfit parameter were calculated from the atomic radii data, where some information regarding the vacancy misfit was given by [6]. The details of the exploitation of the equilibrium conditions are found to be in [4].

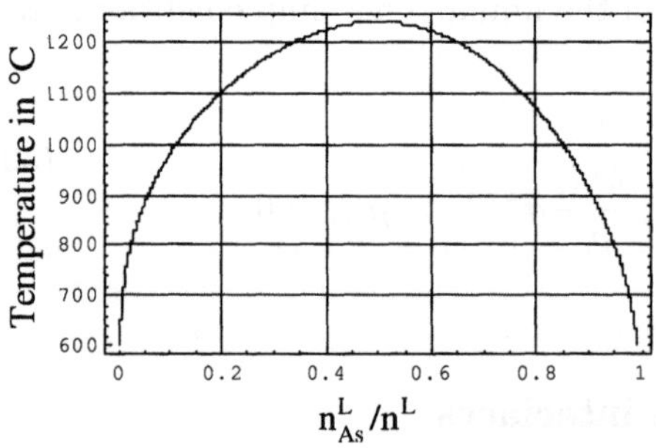

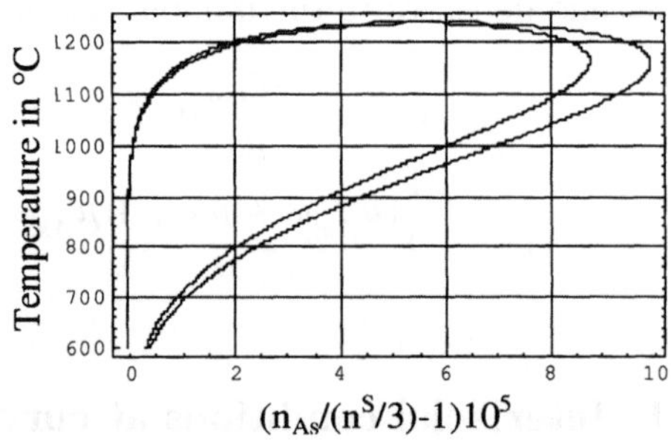

The two graphs were calculated from the solutions of the algebraic system (40) through (50). The graph on the left shows the liquidus line of the GaAs phase diagram, and turns out to be almost independent of mechanical effects. The graph on the right represents the homogeneity region of the phase diagram, where the defects are homogeneously distributed over the lattice sides. The smaller region results if mechanical effects are ignored. If mechanical effects are taken into account, the homogeneity region is affected by the outer pressure and increases if the outer pressure is increased. The graph corresponds to an outer pressure of 500 bar. A similar calculation for the situation with a spherical droplet within a solid matrix, as it is explained in Section 3, has exhibited a much more drastic mechanical effect that does not need such high external pressure as in the current case with a plane interface, see [4].

The concentration of the charged antisite $As_\alpha^{(+)}$ is of most importance regarding the semi-insulating behaviour of GaAs. The next Figure shows the influence of mechanical effects on the concentration of $As_\alpha^{(+)}$.

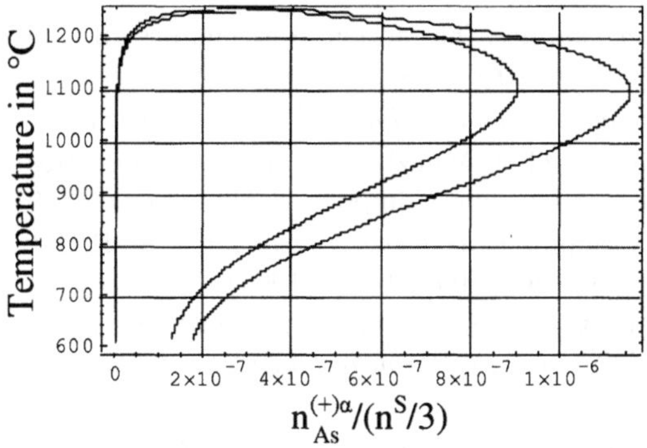

Right from the melting point, the concentration of $As_{\alpha}^{(+)}$ increases drastically with increasing outer pressure.

13. Synopsis and outlook

The current paper sets the beginning of a series of studies on the formation and growth of liquid arsenic droplets in solid GaAs. In the first paper we have developed a thermodynamic model which, in particular, takes local mechanical stress fields into account. Some simple examples show their significance on the laws of mass action and on interfacial equilibrium conditions.
The next step of our investigations will be the study of the diffusion problem including the coupling of the Becker/Döring theory to the liquid/solid interface conditions.

References

[1] Brice, J.C., *Physical and Thermal Properties, Properties of Gallium Arsenide*, EMIS Datareviews Serie No. **2**, (1990).

[2] Clausius, R., *Über eine veränderte Form des zweiten Hauptsatzes der mechanischen Wärmetheorie*, Poggendorf's Annalen der Physik, **93**, (1850).

[3] Dreyer,W., *On jump conditions of moving interfaces in crystals*, WIAS Preprint, (2003).

[4] Dreyer,W & F. Duderstadt, *Modeling of the appearance of arsenic droplets in galliumarsenide*, WIAS Preprint (2003).

[5] Dreyer,W. & F. Duderstadt, On the Becker/Döring Theory of nucleation of liquid droplets in solids, WIAS Preprint (2003).

[6] Eichler, St., private communication, Freiberg, (2002).

[7] Eshelby, J.D., *The Elastic Energy Momentum Tensor*, Journal of Elasticity, **5** (1975).

[8] Gokcen, *The Arsenic System*, Bulletin of Alloy Phase Diagrams Vol. **10**, No 1, (1989).

[9] Jurisch, M., private communication, Freiberg, (2001).

[10] Müller, I., *Thermodynamics*, Interaction of Mechanics and Mathematics Series, Pitman Advanced Publishing Program, Boston, (1985).

[11] Müller, I., *Grundzüge der Thermodynamik*, 3. Auflage, Springer Berlin, (2001).

[12] Oates, W. A., Erikson, G. & Wenzl. H., *Thermodynamic modelling of solid gallium arsenide*, Journal of Alloys and Compounds **220**, Elsevier, (1995).

[13] Truesdell,C. & R. Toupin, *The Classical Field Theories*, Handbuch der Physik, III/1, Springer, Berlin, Göttingen, Heidelberg, (1960).

[14] Wenzl, H., Mika, K. & D. Henkel, *Thermochemical Analysis of Native Point Defects ans Substitutional Carbon and Baron in GaAs Crystal Growth*, Cryst. Res. Technol., (1990).

[15] Wenzl, H., Oates, W. A. & K. Mika, *Defect Thermodynamics and Phase Diagrams in Compound Crystal Growth Process*, Handbook of Crystal Growth, Vol 1, Elsevier, (1993).

(W. Dreyer and F. Duderstadt) WIAS, Berlin

International Series of Numerical Mathematics, Vol. 147, 131–138

On the Intermediate Surface Diffusion Flow

Joachim Escher and Kazuo Ito

Abstract. The intermediate surface diffusion flow is a nonlocal geometric surface evolution law with two physical parameters. We survey known results of this flow, such as existence results, the limiting motion concerning a parameter, and self-intersections of solution surfaces. Some new results on a loss of convexity and singularities of solutions are also announced.

1. Introduction

We study the following nonlocal geometric evolution equation:

$$V(t) = -\Delta_{\Gamma(t)} \left(\delta - \mu\Delta_{\Gamma(t)}\right)^{-1} H(t) \quad \text{on } \Gamma(t) \text{ for } t > 0. \tag{1}$$

Here $t > 0$ denotes the time variable, and $\Gamma(t)$ is an unknown, evolving closed compact oriented connected hypersurface immersed in $\boldsymbol{R}^n$, with $n \geq 2$. We write $\Delta_{\Gamma(t)}$ for the Laplace–Beltrami operator on $\Gamma(t)$ with respect to the Euclidean metric. Since $\Gamma(t)$ is oriented, its unit normal vector field $\{N(t,x); x \in \Gamma(t)\}$ is well-defined. We have two possible directions for the unit normal vector but we fix arbitrarily one of them. Then, in the direction of $N(t)$, $H(t)$ denotes the mean curvature of $\Gamma(t)$ multiplied by $(n-1)$ and $V(t)$ stands for the normal velocity of the family $\{\Gamma(t)\,;\, t > 0\}$. We mention that the evolution equation (1) does not depend on the choice of orientation. Finally, in the setting $D := 1/\delta > 0$ and $M := 1/\mu > 0$, D is the diffusion coefficient and M is the mobility constant. Following J. W. Cahn and J. E. Taylor [2, 20], we call (1) the *intermediate surface diffusion flow.*

The evolution equation (1) has to be supplemented by an appropriate initial condition

$$\Gamma(0) = \Gamma_0, \tag{2}$$

where Γ_0 is a given closed compact oriented connected hypersurface immersed in $\boldsymbol{R}^n$.

2000 *Mathematics Subject Classification.* Primary 35R35; Secondary 53C44.
Key words and phrases. Intermediate surface diffusion flow, Self-intersection, Loss of convexity, Singularity.
The second author was partially supported by the 2001 Canon Foundation Research Fellowship. He is also grateful to Institute of Applied Mathematics, University of Hannover for the warm hospitality during his stay.

The intermediate surface diffusion flow (1) was introduced to describe a geometric growth law for a moving interface where surface diffusion is the only transport mechanism and the reduction of total surface energy is the only driving force for surface motion, cf. [2, 20]. The geometric features of this flow are the signed volume preserving and the area decreasing properties. To see this, we define the signed volume $Vol(t)$ for $\Gamma(t)$ by

$$Vol(t) := -\frac{1}{n}\int_{\Gamma(t)} (\gamma(t)|N(t))d\sigma(t),$$

and the surface area $Area(t)$ of $\Gamma(t)$ by

$$Area(t) := \int_{\Gamma(t)} 1 d\sigma(t).$$

Here $\gamma(t)$ denotes the position vector of $\Gamma(t)$ and $d\sigma(t)$ stands for the surface measure on $\Gamma(t)$ induced by the Euclidean metric in the ambient space $\boldsymbol{R}^n$, and $(\cdot|\cdot)$ is the usual inner product in $\boldsymbol{R}^n$. If $\Gamma(t)$ is embedded and the direction of $N(t)$ points inward, then the divergence theorem ensures that $Vol(t)$ coincides with the usual volume of the domain enclosed by $\Gamma(t)$. It is known that the following variational identities of $Vol(t)$ and $Area(t)$ for general moving surfaces $\Gamma(t)$ hold true:

$$\frac{d}{dt}Vol(t) = \int_{\Gamma(t)} V(t)d\sigma(t),$$
$$\frac{d}{dt}Area(t) = -\int_{\Gamma(t)} H(t)V(t)d\sigma(t).$$

Then one has

$$\frac{d}{dt}Vol(t) = -\int_{\Gamma(t)} \Delta_{\Gamma(t)}\left(\delta - \mu\Delta_{\Gamma(t)}\right)^{-1} H(t)d\sigma(t) = 0,$$

$$\begin{aligned}\frac{d}{dt}Area(t) &= \int_{\Gamma(t)} H(t)\Delta_{\Gamma(t)}\left(\delta - \mu\Delta_{\Gamma(t)}\right)^{-1} H(t)d\sigma(t) \\ &= -\int_{\Gamma(t)} \left|\nabla_{\Gamma(t)}\left(\delta - \mu\Delta_{\Gamma(t)}\right)^{-1/2} H(t)\right|^2 d\sigma(t) \leq 0,\end{aligned}$$

where $\nabla_{\Gamma(t)}$ stands for the gradient on $\Gamma(t)$ induced by the Euclidean metric in the ambient space $\boldsymbol{R}^n$. Hence one sees that the flow (1) is indeed signed volume preserving and area decreasing.

J. W. Cahn and J. E. Taylor [2, 20] expected (1) to play an interesting role as a link between two known signed volume preserving and area decreasing flows. One is the *volume preserving mean curvature flow* (also called the *averaged mean curvature flow*)

$$V(t) = \mu^{-1}\left(H(t) - \overline{H}(t)\right) \quad \text{on } \Gamma(t) \text{ for } t > 0, \tag{3}$$

where $\overline{H}(t)$ is the spatial average of the mean curvature $H(t)$, i.e.,

$$\overline{H}(t) := \int_{\Gamma(t)} H(t) d\sigma(t) \Big/ \int_{\Gamma(t)} 1 d\sigma(t) .$$

The other one is the *surface diffusion flow*

$$V(t) = -\delta^{-1}\Delta_{\Gamma(t)} H(t) \quad \text{on } \Gamma(t) \text{ for } t > 0. \tag{4}$$

Similarly as above, one can check that both flows (3) and (4) are signed volume preserving and area decreasing. It was conjectured in [2, 20] that the limiting motions of (1) as $\delta \to 0$ and $\mu \to 0$ are (3) and (4), respectively. This is the reason why (1) is called the intermediate surface diffusion flow.

2. Existence results

The first mathematical results for (1) was presented by C. M. Elliott and H. Garcke [4]. They proved for $n = 2$ that there exist local-in-time solutions and that solutions starting from initial data close to a circle exist globally and converge to a (possibly another) circle as $t \to \infty$. These results are extended by J. Escher and G. Simonett [10] to the higher dimensional case. Additionally, a general uniqueness result is obtained in [10]. We note that every compact surface of constant mean curvature is an equilibrium for (1), and the converse is also true by Liouville's theorem. In particular, in the embedded case this leaves only the spheres, cf. [1]. Recently, J. Escher [5] has proved that solutions of (1) are real analytic in space and time, provided the initial surface is suitably diffeomorphic to an Euclidean sphere. In [4, 10] there is no rigorous investigation to clarify the dependency of (1) on the parameters δ and μ. Recently, J. Escher, Y. Giga and K. Ito [6, 7] established a uniform local existence result of (1) for $\delta \to 0$, which is applied to prove that the limiting motion of (1) as $\delta \to 0$ is in fact the volume preserving mean curvature flow (3). For this we mention the precise result in Section 3. Up to now the question of the limit of (1) as $\mu \to 0$ is still open.

3. Limiting motion

In this section we are interested in the limiting motion of (1) as $\delta \to 0$. So we are going to restrict ourselves to the case $\delta \in (0, 1]$. Also μ is regarded to be a fixed constant.

Theorem 3.1. *(Limiting motion [6, 7]). Let Γ_0 be of class $C^{2+\alpha}$ with fixed $\alpha \in (0, 1)$. Then the following hold true.*

(i) The problem (1)–(2) admits a unique local solution $\Gamma^\delta := \{\Gamma^\delta(t); t \in [0, T]\}$ of class $C^{1,2+\alpha}$ on a common time interval $[0, T]$, i.e., T depends on $\|\Gamma_0\|_{C^{2+\alpha}}$, but not on $\delta \in (0, 1]$.

(ii) The solution Γ^δ converges, as $\delta \to 0$, to the unique solution of the volume preserving mean curvature flow (3) *supplemented with* (2) *— with the same Γ_0 — in the $C^{1,2+\alpha}$-topology on $[0,T]$.*

4. Self-intersections

In contrast to the mean curvature flow $V(t) = H(t)$, the flows (3) and (4) do not enjoy the maximum principle, cf. [11, 12, 18]. This is understood from the fact that the solutions of (3) and (4) can develop self-intersections even if the initial data are embedded.

As an application of Theorem 3.1, we show that this is also the case for the flow (1), i.e., embedded hypersurfaces can be driven under (1) into self-intersections in finite time for $\delta > 0$.

Theorem 4.1. *(Self-intersections [6, 7]). Let $\delta > 0$ be fixed. There is an embedded closed compact oriented hypersurface Γ_0^δ of class $C^{2+\alpha}$, with $\alpha \in (0,1)$, such that the unique solution Γ^δ of* (1) *starting from Γ_0^δ develops a self-intersection at a time $t_\delta \in (0, T_\delta]$, although Γ^δ stays in $C^{1,2+\alpha}$ on $[0, T_\delta]$. Here T_δ is an existence time of the solution Γ^δ.*

We mention that T_δ is eventually away from 0 also when $\delta \to 0$, thanks to Theorem 3.1 (i).

We are now going to sketch the crucial steps of the proof. Assume that (1) has the desired property for some $\delta_* > 0$. Then the same is true for any $\delta > 0$. To see this, we introduce a scaling both in time and in the ambient space given by $\tilde{t} = \lambda^{-2}t$ and $\tilde{x} = \lambda^{-1}x$ for $\lambda > 0$. We then have the scaling relations $\tilde{V} = \lambda V$, $\tilde{H} = \lambda H$, and $\tilde{\Delta} = \lambda^2 \Delta$. Thus (1) and (2) are rescaled to

$$\tilde{V}(\tilde{t}) = -\tilde{\Delta}_{\tilde{\Gamma}(\tilde{t})}\left(\lambda^2\delta - \mu\tilde{\Delta}_{\tilde{\Gamma}(\tilde{t})}\right)^{-1}\tilde{H}(\tilde{t}),\ \tilde{t} > 0;\quad \tilde{\Gamma}(0) = \tilde{\Gamma}_0. \tag{5}$$

Choosing λ to be $\lambda^2\delta = \delta_*$, one sees that (5) has the property to develop a self-intersection in finite time for an embedded initial datum. Note that this self-intersection is invariant under the present scaling. Therefore (1) must have this property as well.

Thanks to Theorem 3.1, the solutions of (1) and (3) with the same initial data are close to each other in the $C^{1,2+\alpha}$-topology when $\delta_* > 0$ is sufficiently small. Since (3) can develop a self-intersection, cf. [11, 18], the same must be true for (1) with $\delta_* > 0$ small enough. Therefore it must be also true for (1) with any $\delta > 0$ as well, as pointed out previously.

5. Loss of convexity

It is known that the flow (3) preserves the convexity of its solutions for all times, cf. [11, 15], whereas a solution of the flow (4) can lose its convexity in finite time, cf. [13, 16].

We announce here a new result stating that the flow (1) has the latter property. In the following statement, a surface is said to be strictly convex if all the principal curvatures of the surface are positive.

Theorem 5.1. *(Loss of convexity* [8]*). There is a smooth ($C^{6+\alpha}$ with $\alpha \in (0,1)$ is in fact enough), closed, compact, oriented, strictly convex hypersurface Γ_0 in $\boldsymbol{R}^n$ such that the unique solution of* (1) *starting from Γ_0 loses convexity in finite time, although it stays smooth.*

6. Singularities

For the flow (4) in $n = 2$, J. Escher, U. F. Mayer and G. Simonett [9] presented numerical experiments for formations of singularities of solution curves with self-intersections. Recently, K. S. Chou [3] gave an analytical proof for the above mentioned experiments in a certain class of initial curves.

In this section we announce some new results on the formation of singularities of solution curves of the flow (1). When we discuss this problem, we must take into account that two invariant quantities of the solution curves play important roles, namely the rotation index and the signed area. To define them suitably for our purpose, let us consider a general smooth connected closed curve Γ immersed in the plane. Let s and γ be respectively the arclength parameter and the position vector of Γ. When γ is described by s and we set $T := \partial\gamma/\partial s$, T is the unit tangent vector of Γ. Let N be the unit normal vector of Γ so that the frame $\{T, N\}$ obeys the Frenet convention. Then it fulfills the relation $N = R(\pi/2)T$, where $R(\pi/2)$ denotes the matrix which rotates the operand vector by the angle $+\pi/2$. Let us denote by κ the curvature of Γ; κ is computed in the direction of N so that Frenet's formulae $T_s = \kappa N$ and $N_s = -\kappa T$ hold. Then we define the rotation index m of Γ by

$$m := \frac{1}{2\pi} \int_\Gamma \kappa ds.$$

It is known that m has to be an integer. We also define the signed area A for Γ by

$$A := -\frac{1}{2} \int_\Gamma (\gamma|N) ds.$$

Remark 6.1. (i) When Γ is embedded, the above definition yields that $A > 0$ when $m = 1$ and $A < 0$ when $m = -1$. In fact, this is an easy consequence of the divergence theorem.

(ii) Given Γ, the signs of m and A are only the matter of convention. In fact, we are free to choose arbitrary signs of m and A by adopting alternative definitions for the rotation index and the signed area.

In this paper, given connected closed curves, we always adopt the sign convention so that $m \geq 0$. Using the notations mentioned above, we now rewrite (1) in the following form:

$$V(t) = -\partial_s^2(\delta - \mu\partial_s^2)^{-1}\kappa(t) \quad \text{on } \Gamma(t). \tag{6}$$

In (6) ∂_s^2 and $\kappa(t)$ are used instead of $\Delta_{\Gamma(t)}$ and $H(t)$. Equation (6) has also to be supplemented with the initial condition (2). In the sequel, we will always impose (2) without further mentioning.

We say that the solution to (6) develops a singularity (or blows up) in finite time if a smooth global-in-time solution does not exist.

Let us denote by T_{max} the maximal existence time of the solution $\Gamma = \{\Gamma(t); t \in [0, T_{max})\}$ to (6). Thus the statement $T_{max} < \infty$ means that the solution of (6) develops a singularity (or blows up) in finite time.

The following result shows the blow-up rate of the L_2-norm of the curvature $\|\kappa(t)\|_2$ of the solution of (6).

Proposition 6.2. *(Blow-up rate [8]). Let $m \geq 0$ for Γ_0. Assume that the signed area A_0 for Γ_0 is nonzero and that T_{max} is finite. Then the following blow-up rate for the solution of* (6) *holds true:*

$$\|\kappa(t)\|_2 \geq C(T_{max} - t)^{-1/4} \tag{7}$$

for $t < T_{max}$ sufficiently close to T_{max}.

Remark 6.3. The same blow-up rate as in (7) is also obtained for the flow (3) in [8]. K. S. Chou [3] obtained the blow-up rate for the flow (4), where the power in (7) is $-1/8$.

Now we give some sufficient conditions for $T_{\max} < \infty$. The first one is stated in the following result.

Theorem 6.4. *(A sufficient condition for singularities [8]). Set $u := (\delta - \mu\partial_s^2)^{-1/2}\kappa$. Assume that for any $t \in [0, T_{max})$, there is a $\sigma(t) \in [0, L(t)]$ such that $u(t, \sigma(t)) = 0$, where $L(t)$ denotes the total length of $\Gamma(t)$. Then we have $T_{max} < \infty$.*

Corollary 6.5. *([8]) Assume that $m = 0$ for Γ_0. Then $T_{max} < \infty$.*

Remark 6.6. In [8] similar sufficient conditions for singularities of solutions to the flows (3) and (4) are also obtained.

For example, Corollary 6.5 tells us that if the initial data Γ_0 has a configuration of figure-eight, then the solution develops a singularity in finite time. For the flows (3) and (4), this kind of results have been already obtained by M. A. Grayson [14] and A. Polden [19], respectively. See also [3].

Even if the assumption in Theorem 6.4 is not fulfilled, a solution of (6) can develop a singularity, as shown below. For example, locally convex curves, i.e., curves with $\kappa > 0$, do not fulfil the assumption in Theorem 6.4 since the operator $(\delta - \mu\partial_s^2)^{-1/2}$ is positivity preserving, but a class of locally convex curves can still develop a singularity. The next theorem gives a sufficient condition for this case.

Theorem 6.7. *(The second sufficient condition for singularities* [8]*). Assume for* Γ_0 *either*

(i) $m \geq 1$ *and* $A_0 < 0$

or

(ii) $m \geq 2$ *and* $L_0^2 < 4\pi m A_0$.

Then T_{max} *is finite for the flow* (6).

Remark 6.8. (a) By (ii) we can find a limaçon to develop a singularity under the flow (6).

(b) The condition (ii) has been first found by K. S. Chou [3] for the flow (4). Theorem 6.7 holds also for the flows (3) and (4), see [8], and hence it is an extension of the result in [3].

(c) Theorem 6.7 does not consider the case in which the initial curves Γ_0 are embedded. But we would like to mention an interesting numerical experiment presented by U. F. Mayer [17], suggesting that there is an embedded initial closed curve Γ_0 such that the solution of (3) starting from Γ_0 first loses embeddedness and then develops a singularity in finite time.

(d) Theorem 6.7 does not treat the case $A_0 = 0$. But we have a result for the flow (4) saying that if $m \geq 1$ and $A_0 = 0$, then $T_{max} < \infty$, see [8].

References

[1] A. D. Alexandrov, *Uniqueness theorems for surfaces in the large*, Vestnik Leningrad Univ. **11** (1956), 5–7.

[2] J. W. Cahn and J. E. Taylor, *Surface motion by surface diffusion*, Acta Metallurgica **42** (1994), 1045–1063.

[3] K. S. Chou, *A blow-up criterion for the curve shortening flow by surface diffusion*, to appear in Hokkaido Math. J.

[4] C. M. Elliott and H. Garcke, *Existence results for diffusive surface motion laws*, Adv. Math. Sci. Appl. **7** (1997), 467–490.

[5] J. Escher, *The intermediate surface diffusion flow on spheres*, to appear in J. Nonl. Math. Phys.

[6] J. Escher, Y. Giga and K. Ito, *On a limiting motion and self-intersections of curves moved by the intermadiate surface diffusion flow*, Nonlinear Analysis TMA **47** (2001), 3717–3728.

[7] J. Escher, Y. Giga and K. Ito, *On a limiting motion and self-intersections for the intermadiate surface diffusion flow*, J. Evol. Equ. **2** (2002), 349–364.

[8] J. Escher and K. Ito, *Some dynamical properties of volume preserving curvature driven flows*, in preparation.

[9] J. Escher, U. F. Mayer and G. Simonett, *The surface diffusion flow for immersed hypersurfaces*, SIAM J. Math. Anal. **29** (1998), 1419–1433.

[10] J. Escher and G. Simonett, *Moving surfaces and abstract parabolic evolution equations*, Progress in Nonlinear Differential Equations and Their Applications **35** (1999), 183–212, Birkhäuser.

[11] M. Gage, *On an area preserving evolution equation for plane curves*, Contemporary Mathematics **51** (1986), 51–62.

[12] Y. Giga and K. Ito, *On pinching of curves moved by surface diffusion*, Comm. Appl. Anal. **2** (1998), 393–405.

[13] Y. Giga and K. Ito, *Loss of convexity of simple closed curves moved by surface diffusion*, Progress in Nonlinear Differential Equations and Their Applications **35** (1999), 305–320, Birkhäuser.

[14] M. A. Grayson, *The shape of a figure-eight under the curve shortening flow*, Invent. Math. **96** (1989), 177–180.

[15] G. Huisken, *The volume preserving mean curvature flow*, J. Reine Angew. Math. **382** (1987), 35–48.

[16] K. Ito, *Loss of convexity of compact hypersurfaces moved by surface diffusion*, Comm. Appl. Anal. **6** (2002), 105–126.

[17] U. F. Mayer, *A singular example for the averaged mean curvature flow*, Exponential Math. **10** (2001), 103–107.

[18] U. F. Mayer, G. Simonett, *Self-intersections for the surface diffusion and the volume preserving mean curvature flow*, Diff. Int. Eqs. **13** (2000), 1189–1199.

[19] A. Polden, *Curves and surfaces of least total curvature and fourth order flows*, Ph.D. dissertation, Universität Tübingen, Germany (1996).

[20] J. E. Taylor and J. W. Cahn, *Linking anisotropic sharp and diffuse surface motion laws via gradient flows*, J. Stat. Phys. **77** (1994), 183–197.

(J. Escher) Institute of Applied Mathematics, University of Hannover, Welfengarten 1, D-30167 Hannover, Germany
E-mail address: escher@ifam.uni-hannover.de

(K. Ito) Graduate School of Mathematics, Kyushu University, Ropponmatsu, Fukuoka 810-8560, Japan
E-mail address: k-ito@math.kyushu-u.ac.jp

International Series of Numerical Mathematics, Vol. 147, 139–149

Solid Core Revisited

Antonio Fasano, Alberto Mancini and Riccardo Ricci

Abstract. We present a hyperbolic model for the growth of a polymer particel in the Zigler–Natta process.

1. Introduction

This paper is devoted to the problem of the growth of a spherulite in the process known as Ziegler–Natta polymerization. This is a rather complex problem and has been the object of a large literature.

The term *polymerization* indicates the process in which a single monomer molecule joins a *chain* of similar molecules to form the polymer. The main peculiarity of the Ziegler–Natta polymerization is the presence of a catalyst, which is necessary to the polymerization. The catalyst is a metal molecule (titanium allide), spread on the the surface of very small particles (supported catalysis).

A well accepted approach for describing the process in the proximity of a single catalyst particle consists in the so called "solid core model" for polymerization. Let us quote from [6]: "The solid core model for polyolefin polymerization is simply based on a spherical catalyst particle with a spherical shell of polymer growing around it. ... Growth of the polymer shell results directly from the rate at which polymer is produced. The kinetics depends on both the temperature at the surface of the catalyst and the monomer concentration at that point".

According to this description the newly formed polymer surrounds the catalyst, and the polymer shell has to expand to accommodate the new inner layer. This picture however poses a number of non trivial problems when we try to translate it into a coherent mathematical model.

The very first question is how the monomer can reach the catalyst surface after a polymeric shell has been formed around it. It is generally assumed that the polymeric shell is a *porous medium* and the monomer fluid reaches the surface through a diffusion mechanism (or a Darcy type flux, depending on the pressure conditions.)

A more intriguing question is "how does the polymer shell move in its expansion?". Even if we assume *a priori* a simple spherically symmetrical geometry, i.e. that the catalyst particle is a sphere and that the polymer shell is a growing sphere at any time, the equations determining the velocity field should include the structural assumptions about the mechanical behavior of the polymer shell itself.

A simple way to write the equation for the velocity field, under the assumption of spherical symmetry, is to assume that the expansion of the shell is "porosity preserving", i.e. that the spherical shell grows rearranging the polymer chain in such a way that the porous matrix formed by the interlaced chain has constant porosity. This allows to write a simple scalar equation for the velocity field, which is sufficient to determine the unique non trivial (radial) component of the field.

In such a way a coherent mathematical model can be derived. A mathematical analysis of it has been performed in [1].

However this is not the only possible assumption, even in the spherical symmetric case, and it fails to give a sufficient number of equations if the symmetry assumption is not satisfied.

2. The elastic model

In order to overcome the geometrical restriction of the solid core model, we have to introduce continuum mechanics into the model.

In fact we must specify how the porous matrix formed by the growing polymer shell reacts to the expansion. This means that we have to introduce a stress tensor as a new unknown quantity and a constitutive assumption relating the stress to the deformation.

The simplest assumption, at least from the conceptual point of view, is that there is a linear dependence between the strain rate and the deformation rate (hypoelasticity).

This means that at any point in the polymer shell we assume the hypoelastic constitutive relation between the increments of the stress tensor (σ_{ij}) and the strain e_{ij}, given by

$$D\sigma_{ij} = E_{ijkl} De_{kl}, \qquad E_{ijkl} = (K - \frac{2}{3}G)\delta_{ij}\delta_{kl} + 2G\delta_{ik}\delta_{lj}, \tag{1}$$

where K and G are respectively the bulk and the shear moduli, [5].

The strain is then expressed by

$$De_{ij} = \dot{e}_{ij} Dt, \qquad \dot{e}_{ij} = \frac{1}{2}\left(\frac{\partial v_i}{\partial x_j} + \frac{\partial v_j}{\partial x_i}\right), \tag{2}$$

where $\dot{e}_{ij}$ is the strain rate of the velocity field $\mathbf{v}$.

The increment operator D must be, in principle, expressed through the Oldroyd derivative. However, in practice the use of the material derivative

$$\frac{d}{dt} = \frac{\partial}{\partial t} + v_k \frac{\partial}{\partial x_k}$$

(summation over the index k) is sufficient.

With this simplifications, the constitutive relation becomes

$$\frac{d\sigma_{ij}}{dt} = E_{ijkl}\dot{e}_{kl}\,. \tag{3}$$

Once the moduli K and G have been determined, equation (3) tells how the material forming the porous matrix reacts to deformation of the matrix itself.

3. The two phase system

In order to give a complete description of the expansion we have to take into account that our system is a *two phases* system composed by the matrix and the monomer inside the pores. This implies that we have to write appropriate equations expressing the following facts: conservation of mass for both phases, the solid one(i.e. the porous matrix) and the fluid one (i.e. the monomer inside the pores); conservation of momentum for both phases; constitutive relations, including (3); appropriate boundary conditions.

The basic form of the equations for a two-phases porous-fluid system can be found in the book of Nikolaevskij, [5].

Conservation of mass for the solid gives

$$\frac{\partial}{\partial t}(\rho_s(1-n)) + \text{div}\,[\rho_s(1-n)\mathbf{V}_s] = 0\,, \tag{4}$$

and conservation of momentum for the solid phase gives

$$(1-n)\rho_s\left[\frac{\partial}{\partial t}\mathbf{V}_s + \mathbf{V}_s\cdot\nabla\mathbf{V}_s\right] = \text{div}\,\underline{\underline{\sigma}}^f - (1-n)\nabla p - \underline{R}\,, \tag{5}$$

where ρ_s is the density of the solid, $\mathbf{V}_s$ is the solid velocity, n is the porosity, p is the fluid pore pressure and $\underline{\underline{\sigma}}^f$ denotes the Terzaghi effective stress tensor, [5], which can be expressed in terms of the pressure and the solid matrix stress tensor $\underline{\underline{\sigma}}$ by $\underline{\underline{\sigma}}^f = (1-n)\left(\underline{\underline{\sigma}} + p\,\underline{\underline{\text{id}}}\right)$.

The vector $\underline{R}$ is the fluid-solid interaction force given by $\underline{R} = r\,(\mathbf{V}_s - \mathbf{V}_f)\,,$ where $\mathbf{V}_f$ is the fluid velocity (with respect to a fixed frame of reference), and r is the resistence coefficient,... [5, (4.36), p. 192],.

Similar equations can be written for the fluid phase. Here we assume that the pressure conditions are such that the monomer can be treated as a liquid, see [3]. For slow motion we can assume that the fluid motion inside the (moving) porous medium is described by Darcy's law, see Bear [2],

$$\mathbf{q}_r = -k\nabla p, \tag{6}$$

where k is the hydraulic conductivity, $\mathbf{q}_r = (\mathbf{q} - n\mathbf{V}_s)$ is the specific discharge with respect to the moving solid particles, and $\mathbf{q} = n\mathbf{V}_f$ is the specific discharge with respect to the fixed coordinates (here we can disregard the gravity in Darcy's law), [2]. Assuming the fluid to be incompressible, the fluid mass conservation becomes

$$\rho_f\frac{\partial n}{\partial t} + \text{div}\,[\rho_f\,(\mathbf{q}_r + n\mathbf{V}_s)] = 0 \tag{7}$$

Another simplification can be made concerning the solid matrix, see [2]. It consists in assuming that, although the medium as a whole undergoes deformation (reflected in the change of porosity), the solid grains themselves are rigid, which implies that $\rho_s = \text{const}$.

Under this assumption, equation (4) reduces to

$$\frac{\partial}{\partial t}(1-n) + \text{div}\,[(1-n)\mathbf{V}_s] = 0\,, \tag{8}$$

which, together with the assumption of incompressible fluid, allows to cancel the porosity term in the fluid mass conservation equation and to reduce it to

$$\text{div}\,[\mathbf{q}_r + \mathbf{V}_s] = 0 \tag{9}$$

4. Boundary conditions

The boundary conditions are one of the most delicate points in the modelling of the growth of the polymer shell.

According to [6]: "It is thought that the growing polymer is directly attached to an active titanium atom on the catalyst surface, and propagation occurs by insertion of monomer between the metal atom and the growing chain."

Suppose than that at time t the polymer shell (the solid matrix) occupies the region $\Omega(t)\backslash\Omega_0$ surrounding the catalyst particle (solid core) Ω_0 (for simplicity we assume that both $\Omega(t)$ and Ω_0 are simply connected regions). The growth mechanism can be modelled assuming that a reaction takes place at the catalyst surface $\partial\Omega_0$ transforming the monomers in the fluid phase into parts of the chains forming the polymer shell. The conservation of mass in this reaction will be one of the boundary conditions.

In turn, the growth of the chains pushes the shell and this can be accounted for prescribing that the normal velocity of the solid matrix at points on Ω_0 is proportional to the quantity of monomer entering the reaction.

So, if $\mathbf{V}_s$ is the solid velocity at the surface of the catalyst, and $\underline{\nu}$ denotes the outward normal to $\partial\Omega_0$, the quantity $\rho_s(1-n)\mathbf{V}_s \cdot \underline{\nu}$ is the quantity of newly created polymer for unit time, and $\rho_f n \mathbf{V}_s \cdot \underline{\nu}$ is the quantity of monomer which has been trapped into the newly formed porous matrix. The sum of these two terms must equal the quantity of monomer which has reached the inner boundary, namely $-\rho_f \mathbf{q}_r \cdot \underline{\nu}$, so that we can write the first boundary condition

$$\rho_s(1-n)\mathbf{V}_s \cdot \underline{\nu} + \rho_f n \mathbf{V}_s \cdot \underline{\nu} = -\rho_f \mathbf{q}_r \cdot \underline{\nu}, \tag{10}$$

or, using (6),

$$\left(\left[n + \frac{\rho_s}{\rho_f}(1-n)\right]\mathbf{V}_s - k\nabla p\right) \cdot \underline{\nu} = 0. \tag{11}$$

We assume that the speed of reaction is proportional to the pressure of the monomer fluid, in agreement with the microscopical interpretation of the pressure. Thus we can write the second boundary condition as

$$\rho_s(1-n)\mathbf{V}_s \cdot \underline{\nu} = \lambda p \tag{12}$$

Finally one has to decide how to fix the value of the porosity of the newly formed polymer matrix. The simplest thing to do is to assume that this porosity is constant (i.e. depends only on the characteristics of the catalyst surface)

$$n = n_0 \tag{13}$$

The first boundary condition on the external boundary $\partial\Omega(t)$ is simply deduced from the fact that $\partial\Omega(t)$ is a material surface whose point are moving with speed $\mathbf{V}_s$. Then, if $\partial\Omega(t)$ is the set of the points satisfying an equation of the form $f(\mathbf{x}, t) = 0$, the boundary condition is

$$\mathbf{V}_s \cdot \underline{\nu} = -\frac{\frac{df}{dt}}{|\nabla f|} \tag{14}$$

Since the particle is surrounded by the monomer fluid, it is reasonable to assume that at $\partial\Omega(t)$ the pore pressure of the fluid equals the exterior pressure

$$p = p_0. \tag{15}$$

Finally, since there is no extra load on the external surface, but for the fluid pressure, the effective stress vanishes on the external boundary,

$$\left(\underline{\underline{\sigma}} + p\,\underline{\underline{\mathrm{id}}}\right)_{ij} \underline{\nu}_j = 0. \tag{16}$$

5. Equation for the one dimensional case

To investigate the well-posedness of the complete set of equations is a very hard task because of the lack of a good theory for hyperbolic systems in $\mathbf{R}^3$. In order to illustrate the kind of mathematics involved in the problem, we devote this section to the one-dimensional problem

$$(v - Kp_x)_x = 0, \tag{17}$$

$$\varphi_t + (v\varphi)_x = 0, \tag{18}$$

$$v_t + vv_x - \frac{1}{\rho}\sigma_x - \frac{\sigma + p}{\rho\varphi}\varphi_x + \frac{1}{\rho\varphi}R = 0, \tag{19}$$

$$\sigma_t + v\sigma_x - \alpha v_x = 0, \tag{20}$$

where v is the velocity, $\varphi = (1 - n)$ is the local solid fraction, ρ is the solid density and σ is the solid stress. The reaction term R is given by

$$R = \frac{rkp_x}{1 - \varphi}, \tag{21}$$

where r is the resistance constant, see [5].

Equations (17)–(20) must be solved in a domain

$$D(\delta) = \{0 < x < x(t), 0 < t < \delta\}, \tag{22}$$

whose right boundary $x = x(t)$ is a free boundary defined by the differential equation

$$\frac{dx}{dt}(t) = v(t, x(t)), \quad x(0) = 0. \tag{23}$$

Boundary conditions for (17)–(20) are given by

$$
\begin{aligned}
p(t,x(t)) &= p_0\,, & (24)\\
v(t,0) - Kp_x(t,0) &= 0\,, & (25)\\
\sigma(t,x(t)) + p(t,x(t)) &= 0\,, & (26)\\
\varphi(t,0) &= \varphi_0\,, & (27)\\
v(t,0) &= \mu p(t,0)\,, & (28)
\end{aligned}
$$

where p_0, φ_0 and μ are given positive constants[1]

For any given pressure $p(x,t)$, we can regard the last three equations of the system (17)–(20) as a (possibly) hyperbolic system, which can be written in the vector form

$$\mathbf{u}_t + B(p,\mathbf{u})\mathbf{u}_x = F(p,p_x,\mathbf{u}), \tag{29}$$

with

$$\mathbf{u} = \begin{pmatrix} \varphi \\ v \\ \sigma \end{pmatrix}, \quad B = \begin{pmatrix} v & \varphi & 0 \\ -\frac{\sigma+p}{\rho\varphi} & v & -\frac{1}{\rho} \\ 0 & -\alpha & v \end{pmatrix}, \quad F = \begin{pmatrix} 0 \\ -\frac{1}{\rho\varphi}R \\ 0 \end{pmatrix}. \tag{30}$$

The eigenvalues of B are given by

$$v - \sqrt{\frac{\alpha-(\sigma+p)}{\rho}}\,, \quad v\,, \quad v + \sqrt{\frac{\alpha-(\sigma+p)}{\rho}}\,, \tag{31}$$

which implies that the system (29) is hyperbolic for any given pressure p and stress σ satisfying

$$\sigma + p \le \alpha. \tag{32}$$

It is worth noticing that condition (32) requires the *effective stress* to be bounded by the elastic modulus of the porous matrix. Since v is an eigenvalue of B and in view of the definition of $D(\delta)$, it results that the hyperbolic problem is a *characteristic free boundary problem*, i.e. the free boundary is one of the characteristics of the system.

We observe that, whereas equations (17)–(19), and in general (5)–(7), can be written in conservative form, this is not true for the stress-strain relation (19). So we cannot rely on most of the results in the theory of hyperbolic systems, which requires conservative systems.

We can consider (26)–(28) as the boundary conditions for the hyperbolic system, and regard (24) and (25) as the two boundary conditions for the elliptic equation for the pressure (with given velocity field $v(t,x)$). This can be easily done, formally, by expressing the pressure and its spatial derivative as functions of the

[1]We used in (25) the simplifying assumption $\rho_f = \rho_s$; we shall explain how to deal with the general case at the end of the existence proof.

solution of the hyperbolic problem,.

$$v(t,x) - Kp_x(t,x) = 0\,, \tag{33}$$

$$p(t,x) = p_0 - \frac{1}{K}\int_x^{x(t)} v(t,\xi)\,\mathrm{d}\xi\,. \tag{34}$$

Notice that in (34) the upper limit of the integral is the free boundary itself.

6. Local existence and Uniqueness

In order to perform the mathematical analysis of the hyperbolic system (29) we write it in *characteristic form* by multiplying equations by the left eigenvectors of the matrix B and summing them up. The resulting system is

$$\left(\frac{\sigma+p}{\varphi}, \sqrt{\rho(\alpha-(\sigma+p))}, 1\right) \cdot \left[\mathbf{u}_t + \left(v - \sqrt{\tfrac{\alpha-(\sigma+p)}{\rho}}\right)\mathbf{u}_x\right] = -\frac{\sqrt{\frac{\alpha-(\sigma+p)}{\rho}}}{\varphi}R\,, \tag{35}$$

$$\left(1, 0, \frac{\varphi}{\alpha}\right) \cdot [\mathbf{u}_t + v\mathbf{u}_x] = 0\,, \tag{36}$$

$$\left(\frac{\sigma+p}{\varphi}, -\sqrt{\rho(\alpha-(\sigma+p))}, 1\right) \cdot \left[\mathbf{u}_t + \left(v + \sqrt{\tfrac{\alpha-(\sigma+p)}{\rho}}\right)\mathbf{u}_x\right] = \frac{\sqrt{\frac{\alpha-(\sigma+p)}{\rho}}}{\varphi}R. \tag{37}$$

In (35)–(37), we express the source term as a function of the solution itself, namely

$$R = \frac{rk}{1-\varphi}v, \tag{38}$$

and we express $p = p[\mathbf{u}]$ as a function of the solution $\mathbf{u}$ given by (34). The system must again be solved in the domain $D(\delta)$ (for, at least, a suitably small δ) with the boundary conditions (24)–(28).

According to the classification in [4], problem (35)–(37), together with boundary conditions (24)–(28), is both a *free boundary problem with characteristic boundary* and a problem *in functional form.*

The strategy proposed in [4] for solving the free boundary problem consists in transforming it into a problem in functional form in the fixed domain

$$R(\delta) = \{(\overline{x},\overline{t})\,|\,0<\overline{x}<\overline{t},\ 0<\overline{t}<\delta\}, \tag{39}$$

by means of the domain transformation

$$\overline{t} = t\,, \quad \overline{x} = \frac{t}{x(t)}x\,. \tag{40}$$

We can use the results in [4] about the *free boundary problem* once we have checked that the extra functional dependence introduced by the term $p[\mathbf{u}]$ in the coefficients and the eigenvalues of the hyperbolic system satisfies the hypotheses ... required in [4, 2.6]. Further, we have to control that the functional dependence in the last boundary condition does not affect the *characterizing matrix*, so that definitions [4, 2, (6.29)] and [4, 3, (1.11)] coincide.

Since the coefficients are regular functions of p (as long as the hyperbolicity condition (32) is satisfied), the regularity requirements on coefficients and eigenvalues are satisfied if we prove that the sup norm of $p[\mathbf{u}]$ is

Lipschitz continuous with respect to the sup norm of the solution $\mathbf{u}$. This must be done in terms of the transformed variables: let us denote by

$$t = \overline{t}, \quad x = X(\overline{t}, \overline{x}) = \frac{x(t)}{\overline{t}} \overline{x}, \tag{41}$$

the inverse transformation of (40), and by $\overline{v}(\overline{t}, \overline{x}) := v(\overline{t}, X(\overline{t}, \overline{x}))$, and analogously for $\overline{\varphi}$, $\overline{\sigma}$. The pressure $\overline{p}(\overline{t}, \overline{x}) = p(\overline{t}, X(\overline{t}, \overline{x}))$ can be computed directly from the transformed variables

$$\overline{p}(\overline{t}, \overline{x}) = p_0 - \frac{x(\overline{t})}{K\overline{t}} \int_{\overline{x}}^{\overline{t}} \overline{v}(\overline{t}, \xi)\, \mathrm{d}\xi, \quad \text{with} \quad x(\overline{t}) = \int_0^{\overline{t}} \overline{v}(\tau, \tau)\, \mathrm{d}\tau. \tag{42}$$

Now, for any given $\mathbf{u}_1$ and $\mathbf{u}_2$, let $\overline{p}_1$ and $\overline{p}_2$ be the corresponding pressures defined by (42) (with $\overline{v}_i$ instead of $\overline{v}$). We have

$$\begin{aligned} \|\overline{p}_1 - \overline{p}_2\| &= \sup \frac{1}{K\overline{t}} \left| x_1(\overline{t}) \int_{\overline{x}}^{\overline{t}} \overline{v}_1(\overline{t}, \xi)\, \mathrm{d}\xi - x_2(\overline{t}) \int_{\overline{x}}^{\overline{t}} \overline{v}_2(\overline{t}, \xi)\, \mathrm{d}\xi \right| \\ &\le \sup \frac{x_1(\overline{t})}{K\overline{t}} \left| \int_{\overline{x}}^{\overline{t}} (\overline{v}_1(\overline{t}, \xi) - \overline{v}_2(\overline{t}, \xi))\, \mathrm{d}\xi \right| \\ &+ \sup \frac{1}{K\overline{t}} \left| \int_{\overline{x}}^{\overline{t}} \overline{v}_2(\overline{t}, \xi)\, \mathrm{d}\xi \right| \left| \int_0^{\overline{t}} (\overline{v}_1(\tau, \tau) - \overline{v}_2(\tau, \tau))\, \mathrm{d}\tau \right|, \end{aligned} \tag{43}$$

and the Lipschitz continuity follows immediately, because both $\frac{x_1(\overline{t})}{K\overline{t}}$ and $\sup \frac{1}{K\overline{t}} \left| \int_{\overline{x}}^{\overline{t}} \overline{v}_1(\overline{t}, \xi)\, \mathrm{d}\xi \right|$ are uniformly bounded.

We now write the boundary conditions for the characteristic variables, namely

$$u_1^* = \sqrt{\rho\alpha}\, v + \sigma = -p_0 + \sqrt{\rho\alpha}\, v := G_1 \tag{44}$$

$$u_2^* = \varphi + \tfrac{\varphi_0}{\alpha} \sigma = \varphi_0 + \tfrac{\varphi_0}{\alpha} \sigma := G_2 \tag{45}$$

$$u_3^* = -\sqrt{\rho\alpha}\, v + \sigma = -\sqrt{\rho\alpha}\, \mu\, p[\mathbf{u}](t, 0) + \sigma := G_3. \tag{46}$$

To prove that our problem has a solution, it remains to check the resolubility conditions on the *characterizing matrix* of the hyperbolic problem.

To this aim, let $\Sigma(\delta, \Omega_1)$ be the space of the C^1 vector valued functions $\mathbf{v} = (v_1, v_2, v_3)$ defined on the angular region $R(\delta)$ and satisfying the inequality

$$\max_k \|q_k\| \le \Omega_1, \tag{47}$$

where (for simplicity, we omit the $^-$ superscript in the following computation)

$$q_i = \frac{\partial v_i}{\partial t}, \quad q_{3+i} = \frac{\partial v_i}{\partial t} + \frac{\partial v_i}{\partial x}, \quad i = 1, 2, 3, . \tag{48}$$

and $\| \ \|$ denotes the C^1 norm in $R(\delta)$.

Let $\mathbf{v} \in \Sigma(\delta, \Omega_1)$, and

$$\tilde{G}_1'(t) = \frac{d}{dt} G_1(t; \mathbf{v}(t,0)), \ \ \tilde{G}_2'(t) = \frac{d}{dt} G_2(t; \mathbf{v}(t,t)), \ \ \tilde{G}_3'(t) = \frac{d}{dt} G_3(t; \mathbf{v}(t,t)), \tag{49}$$

where the G_i are defined by (44)–(46).

By definition, the *characterizing matrix* of the problem is a matrix (θ_{lk}) satisfying the inequality

$$\|\tilde{G}_l'(t)\| \leq \sum_{k=1}^{3} (\theta_{lk} + \varepsilon(\delta, \Omega_1)) \max(\|q_k^\star\|, \|q_{3+k}^\star\|) + R_2, \tag{50}$$

where R_2 is a constant independent from δ and Ω_1,

$$\varepsilon(\delta, \Omega_1) \to 0 \text{ as } \delta \to 0 \quad \text{ for any } \Omega_1, \tag{51}$$

and $q_i^\star = \zeta_{ik}^0 q_k$, where (ζ_{ik}^0) is the matrix of the left eigenvectors of the matrix B defined by (30), computed for $x = 0$ and $t = 0$.

In order to determine the matrix (θ_{lk}), we need to give upper bounds for the $\tilde{G}_l'$. The first two derivatives $\tilde{G}_1'$ and $\tilde{G}_2'$ are trivially bounded choosing the matrix $\theta = H$ with

$$H = \left(\frac{\partial G_l}{\partial u_j} (0, 0, \mathbf{u}(0,0)) \right) \left(\zeta^{jk}(0, 0, \mathbf{u}(0,0)) \right), \tag{52}$$

where $\left(\zeta^{jk}(0,0,\mathbf{u}(0,0))\right)$ is the inverse matrix of the matrix of (ζ_{jk}^0).

It remains to show that $\tilde{G}_3'$ can be bounded by the same matrix H, i.e.

$$\begin{aligned} \tilde{G}_3'(t) &= \frac{d}{dt} G_3(\mathbf{v}(t, x(t))) = \\ &\frac{\partial \sigma}{\partial t} - \frac{\mu}{K}\sqrt{\rho\alpha} \left[\frac{d}{dt}\frac{x(t)}{t} \int_0^t v_2(t,\xi)\, d\xi + v_2(t,t)\frac{x(t)}{t} + \frac{x(t)}{t}\int_0^t \frac{\partial}{\partial t} v_2(t,\xi)\, d\xi \right] \\ &\leq \max\{q_3, q_{3+3}\}\, (1 + \varepsilon(\delta, \Omega_1)) + R_2. \end{aligned} \tag{53}$$

In fact, the estimates

$$\left| v_2(t,t)\frac{x(t)}{t} \right| \leq v_0^2 + (2v_0 q_{2+3} + q_{2+3}^2 t)\, t, \tag{54}$$

$$\left| \frac{x(t)}{t} \int_0^t \frac{\partial}{\partial t} v_2(t,\xi)\, d\xi \right| \leq (v_0 + q_{2+3} t)\, q_2 t \tag{55}$$

are trivial.

On the other hand,

$$|v_2(t,\xi)| \leq v_0 + t \max_{\tau < t} \left| \frac{\partial v}{\partial t}(\tau, 0) \right| + t \max_{\mathrm{x} < \xi} \left| \frac{\partial v}{\partial x}(t, x) \right| \leq v_0 + 2t\Omega_1, \tag{56}$$

and

$$\left|\frac{x'(t)t - x(t)}{t^2}\right| \leq 2\frac{v_0}{t} + 2q_{2+3}, \tag{57}$$

whence there follows

$$\frac{d}{dt}\frac{x(t)}{t}\int_0^t v_2(t,\xi)\,d\xi \leq (2v_0 + 2q_{2+3}t)(v_0 + 2t\Omega_1). \tag{58}$$

Substituting q_i by $q_i^\star$, we finally get that the matrix H given by (52), namely

$$H = \begin{pmatrix} \frac{1}{2} & 0 & -\frac{1}{2} \\ \frac{1}{2}\varphi_0/\alpha & 0 & \frac{1}{2}\varphi_0/\alpha \\ \frac{1}{2} & 0 & \frac{1}{2} \end{pmatrix}, \tag{59}$$

is the characterizing matrix.

We have now to compute the norm defined according to [4][2,(4.12)], namely

$$|H|_{\min} = \inf_{\gamma=\mathrm{diag}(\gamma_i),\gamma_i\neq 0} |\gamma H \gamma^{-1}|, \tag{60}$$

where, denoting $H = (\theta_{lk})$, $|H| = \max_l \sum_{k=1}^3 |\theta_{lk}|$.

It is a matter of straightforward computation to show that H just fails to satisfy the sufficient condition for the C^1 resolubility of the problem, since the above norm is exactly equal to 1 (the sufficient condition being $|H|_{\min} < 1$.)

Then we have to relay on the solvability results in the higher regularity classes C^{N+1} given by the conditions in Chap.7 of [4].

We want to show that these conditions are satisfied for $N = 1$.

In order to do this, we rewrite the boundary conditions as

$$u_1^\star(t,x(t)) = -\frac{1}{2}u_3^\star - p_0 = \Gamma_1, \tag{61}$$

$$u_2^\star(t,0) = \varphi_0 + \frac{\varphi_0}{\alpha}\left(u_1^\star - \sqrt{\rho\alpha}\,\mu\, p(t,0)\right) = \Gamma_2, \tag{62}$$

$$u_3^\star(t,0) = u_1^\star - 2\sqrt{\rho\alpha}\,\mu\, p(t,0) = \Gamma_3. \tag{63}$$

(64)

Then the C^2 resolubility condition consists in showing that the matrix

$$H_1 = H\varsigma = \begin{pmatrix} 0 & \frac{\partial\Gamma_1}{\partial u_2^\star}\varsigma_2 & \frac{\partial\Gamma_1}{\partial u_3^\star}\varsigma_3 \\ \frac{\partial\Gamma_2}{\partial u_1^\star}\varsigma_1 & 0 & 0 \\ \frac{\partial\Gamma_3}{\partial u_1^\star}\varsigma_1 & 0 & 0 \end{pmatrix}, \tag{65}$$

where

$$\varsigma_1 = \frac{\lambda_1^0}{\lambda_1^0 - \lambda_3^0} = \frac{v - \sqrt{\frac{\alpha}{\rho}}}{-\sqrt{\frac{\alpha}{\rho}}},\ \varsigma_2 = \frac{\lambda_2^0 - x_0'}{\lambda_2^0} = 0,\ \varsigma_3 = \frac{\lambda_3^0 - \lambda_2^0}{\lambda_3^0} = \frac{\sqrt{\frac{\alpha}{\rho}}}{v + \sqrt{\frac{\alpha}{\rho}}}. \tag{66}$$

satisfies $|H_1|_{\min} < 1$.

A straightforward computation shows that $|H_1|_{\min} = \frac{1}{2}$ whence we obtain the following result:

Theorem 6.1. *There exists a sufficiently small δ such that problem* (35)–(37), (26)–(28) *and* (34) *admits a unique C^2 solution in $D(\delta)$.*

The general case $\rho_s \neq \rho_f$ can be handled by substituting (25) with the more general boundary condition

$$v(t,0) - Kp_x(t,0) = \left(1 - \frac{\rho_s}{\rho_f}\right)\varphi_0 v(t,0), \tag{67}$$

which gives

$$p(t,x) = p_0 - \frac{1}{K}\int_x^{x(t)} v(\xi,t)\,dt + \frac{\beta}{K}\mu p(t,0), \tag{68}$$

$$p(t,0)\left[1 - \frac{\beta}{K}\mu x(t)\right] = p_0 - \frac{1}{K}\int_0^{x(t)} v(\xi,t)\,dt, \tag{69}$$

where $\beta = \left(1 - \frac{\rho_s}{\rho_f}\right)\varphi_0$.

A straightforward computation shows that the functional p defined in (68) is still a Lipschitz continuous functional of the solution, so that Theorem (6.1) applies in this case as well.

References

[1] D. Andreucci, A. Fasano, R. Ricci, *On the growth of a polymer layer around a catalytic particle: a free boundary problem*, NoDEA **4** (1997), 511–520.

[2] J. Bear, *Dynamics of Fluids in Porous Media*, Elsevier P., Toronto, 1972.

[3] A. Fasano, *Mathematical models in polymer processing*, MECCANICA J. 35 (2000) 163–198.

[4] Li Ta-tsien, Yu Wen-ci *Boundary Value Problems for Quasilinear Hyperbolic Systems*, Duke Univ. Maths Series, 1985.

[5] V. N. Nikolaevskij, *Mechanics of Porous and Fractured media*, World Scientific, Singapore, 1990.

[6] E.J. Nagel, V.A. Kirillov, W.H. Ray, *Prediction of Molecular Weight Distribution for High-Desity Polyolefins*, Ind. Eng. Prod. Res. Dev. **19** (1980), 372–379.

Department of Mathematics "U.Dini",
University of Florence,
42/a, viale Morgagni
50134 Firenze, Italy
E-mail address: ricci@math.unifi.it

International Series of Numerical Mathematics, Vol. 147, 151–165

Transmission-Stefan Problems Arising in Czochralski Process of Crystal Growth

Takesi Fukao, Nobuyuki Kenmochi, and Irena Pawłow

Abstract. We consider a transmission problem of two parabolic partial differential equations as a mathematical model of the Czochralski crystal growth process. Originally this process is a three-phase system and involves the three interfaces of solid-liquid, liquid-gas, and solid-gas, which should be determined as a part of solution from the mathematical point of view. Nevertheless, assuming that the solid-gas and liquid-gas interfaces are prescribed, we propose in this paper a transmission problem composed of a linear heat equation in the gas region and a nonlinear degenerate parabolic equation in the material region as a simplified mathematical model of the Czochralski process.

1. Introduction

The Czochralski pulling method is widely used for the production of a column of single silicon crystal from the melt. Mathematically, this is a free boundary problem with three phases, solid, liquid, and gas. These locations should be determined by some balance laws on their interfaces as a part of the solution of the problem. This process has many interesting physical aspects; for instance, there is neither a phase transition between gas and liquid nor between gas and solid, and the solid-liquid phase transition occurs on its interface, which means that the balance equation on the solid-liquid interface is of the Stefan type and those on the gas-liquid or gas-solid interface are of the usual transmission type. Many authors have studied this process from various mathematical and mechanical points of view (cf. Crowley [2], Francu [5], Kenmochi and Pawłow [11], Pawłow [12], and so on). In connection with this process, a transmission problem between two kinds of degenerate parabolic equations has been recently proposed by Fukao, Kenmochi, and Pawłow [7], in which the main idea is to consider a transmission condition on the interface of the material and gas regions. This problem has been solved by the abstract theory of evolution equations governed by time-dependent subdifferential operators. But

2000 *Mathematics Subject Classification.* Primary 35K65; Secondary 80A22.
Key words and phrases. Stefan problems, Transmission problems.
This work has been carried out during a staying of the first author at the Department of Mathematics, Pavia University, under the sponsorship of the Italian COFIN 2000 Research Program "EQUAZIONI VARIAZIONALI D'EVOLUZIONE ASSOCIATE A DINAMICHE DI FASE". .

this is incomplete as a model of the crystal growth process from some physical point of view. For example [7; Remark 2.1], and so on. Therefore, in this paper we propose a new model for the Czochralski pulling process, which is physically more acceptable; in our new model, the transmission condition on the gas-material interface requires the continuity of the temperature and the balance of the heat flux, and we still suppose, just as in [6, 7, 8], that the movement of the gas-material interface is prescribed smoothly in time. The last assumption is based on the fact that the crystal radius (hence the shape of the crystal) can be technologically controlled by a suitable choice of the pulling velocity and of the thermal situation in the gas region.

We use the following notation (see Fig.1). For $0 < T < \infty$ and $t \in [0, T]$, we let $\Omega \subset \mathbf{R}^3$ be a bounded domain with smooth boundary $\Gamma := \partial\Omega$ and

$$\Omega := \Omega_m(t) \cup \Omega_g(t) \cup \Gamma_{gm}(t).$$

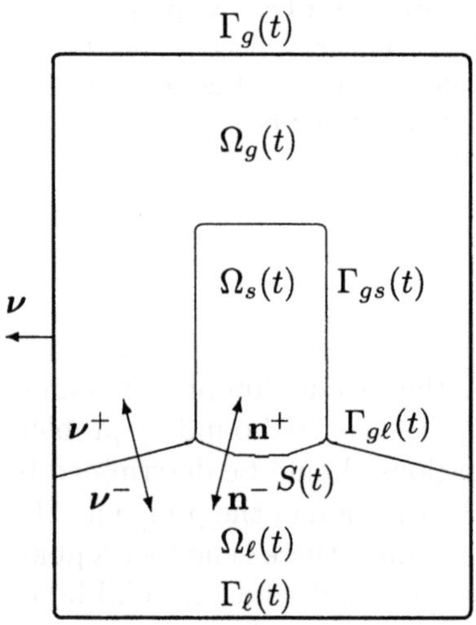

Fig. 1

$\Omega_m(t) := \Omega_\ell(t) \cup \Omega_s(t) \cup S(t)$ is the material region, in which the liquid region $\Omega_\ell(t)$ is separated from the solid region $\Omega_s(t)$ by the solid-liquid interface $S(t)$.

$\Omega_g(t)$ is the gas region.

$\Gamma_{gm}(t) := \Gamma_{g\ell}(t) \cup \Gamma_{gs}(t)$ is the interface between $\Omega_m(t)$ and $\Omega_g(t)$, which consists of the gas-liquid interface $\Gamma_{g\ell}(t)$ and the gas-solid interface $\Gamma_{gs}(t)$.

$\Gamma := \Gamma_\ell(t) \cup \Gamma_g(t)$, where $\Gamma_\ell(t)$ (resp. $\Gamma_g(t)$) is the subset of Γ on which the liquid (resp. gas) touches Γ; hence $\Gamma_m(t) := \partial\Omega_m(t) = \Gamma_\ell(t) \cup \Gamma_{g\ell}(t) \cup \Gamma_{gs}(t)$.

Note here that our geometrical assumptions ensure that in our model the solid region is detached from the crucible boundary Γ. Moreover we use the following notation:

$\boldsymbol{\nu} = \boldsymbol{\nu}(x)$: 3-dimensional unit vector normal to Γ at $x \in \Gamma$,
$\boldsymbol{\nu}^+ := \boldsymbol{\nu}^+(t,x)$: 3-dimensional unit vector normal to $\Gamma_{gm}(t)$, pointing to $\Omega_g(t)$,
$\mathbf{n}^+ := \mathbf{n}^+(t,x)$: 3-dimensional unit vector normal to $S(t)$, pointing to $\Omega_s(t)$,
$\boldsymbol{\nu}^- := \boldsymbol{\nu}^-(t,x) = -\boldsymbol{\nu}^+$, $\mathbf{n}^- := \mathbf{n}^-(t,x) = -\mathbf{n}^+$,
$Q := (0,T) \times \Omega$, $\Sigma := (0,T) \times \Gamma$,

$$Q_i := \bigcup_{t \in (0,T)} \{t\} \times \Omega_i(t), \quad i = m, \ell, s, g,$$

$$\Sigma_i := \bigcup_{t \in (0,T)} \{t\} \times \Gamma_i(t), \quad i = m, \ell, g, gm, g\ell, gs, \quad S := \bigcup_{t \in (0,T)} \{t\} \times S(t).$$

Next, we denote by $v_{\Sigma_m} := v_{\Sigma_m}(t,x)$ and $v_S := v_S(t,x)$ the normal speed of $\Gamma_m(t)$ at $(t,x) \in \Sigma_m$ and of $S(t)$ at $(t,x) \in S$, respectively. Then 4-dimensional (with respect to the (t,x)-space) unit normal vectors $\vec{\boldsymbol{\nu}}^+$ and $\vec{\mathbf{n}}^+$, outward from Q_m to Q_g and Q_ℓ to Q_s are given by

$$\vec{\boldsymbol{\nu}}^+ = \frac{1}{(|v_{\Sigma_m}|^2+1)^{\frac{1}{2}}}(-v_{\Sigma_m}, \boldsymbol{\nu}^+), \quad \vec{\mathbf{n}}^+ = \frac{1}{(|v_S|^2+1)^{\frac{1}{2}}}(-v_S, \mathbf{n}^+). \tag{1.1}$$

In this paper we assume that the material domain $\Omega_m(t)$ depends smoothly on time t in the following sense:

(A1) There is a transformation $y = \mathbf{y}(t,x) := (y_1(t,x), y_2(t,x), y_3(t,x))$ which is a function of C^2-class from $\overline{Q}$ into $\mathbf{R}^3$, satisfying that $y = \mathbf{y}(t,x)$ is a C^2-diffeomorphism from $\overline{\Omega}$ onto itself for each $t \in [0,T]$ and

$$\mathbf{y}(t, \overline{\Omega_i(t)}) = \overline{\Omega_i(0)} \quad \text{for all } t \in [0,T] \text{ and } i = m, g, \quad \mathbf{y}(0,\cdot) = I \text{ (identity) in } \overline{\Omega}.$$

Also assume that the convective fields $\mathbf{v}_m$ and $\mathbf{v}_g$ which occur by the deformation of the domains $\Omega_m(t)$ and $\Omega_g(t)$ are prescribed, and they satisfy (A2):

(A2) $\mathbf{v}_i \in C^1(\overline{Q_i})^3$ and the following properties are satisfied.

$$\mathrm{div}\mathbf{v}_i(t,\cdot) = 0 \quad \text{in } \Omega_i(t) \quad \text{for all } t \in (0,T), \ i = m, g, \tag{1.2}$$

where div means the divergence with respect to the space variable x,

$$\mathbf{v}_m \cdot \boldsymbol{\nu}^+ = \mathbf{v}_g \cdot \boldsymbol{\nu}^+ = v_{\Sigma_m} \quad \text{on } \Sigma_{gm}, \tag{1.3}$$

$$\mathbf{v}_m \cdot \boldsymbol{\nu} = \mathbf{v}_g \cdot \boldsymbol{\nu} = v_{\Sigma_m} := 0 \quad \text{on } \Sigma. \tag{1.4}$$

For the convective vector fields $\mathbf{v}_i$ on Q_i, $i = m, g$, the assumptions (1.2)–(1.4) are quite natural from the physical point of view. Under these conditions, we consider the following system, where $\theta_i := \theta_i(t,x)$ is the temperature on Q_i, $i = \ell, s, g$,

$$c_i(\theta_{i,t} + \mathbf{v}_i \cdot \nabla\theta_i) - k_i\Delta\theta_i = f \quad \text{in } Q_i, \ i = \ell, s, g, \tag{1.5}$$

$$\theta_\ell = \theta_s = 0, \quad k_\ell \frac{\partial\theta_\ell}{\partial\mathbf{n}^+} + k_s \frac{\partial\theta_s}{\partial\mathbf{n}^-} = L(\mathbf{v}_m \cdot \mathbf{n}^+ - v_S) \quad \text{on } S, \tag{1.6}$$

where and hereafter $\theta_{i,t} := (\partial\theta_i/\partial t)$, $\nabla\theta_i := (\partial\theta_i/\partial x_1, \partial\theta_i/\partial x_2, \partial\theta_i/\partial x_3)$, $\Delta\theta_i := \sum_{j=1}^3 (\partial^2\theta_i/\partial x_j^2)$, $\partial\theta_i/\partial\mathbf{n}^+ := \nabla\theta_i \cdot \mathbf{n}^+$ for $i = \ell, s, g$. In (1.5) and (1.6), we take $\mathbf{v}_i = \mathbf{v}_m$ for $i = s, \ell$. Moreover, the positive constants c_i, k_i $(i = \ell, s, g)$, and L are positive constants which are the specific heats, heat conductivities and latent heats in Q_i. A heat source f is given on Q. Note that (1.6) is the Stefan conditions taking into account the convection $\mathbf{v}_m$, where the phase change temperature is supposed to be 0 for simplicity. Moreover, on the gas-material boundary Σ_{gm} we require the following transmission condition:

$$\theta_i = \theta_g, \quad k_i \frac{\partial\theta_i}{\partial\boldsymbol{\nu}^+} + k_g \frac{\partial\theta_g}{\partial\boldsymbol{\nu}^-} = 0 \quad \text{on } \Sigma_{gi}, \ i = \ell, s. \tag{1.7}$$

The boundary condition on Σ is given by

$$k_i \frac{\partial \theta_i}{\partial \nu} + n_0 \theta_i = p \quad \text{on } \Sigma_i, \; i = \ell, g, \tag{1.8}$$

where n_0 is a positive constant, p is the boundary datum prescribed on Σ, and the initial conditions are prescribed as follows:

$$\theta_i(0, \cdot) = \theta_{i0} \quad \text{on } \Omega_i(0), \; i = \ell, s, g, \quad S(0) = S_0, \tag{1.9}$$

where S_0 is the initial location of the solid-liquid interface and θ_{i0}, $i = \ell, s, g$, are the initial temperature on $\Omega_i(0)$ which satisfy the compatibility condition

$$\theta_{\ell 0} > 0, \quad \theta_{s0} < 0 \text{ and } \theta_{\ell 0} = \theta_{s0} = 0 \quad \text{on } S_0. \tag{1.10}$$

As a remark, when $\theta_i = \theta_g$ is replaced by $k_i \theta_i = k_g \theta_g$, for $i = \ell, s$, in (1.7), the problem was treated in Fukao, Kenmochi and Pawłow [7] and an existence-uniqueness result was established in the framework of the abstract theory of evolution equations of time-dependent subdifferentials. However, the same approach cannot be applied to the system (1.5)–(1.9) proposed above, and one of the difficulties in its mathematical treatment comes from the transmission condition (1.7). The uniqueness argument, based on a dual method [7; Section 5], does not seem to work in this context, either.

2. Weak Formulation and Existence Result

In this section we give a weak formulation of system (1.5)–(1.9) and state an existence result for it. We begin with defining the enthalpy function u on Q and the heat conductivity function $k(r)$ on $\mathbf{R}$ by

$$u := u(t,x) \begin{cases} = c_s \theta & \text{if } x \in \Omega_m(t), \; \theta < 0, \\ \in [0, L] & \text{if } x \in \Omega_m(t), \; \theta = 0, \\ = c_\ell \theta + L & \text{if } x \in \Omega_m(t), \; \theta > 0, \\ = c_g \theta & \text{if } x \in \overline{\Omega_g(t)}, \end{cases} \tag{2.1}$$

$$k(r) := \begin{cases} k_s & \text{if } r < 0, \\ k_\ell & \text{if } r \geq 0, \end{cases} \tag{2.2}$$

where for each $t \in [0, T]$ we put $\theta(t, x) := \theta_i(t, x)$ if $x \in \overline{\Omega_i(t)}$ for $i = \ell, s, g$. By using these functions u, θ and k our problem (1.5)–(1.9) is reformulated in the following weak sense (cf. [7]).

$$\begin{aligned} &-\int_Q u(\eta_t + \mathbf{v} \cdot \nabla \eta) dx dt + \int_Q \left\{ k(\theta) \chi_{\Omega_m(t)} + k_g \chi_{\overline{\Omega_g(t)}} \right\} \nabla \theta \cdot \nabla \eta dx dt \\ &+ n_0 \int_\Sigma \theta \eta d\Gamma dt = \int_Q f \eta dx dt + \int_\Sigma p \eta d\Gamma dt + \int_\Omega u_0 \eta(0) dx \quad \text{for all } \eta \in W, \end{aligned} \tag{2.3}$$

where $W := \{ \eta \in H^1(Q); \; \eta(T, \cdot) = 0 \text{ on } \Omega \}$, and

$\mathbf{v}(t) := \mathbf{v}_m(t) \chi_{\Omega_m(t)} + \mathbf{v}_g(t) \chi_{\overline{\Omega_g(t)}}$, $u_0 := c_s \theta_0 \chi_{\Omega_s(0)} + (c_\ell \theta_0 + L) \chi_{\Omega_\ell(0)} + c_g \theta_0 \chi_{\overline{\Omega_g(0)}}$,

$\chi_{\Omega_i(t)}, \chi_{\overline{\Omega_g(t)}}$ are the characteristic functions of $\Omega_i(t)$ for $i = \ell, s, m$ and of $\overline{\Omega_g(t)}$ for each $t \in [0, T]$.

In fact, the variational identity (2.3) is derived in the following way. Assume that θ_i, $i = \ell, s, g$, satisfy system (1.5)–(1.9) and $\theta_\ell > 0$ on Q_ℓ, $\theta_s < 0$ on Q_s, and let η be any test function in W. First, with the help of the Green–Stokes' formula and the relations $d\Sigma_i = (|v_{\Sigma_m}|^2 + 1)^{1/2} d\Gamma_i(t)dt$ for each i, we see from conditions (1.2)–(1.4) and (1.6) with expression (1.1) of $\vec{\boldsymbol{\nu}}^+$ that

$$\begin{aligned}
&\int_{Q_\ell} c_\ell(\theta_{\ell,t} + \mathbf{v}_\ell \cdot \nabla\theta_\ell)\eta dxdt \\
&= -\int_{Q_\ell} c_\ell\theta_\ell(\eta_t + \mathbf{v}_\ell \cdot \nabla\eta)dxdt + \int_{\Sigma_{g\ell}} c_\ell\theta_\ell\eta(\vec{\boldsymbol{\nu}}^+)^t d\Sigma_{g\ell} + \int_{\Omega_\ell(0)} c_i\theta_{\ell 0}\eta(0)(-1)dx \\
&\quad + \int_0^T \int_{\Gamma_{g\ell}(t)} c_\ell\theta_\ell\eta(\mathbf{v}_\ell \cdot \boldsymbol{\nu}^+)d\Gamma_{g\ell}(t)dt \\
&= -\int_{Q_\ell} c_\ell\theta_\ell(\eta_t + \mathbf{v}_\ell \cdot \nabla\eta)dxdt - \int_{\Omega_\ell(0)} c_\ell\theta_{\ell 0}\eta(0)dx,
\end{aligned}$$

where $(\vec{\boldsymbol{\nu}}^+)^t$ denotes the t-component of $\vec{\boldsymbol{\nu}}^+$; in order to get the last equality we used

$$(\vec{\boldsymbol{\nu}}^+)^t d\Sigma_{g\ell} = -v_{\Sigma_m} d\Gamma_{g\ell}(t)dt = (\mathbf{v}_\ell \cdot \boldsymbol{\nu}^+)d\Gamma_{g\ell}(t)dt \quad \text{on } \Sigma_{g\ell}.$$

We have a similar result for $i = s, g$, so that for $i = \ell, s, g$,

$$\int_{Q_i} c_i(\theta_{i,t} + \mathbf{v}_i \cdot \nabla\theta_i)\eta dxdt = -\int_{Q_i} c_i\theta_i(\eta_t + \mathbf{v}_i \cdot \nabla\eta)dxdt - \int_{\Omega_i(0)} c_i\theta_{i0}\eta(0)dx. \quad (2.4)$$

Therefore, summing up (2.4) for $i = \ell, s, g$, and using the enthalpy u, we have

$$\begin{aligned}
\sum_{i=\ell,s,g} \int_{Q_i} c_i(\theta_{i,t} + \mathbf{v}_i \cdot \nabla\theta_i)\eta dxdt &= -\int_Q u(\eta_t + \mathbf{v} \cdot \nabla\eta)dxdt \\
- \int_\Omega u_0\eta(0)dx + \int_{Q_\ell} L(\eta_t &+ \mathbf{v}_m \cdot \nabla\eta)dxdt + \int_{\Omega_\ell(0)} L\eta(0)dx.
\end{aligned} \quad (2.5)$$

Next, by (1.6)–(1.8) we have

$$\begin{aligned}
&-\int_{Q_\ell} k_\ell\Delta\theta_\ell\eta dxdt - \int_{Q_s} k_s\Delta\theta_s\eta dxdt - \int_{Q_g} k_g\Delta\theta_g\eta dxdt \\
&= \int_Q \left\{k(\theta)\chi_{\Omega_m(t)} + k_g\chi_{\overline{\Omega_g(t)}}\right\} \nabla\theta \cdot \nabla\eta dxdt + \int_\Sigma (n_0\theta - p)\eta d\Gamma(t)dt \\
&\quad - \int_S L(\mathbf{v} \cdot \mathbf{n}^+ - v_S)\eta dS(t)dt.
\end{aligned}$$

Thus we have

$$
\begin{aligned}
-\sum_{i=\ell,s,g}\int_{Q_i} k_i\Delta\theta_i\eta dxdt = &\int_Q \left\{k(\theta)\chi_{\Omega_m(t)} + k_g\chi_{\overline{\Omega_g(t)}}\right\}\nabla\theta\cdot\nabla\eta dxdt \\
&+\int_\Sigma (n_0\theta - p)\eta d\Gamma(t)dt - \int_S L(\mathbf{v}\cdot\mathbf{n}^+ - v_S)\eta dS(t)dt.
\end{aligned}
\tag{2.6}
$$

Here add (2.5) and (2.6). Then it follows from (1.5) that

$$
\begin{aligned}
&-\int_Q u(\eta_t + \mathbf{v}\cdot\nabla\eta)dxdt + \int_Q \left\{k(\theta)\chi_{\Omega_m(t)} + k_g\chi_{\overline{\Omega_g(t)}}\right\}\nabla\theta\cdot\nabla\eta dxdt + n_0\int_\Sigma \theta\eta d\Gamma dt \\
&+\int_{Q_\ell} L(\eta_t + \mathbf{v}_m\cdot\nabla\eta)dxdt + \int_{\Omega_\ell(0)} L\eta(0)dx - \int_S L(\mathbf{v}_m\cdot\mathbf{n}^+ - v_S)\eta dS(t)dt \\
&= \int_Q f\eta dxdt + \int_\Sigma p\eta d\Gamma dt + \int_\Omega u_0\eta(0)dx.
\end{aligned}
$$

Here, with the help of the Green–Stokes' formula, we see from conditions (1.2)–(1.4), and the expression (1.1) of $\vec{\mathbf{n}}^+$ and the relation $dS = (|v_S|^2+1)^{1/2}dS(t)dt$ that

$$
\int_{Q_\ell} L(\eta_t + \mathbf{v}_m\cdot\nabla\eta)dxdt + \int_{\Omega_\ell(0)} L\eta(0)dx = \int_S L(\mathbf{v}\cdot\mathbf{n}^+ - v_S)\eta dS(t)dt.
$$

Therefore the variational identity (2.3) is obtained.

Here we define a weak solution of our problem (1.5)–(1.9). For simplicity, we denote by V the Sobolev space $H^1(\Omega)$ with the norm $|\cdot|_V$ given by

$$
|u|_V := \left\{\int_\Omega |\nabla u|^2 dx + n_0\int_\Gamma |u|^2 d\Gamma\right\}^{\frac{1}{2}} \quad \text{for all } u\in V,
$$

which is equivalent to the standard norm of $H^1(\Omega)$. The dual space of V is denoted by V^*, with duality pairing denoted by $\langle\cdot,\cdot\rangle_{V^*,V}$. Then we have the usual dense and compact imbeddings:

$$
V \overset{d}{\hookrightarrow} L^2(\Omega) \overset{d}{\hookrightarrow} V^*.
$$

Next, we define a multivalued function $A(t,x,r)$ on $[0,T]\times\Omega\times\mathbf{R}$ into $\mathbf{R}$ by

$$
A(t,x,r) = \begin{cases} c_s r & \text{if } x\in\Omega_m(t),\ r<0, \\ [0,L] & \text{if } x\in\Omega_m(t),\ r=0, \\ c_\ell r + L & \text{if } x\in\Omega_m(t),\ r>0, \\ c_g r & \text{if } x\in\overline{\Omega_g(t)}. \end{cases}
\tag{2.7}
$$

By using this function A and the temperature θ, the enthalpy u can be given as a measurable selection of $A(t,x,r)$, namely

$$
u(t,x)\in A(t,x,\theta(t,x)) =: A(\theta)(t,x);
$$

note here that θ is a single valued function of u, since A has a single valued inverse.

Now we are in a position to give the weak formulation of our problem (1.5)–(1.9).

Definition 2.1. A pair $\{\theta, u\}$ of functions θ and $u : [0,T] \to L^2(\Omega)$ is called a weak solution of (1.5)–(1.9) if $\{\theta, u\}$ satisfy the following conditions:

(D1) $\theta \in L^\infty(0,T;L^2(\Omega)) \cap L^2(0,T;V)$, $u \in W^{1,2}(0,T;V^*) \cap L^\infty(0,T;L^2(\Omega))$.

(D2) $u \in A(\theta)$ and θ and u satisfy the variational identity (2.3).

Our main result is stated as follows.

Theorem 2.2. *Assume that* (A1) *and* (A2) *hold. Let* $f \in L^2(Q)$, $p \in C^1(\overline{\Sigma})$, $\theta_0 \in V$ *and* $u_0 \in L^2(\Omega)$. *Then there exist at least one weak solution* $\{\theta, u\}$ *of* (1.5)–(1.9).

Our assumption (A1) ensures that by the coordinate transformation $y = \mathbf{y}(t,x)$ the time-dependent domains $\Omega_i(t)$, $i = m, g$, are mapped onto time-independent ones. In Section 3 we consider the system (1.5)–(1.9) written in the new coordinate (t,y). In Section 4, we show the existence of a weak solution of the transformed system by the so-called time discretization method. Note here that for each $u \in W^{1,2}(0,T;V^*) \cap L^2(0,T;V)$ with $u(0) = u_0$ in $L^2(\Omega)$ and for $\eta \in V$

$$\int_0^T \langle u'(t), \eta(t)\rangle_{V^*,V} dt = -\int_0^T (u(t), \eta'(t))_{L^2(\Omega)} dt - (u_0, \eta(0))_{L^2(\Omega)},$$

where $(\cdot,\cdot)_{L^2(\Omega)}$ denotes the usual inner product in $L^2(\Omega)$ and $u' = (du/dt)$. By this relation we see that the weak variational form (2.3) is equivalent to

$$\int_0^T \langle u'(t), \eta(t)\rangle_{V^*,V} dt + \int_0^T \langle K(t,\theta(t)), \eta(t)\rangle_{V^*,V} dt + \int_0^T \langle G(t,u(t)), \eta(t)\rangle_{V^*,V} dt$$

$$= \int_0^T \langle f^*(t), \eta(t)\rangle_{V^*,V} dt \quad \text{for all } \eta \in L^2(0,T;V), \tag{2.8}$$

$$u(0) = u_0 \quad \text{in } L^2(\Omega), \tag{2.9}$$

where for each $t \in [0,T]$, $K(t,\cdot) : V \to V^*$ is a nonlinear second order differential operator and $G(t,\cdot) : L^2(\Omega) \to V^*$ is a linear operator given for each z, $z_1 \in V$ and $z_2 \in L^2(\Omega)$ by the following formulae:

$$\langle K(t,z_1), z\rangle_{V^*,V} := \int_\Omega \left\{k(z_1)\chi_{\Omega_m(t)} + k_g \chi_{\overline{\Omega_g(t)}}\right\} \nabla z_1 \cdot \nabla z dx + n_0 \int_\Gamma z_1 z d\Gamma,$$

$$\langle G(t,z_2), z\rangle_{V^*,V} := -\int_\Omega z_2(\mathbf{v}(t)\cdot\nabla z) dx,$$

and $f^* \in L^2(0,T;V^*)$ is defined by

$$\langle f^*(t), z\rangle_{V^*,V} := (f(t), z)_{L^2(\Omega)} + \int_\Gamma p(t) z d\Gamma \quad \text{for all } z \in V.$$

It is clear that (2.8) can be written in the form

$$u'(t) + K(t,\theta(t)) + G(t,u(t)) = f^*(t) \quad \text{in } V^* \text{ for a.e. } t \in [0,T]. \tag{2.10}$$

3. Transformation of (1.5)–(1.9)

In the mathematical treatment of our problem (1.5)–(1.9), one of the difficulties comes from the fact that the domain depends on time. In order to overcome this difficulty, we write the equations (1.5)–(1.8) in the coordinates (t, y), with $y = \mathbf{y}(t, x)$ given by assumption (A1). For convenience, we denote by $x = \mathbf{x}(t, y)$ the inverse of $\mathbf{y}(t, x)$.

Actually assumption (A1) ensures that for each $t \in [0, T]$

$$\mathbf{y}(t, \overline{\Omega_i(t)}) = \overline{\Omega_i(0)}, \quad \mathbf{x}(t, \overline{\Omega_i(0)}) = \overline{\Omega_i(t)}, \quad i = m, g,$$

$$\mathbf{y}(t, \overline{\Gamma_i(t)}) = \overline{\Gamma_i(0)}, \quad \mathbf{x}(t, \overline{\Gamma_i(0)}) = \overline{\Gamma_i(t)}, \quad i = m, g.$$

Moreover we put

$$\bar{\theta}(t, y) := \theta(t, \mathbf{x}(t, y)), \quad \bar{u}(t, y) := u(t, \mathbf{x}(t, y)),$$

and define a time-independent multivalued function $\bar{A} := \bar{A}(y, r)$ on $\Omega \times \mathbf{R}$ by putting

$$\bar{A}(y, r) := \begin{cases} c_s r & \text{if } y \in \Omega_m(0),\ r < 0, \\ [0, L] & \text{if } y \in \Omega_m(0),\ r = 0, \\ c_\ell r + L & \text{if } y \in \Omega_m(0),\ r > 0, \\ c_g r & \text{if } y \in \overline{\Omega_g(0)}. \end{cases} \tag{3.1}$$

Now, let $\{u, \theta\}$ be a weak solution of (1.5)–(1.9) and define $\bar{u}$ and $\bar{\theta}$ as above. Then it follows from (3.1) that

$$\bar{u}(t, y) \in \bar{A}(y, \bar{\theta}(t, y)) \quad \text{for a.e. } (t, y) \in Q. \tag{3.2}$$

Moreover the inverse $\bar{B} := \bar{B}(y, q)$ of $q := \bar{A}(y, r)$ is a singlevalued function given by

$$r := \bar{B}(y, q) = \begin{cases} \dfrac{q}{c_s} & \text{if } y \in \Omega_m(0),\ q < 0, \\ 0 & \text{if } y \in \Omega_m(0),\ q \in [0, L], \\ \dfrac{q - L}{c_\ell} & \text{if } y \in \Omega_m(0),\ q > L, \\ \dfrac{q}{c_g} & \text{if } y \in \overline{\Omega_g(0)}. \end{cases} \tag{3.3}$$

Hereafter we use the abridged forms $\bar{A}(\bar{\theta})$ and $\bar{B}(\bar{u})$ for $\bar{A}(\cdot, \bar{\theta})$ and $\bar{B}(\cdot, \bar{u})$, respectively. Clearly $\bar{A}$ and $\bar{B}$ are bounded and maximal monotone operators in $L^2(\Omega)$ and $\bar{A}$ satisfies the following inequality

$$(z_1^* - z_2^*, z_1 - z_2)_{L^2(\Omega)} \geq C_1 |z_1 - z_2|^2_{L^2(\Omega)}, \tag{3.4}$$

for all $z_i \in L^2(\Omega)$ and all $z_i^* \in \bar{A}(z_i)$ for $i = 1, 2$, where C_1 is a positive constant. Also, we see that the graphs of $\bar{A}$ and $\bar{B}$ in $L^2(\Omega) \times L^2(\Omega)$ are strongly-weakly closed, that is, $\bar{A}$ and $\bar{B}$ are demi-closed. As a remark, we can not identify $V = H^1(\Omega)$ with the space transformed by $y = \mathbf{y}(t, x)$ in the strict sense, although V is invariant under this coordinate transformation. Nevertheless, for simplicity, we use the same notation $V = H^1(\Omega)$ to indicate the space given after transformation, using the symbol x and y separately.

Now for each $t \in [0,T]$, we denote by $J_{\mathbf{x}(t)} := J_{\mathbf{x}(t)}(y)$ (resp. $J_{\mathbf{y}(t)} := J_{\mathbf{y}(t)}(x)$) the Jacobian matrix of the transformation $x = \mathbf{x}(t,y)$ (resp. $y = \mathbf{y}(t,x)$). Also, we denote by $x = \tilde{\mathbf{x}}(t,y)$ the restriction of $x = \mathbf{x}(t,y)$ to the boundary Γ. Further, we denote by $dx = ||J_{\mathbf{x}(t)}(y)||dy$, $d\Gamma(x) = ||J_{\tilde{\mathbf{x}}(t)}(y)||d\Gamma(y)$ the relations between the volume elements dx and dy in Ω, and the surface elements $d\Gamma(x)$ and $d\Gamma(y)$ on Γ, respectively. Actually $||J_{\mathbf{x}(t)}(y)||$ is the determinant of $J_{\mathbf{x}(t)}(y)$ and $||J_{\tilde{\mathbf{x}}(t)}(y)||$ is similarly expressed by using a suitable 2-dimensional local coordinate system on Γ. Under this setting, (2.8) is transformed to the following form:

$$\int_0^T \langle \bar{u}'(t), \xi(t)\rangle_{V^*,V} dt + \int_0^T \langle \bar{K}(t,\bar{\theta}(t)), \xi(t)\rangle_{V^*,V} dt + \int_0^T \langle \bar{G}(t,\bar{u}(t)), \xi(t)\rangle_{V^*,V} dt$$
$$= \int_0^T \langle \bar{f}^*(t), \xi(t)\rangle_{V^*,V} dt \quad \text{for all } \xi \in L^2(0,T;V), \tag{3.5}$$

subject to the initial condition

$$\bar{u}(0,y) = u_0(y) \quad \text{for a.e. } y \in \Omega, \tag{3.6}$$

where for each $\bar{z}_1 \in V$ the operators $\bar{K}(t,\cdot) : V \to V^*$ is defined by

$$\begin{aligned}\langle \bar{K}(t,\bar{z}_1), \bar{z}\rangle_{V^*,V} &:= \int_\Omega \sum_{i,j=1}^3 \left\{k(\bar{z}_1)\chi_{\Omega_m(0)} + k_g \chi_{\overline{\Omega_g(0)}}\right\} a_{ij}(t) \frac{\partial \bar{z}_1}{\partial y_j}\frac{\partial \bar{z}}{\partial y_i} dy \\ &- \int_\Omega \sum_{i,j=1}^3 \left\{k(\bar{z}_1)\chi_{\Omega_m(0)} + k_g \chi_{\overline{\Omega_g(0)}}\right\} a_{ij}(t) \frac{\partial \bar{z}_1}{\partial y_j}\frac{\partial ||J_{\mathbf{x}(t)}||}{\partial y_i}\frac{\bar{z}}{||J_{\mathbf{x}(t)}||} dy \\ &+ n_0 \int_\Gamma \bar{z}_1 \bar{z} \frac{||J_{\tilde{\mathbf{x}}(t)}||}{||J_{\mathbf{x}(t)}||} d\Gamma(y) \quad \text{for all } \bar{z} \in V,\end{aligned} \tag{3.7}$$

and for each $\bar{z}_2 \in L^2(\Omega)$ the operators $\bar{G}(t,\cdot) : L^2(\Omega) \to V^*$ is defined by

$$\begin{aligned}\langle \bar{G}(t,\bar{z}_2), \bar{z})\rangle_{V^*,V} &:= -\int_\Omega \bar{z}_2(\mathbf{w}(t)\cdot\nabla_y \bar{z}) dy \\ &+ \int_\Omega \bar{z}_2 \left(-\frac{\partial ||J_{\mathbf{x}(t)}||}{\partial t} + \mathbf{w}(t)\cdot\nabla_y(||J_{\mathbf{x}(t)}||)\right)\frac{\bar{z}}{||J_{\mathbf{x}(t)}||} dy \quad \text{for all } \bar{z} \in V.\end{aligned}$$

Moreover $\bar{f}^* \in L^2(0,T;V^*)$ is given by

$$\langle \bar{f}^*(t), \bar{z}\rangle_{V^*,V} := \int_\Omega \bar{f}(t)\bar{z} dy + \int_\Gamma \bar{p}(t)\bar{z} d\Gamma(y) \quad \text{for all } \bar{z} \in V,$$

with

$$\bar{f}(t,y) := f(t,\mathbf{x}(t,y)), \quad \bar{p}(t,y) := p(t,\mathbf{x}(t,y))\frac{||J_{\tilde{\mathbf{x}}(t)}(y)||}{||J_{\mathbf{x}(t)}(y)||},$$

$$a_{ij}(t,y) := \sum_{h=1}^3 \frac{\partial y_i}{\partial x_h}(t,\mathbf{x}(t,y))\frac{\partial y_j}{\partial x_h}(t,\mathbf{x}(t,y)) \quad \text{for } i,j = 1,2,3,$$

$\mathbf{w} := (w_1, w_2, w_3)$ with

$$w_j(t,y) := \sum_{i=1}^{3} \frac{\partial y_j}{\partial x_i}(t,\mathbf{x}(t,y)) \left(-\frac{\partial x_i}{\partial t}(t,y) + v_i(t,\mathbf{x}(t,y)) \right) \quad \text{for all } (t,y) \in Q.$$

In fact, perform the coordinate transformation from (t,x) to (t,y) in (2.3) and take $\eta(t,\mathbf{x}(t,y))||J_{\mathbf{x}(t)}(y)||$ as test function ξ. Then we have (3.5).

As regards the operators $\bar{K}$ and $\bar{G}$, we have the following lemma.

Lemma 3.1. (i) *If $\bar{z}_n \to \bar{z}$ weakly in V and $t_n \to t$ as $n \to +\infty$, then $\bar{K}(t_n, \bar{z}_n) \to \bar{K}(t,\bar{z})$ weakly in V^*. Moreover, there exist positive constants $C_i, i = 2,3,4$, such that*

$$|\bar{K}(t,\bar{z}_1)|_{V^*} \le C_2|\bar{z}_1|_V \quad \textit{for all } \bar{z}_1 \in V \textit{ and } t \in [0,T], \tag{3.8}$$

$$\langle \bar{K}(t,\bar{z}_1), \bar{z}_1\rangle_{V^*,V} \ge C_3|\bar{z}_1|_V^2 - C_4|\bar{z}_1|_{L^2(\Omega)}^2 \quad \textit{for all } \bar{z}_1 \in V \textit{ and } t \in [0,T]. \tag{3.9}$$

(ii) *If $v_n \to v$ weakly in $L^2(0,T;V)$ and in $L^2(0,T;L^2(\Omega))$ as $n \to +\infty$, then $\bar{K}(\cdot, v_n) \to \bar{K}(\cdot, v)$ weakly in $L^2(0,T;V^*)$.*
(iii) *If $\bar{z}_n \to \bar{z}$ weakly in $L^2(\Omega)$ and $t_n \to t$ as $n \to +\infty$, then $\bar{G}(t_n,\bar{z}_n) \to \bar{G}(t,\bar{z})$ weakly in V^*. Moreover, there exists a positive constant C_5 such that*

$$|\bar{G}(t,\bar{z}_2)|_{V^*} \le C_5|\bar{z}_2|_{L^2(\Omega)} \quad \textit{for all } \bar{z}_2 \in L^2(\Omega) \textit{ and } t \in [0,T]. \tag{3.10}$$

Proof. Putting $\hat{k}(r) := \int_0^r k(s)ds$, we reformulate the definition (3.7) of $\bar{K}(t,\bar{z}_1)$ as follows. For each $t \in [0,T]$ and each $\bar{z}_1 \in V$,

$$\begin{aligned}
&\langle \bar{K}(t,\bar{z}_1), \bar{z}\rangle_{V^*,V} \\
&= \int_{\Omega_m(0)} \sum_{i,j=1}^{3} a_{ij}(t) \frac{\partial \hat{k}(\bar{z}_1)}{\partial y_j}\frac{\partial \bar{z}}{\partial y_i} dy + \int_{\Omega_g(0)} k_g \sum_{i,j=1}^{3} a_{ij}(t)\frac{\partial \bar{z}_1}{\partial y_j}\frac{\partial \bar{z}}{\partial y_i} dy \\
&\quad - \int_{\Omega_m(0)} \sum_{i,j=1}^{3} a_{ij}(t)\frac{\partial \hat{k}(\bar{z}_1)}{\partial y_j}\frac{\partial ||J_{\mathbf{x}(t)}||}{\partial y_i}\frac{\bar{z}}{||J_{\mathbf{x}(t)}||} dy \\
&\quad - \int_{\Omega_g(0)} k_g \sum_{i,j=1}^{3} a_{ij}(t)\frac{\partial \bar{z}_1}{\partial y_j}\frac{\partial ||J_{\mathbf{x}(t)}||}{\partial y_i}\frac{\bar{z}}{||J_{\mathbf{x}(t)}||} dy + n_0 \int_\Gamma \bar{z}_1 \bar{z} \frac{||J_{\tilde{\mathbf{x}}(t)}||}{||J_{\mathbf{x}(t)}||} d\Gamma
\end{aligned}$$

for all $\bar{z} \in V$. If $\bar{z}_n \to \bar{z}$ weakly in V, $t_n \to t$ as $n \to +\infty$, then $\bar{z}_n \to \bar{z}$ in $L^2(\Omega)$ by the compact imbedding of V into $L^2(\Omega)$. Also, we have $\hat{k}(\bar{z}_n) \to \hat{k}(\bar{z})$ weakly in V by virtue of the definition of k. This implies that $\bar{K}(t_n,\bar{z}_n) \to \bar{K}(t,\bar{z})$ weakly in V^*. It is easy to deduce (3.8) and (3.9) from (3.7) and the definition of k. Similarly, (ii) is obtained from the expression of $\bar{K}$. Finally ,(iii) is also inferred from the definition $\bar{G}$. □

The system $\{(3.5), (3.6)\}$ is referred to as $(\bar{\mathrm{P}})$ and in the next section its solvability is discussed.

4. Construction of a Weak Solution

In this section we employ the time discretization method in order to solve $(\bar{\mathrm{P}})$. For each fixed $N \in \mathbf{N}$, we put $h := T/N$, $t_i^N := ih$ for $i = 0, 1, \ldots, N$, $\bar{\theta}_0^N := \theta_0, \bar{u}_0^N := u_0$, and derive the following problem. For each i, find the pair of functions $\{\bar{\theta}_{i+1}^N, \bar{u}_{i+1}^N\}$, with $\bar{\theta}_{i+1}^N \in V, \bar{u}_{i+1}^N \in L^2(\Omega)$ and $\bar{u}_{i+1}^N \in \bar{A}(\bar{\theta}_{i+1}^N)$, such that the following equation holds:

$$\left(\frac{\bar{u}_{i+1}^N - \bar{u}_i^N}{h}, \bar{z}\right)_{L^2(\Omega)} + \langle \bar{K}(t_{i+1}^N, \bar{\theta}_{i+1}^N), \bar{z}\rangle_{V^*,V} + \langle \bar{G}(t_{i+1}^N, \bar{u}_{i+1}^N), \bar{z}\rangle_{V^*,V} = \langle \bar{f}_{i+1}^{*N}, \bar{z}\rangle_{V^*,V} \quad \text{for all } \bar{z} \in V, \tag{4.1}$$

that is,

$$\frac{\bar{u}_{i+1}^N}{h} + \bar{K}(t_{i+1}^N, \bar{\theta}_{i+1}^N) + \bar{G}(t_{i+1}^N, \bar{u}_{i+1}^N) = \frac{\bar{u}_i^N}{h} + \bar{f}_{i+1}^{*N} \quad \text{in } V^*. \tag{4.2}$$

Here $\bar{f}_{i+1}^{*N}$ is the discrete approximation of f defined by

$$\bar{f}_{i+1}^{*N} := \frac{1}{h}\int_{t_i^N}^{t_{i+1}^N} \bar{f}^*(s)ds \quad \text{for } i = 0, 1, \ldots, N-1.$$

The following lemma ensures that the above difference approximation scheme $\{\bar{\theta}_i^N, \bar{u}_i^N\}_{i=1}^N$ can be in fact constructed.

Lemma 4.1. *Given $\bar{\theta}_i^N \in V$ and $\bar{u}_i^N \in L^2(\Omega)$, $i \le N-1$, there exist $\bar{\theta}_{i+1}^N \in V$ and $\bar{u}_{i+1}^N \in L^2(\Omega)$ with $\bar{u}_{i+1}^N \in \bar{A}(\bar{\theta}_{i+1}^N)$ such that $\bar{\theta}_{i+1}^N$ and $\bar{u}_{i+1}^N$ satisfy the equation* (4.2) (*hence* (4.1)).

Proof. Fix any $t \in [0, T]$. We define a multivalued operator $\Phi : V \to V^*$ by

$$\Phi\bar{z} := \left\{\frac{\bar{z}_1}{h} + \bar{K}(t, \bar{z}) + \bar{G}(t, \bar{z}_1) \in V^*; \bar{z}_1 \in \bar{A}(\bar{z})\right\} \quad \text{for all } \bar{z} \in V.$$

Then we see from Lemma 3.1 that for each $\bar{z} \in V$, $\Phi\bar{z}$ is a non-empty bounded and closed convex set in V^*, since $\bar{A}$ is maximal monotone and bounded in $L^2(\Omega)$. Also, by Lemma 3.1, its graph is weakly closed in $V \times V^*$. The boundedness and coerciveness of Φ are immediately obtained. Thus (4.2) is solved by virtue of the abstract results on the surjectiveness for nonlinear multivalued operators of Type M (see Brézis [1], Kenmochi [9]). □

Now, let $\{\bar{u}_i^N, \bar{\theta}_i^N\}_{i=0}^N$ be constructed as in Lemma 4.1, and define the piecewise linear function $\bar{u}^N$ on Q generated by $\{\bar{u}_i^N\}$ on $[0, T]$, that is,

$$\bar{u}^N(t, y) := \begin{cases} \bar{u}_i^N(y) & \text{if } t = t_i^N, i = 0, 1, \ldots, N, \\ \dfrac{t - t_{i-1}^N}{h}\bar{u}_{i-1}^N(y) + \left(1 - \dfrac{t - t_{i-1}^N}{h}\right)\bar{u}_i^N(y) & \\ \qquad\qquad \text{if } t \in (t_{i-1}^N, t_i^N), i = 1, 2, \ldots, N. \end{cases}$$

It is clear that $\bar{u}^N \in C([0,T];L^2(\Omega))$ for each $N \in \mathbf{N}$. Also, we can define the step functions $\bar{\bar{\theta}}^N$, $\bar{\bar{K}}^N$, $\bar{\bar{u}}^N$ and $\bar{\bar{G}}^N$ on $[0,T)$ by

$$\bar{\bar{\theta}}^N(t) := \bar{\theta}_i^N, \quad \bar{\bar{K}}^N(t) := \bar{K}(t_i^N, \bar{\theta}_i^N) \quad \text{if } t \in [t_{i-1}^N, t_i^N),\ i = 1,2,\ldots,N,$$

$$\bar{\bar{u}}^N(t) := \bar{u}_i^N, \quad \bar{\bar{G}}^N(t) := \bar{G}(t_i^N, \bar{u}_i^N) \quad \text{if } t \in [t_{i-1}^N, t_i^N),\ i = 1,2,\ldots,N.$$

The properties of these sequences of functions are stated in the following lemmas.

Lemma 4.2. *There exists a positive constant C_6 such that*

$$|\bar{\theta}_i^N|_{L^2(\Omega)} \le C_6 \quad \textit{for all } N \in \mathbf{N} \textit{ and } i = 0,1,\ldots,N. \tag{4.3}$$

Moreover, there exist positive constants $C_i, i = 7,8$, such that

$$|\bar{\bar{\theta}}^N|_{L^2(0,T;V)} \le C_7 \quad \textit{for all } N \in \mathbf{N}, \tag{4.4}$$

$$|\bar{u}^N|_{C([0,T];L^2(\Omega))} \le C_8 \quad \textit{for all } N \in \mathbf{N}. \tag{4.5}$$

Proof. By taking $\bar{\theta}_i^N$ as $\bar{z}$ in (4.1), we get

$$(\bar{u}_i^N - \bar{u}_{i-1}^N, \bar{\theta}_i^N)_{L^2(\Omega)} + h\langle \bar{K}(t_i^N, \bar{\theta}_i^N), \bar{\theta}_i^N\rangle_{V^*,V} + h\langle \bar{G}(t_i^N, \bar{u}_i^N), \bar{\theta}_i^N\rangle_{V^*,V}$$
$$= h\langle \bar{f}_i^{*N}, \bar{\theta}_i^N\rangle_{V^*,V} \quad \text{for } i = 1,2,\ldots,N. \tag{4.6}$$

From the definition of $\bar{B}$, there exists a convex and Lipschitz continuous function b on $L^2(\Omega)$ such that its subdifferential, denoted by ∂b, coincides with $\bar{B}$ in $L^2(\Omega)$. Then we have

$$(\bar{u}_i^N - \bar{u}_{i-1}^N, \bar{\theta}_i^N)_{L^2(\Omega)} \ge b(\bar{u}_i^N) - b(\bar{u}_{i-1}^N), \tag{4.7}$$

$$C_9|\bar{u}_i^N|^2_{L^2(\Omega)} - C_9' \le b(\bar{u}_i^N) \le C_{10}|\bar{u}_i^N|^2_{L^2(\Omega)} + C_{10}', \tag{4.8}$$

since $\bar{\theta}_i^N = \bar{B}(\bar{u}_i^N) = \partial b(\bar{u}_i^N)$ for $i = 0,1,\ldots,N$, where C_i and C_i' for $i = 9,10$, are positive constants independent of N. On the other hand, by virtue of Lemma 3.1, there exist positive constants C_3' and C_4' depending only on C_3, C_4, $\mathbf{w}$, and b such that

$$\langle \bar{K}(t_i^N, \bar{\theta}_i^N) + \bar{G}(t_i^N, \bar{u}_i^N), \bar{\theta}_i^N\rangle_{V^*,V} \ge C_3'|\bar{\theta}_i^N|_V^2 - C_4'|\bar{u}_i^N|^2_{L^2(\Omega)}. \tag{4.9}$$

Substituting (4.7)–(4.9) into (4.6) and summing up the results from $i = 1$ to $i = j \le N$, we obtain

$$b(\bar{u}_j^N) + (C_3' - \varepsilon)h\sum_{i=1}^{j}|\bar{\theta}_i^N|_V^2 \le \frac{C_4'}{C_9}h\sum_{i=1}^{j} b(\bar{u}_i^N) + C_\varepsilon h\sum_{i=1}^{j}|\bar{f}_i^{*N}|^2_{V^*} + \frac{C_4'C_9'}{C_9} + b(u_0),$$

for any $\varepsilon \in (0, C_3')$ and $C_\varepsilon := 1/(4\varepsilon)$. Note here that all constants are independent of f, u_0, and N. Therefore, applying the Gronwall lemma to this inequality, we derive the estimates (4.3)–(4.5). □

Lemma 4.3. $\{\bar{u}^N\}_{N=1}^\infty$ *is equicontinuous as a family of functions from* $[0,T]$ *into* V^*.

Proof. For each s, t with $s \le t$ such that $s \in [t^m_{s'-1}, t^m_{s'}], t \in [t^m_{t'-1}, t^m_{t'}]$ and $s' \le t'$, with the help of the definition of $\bar{u}^N(\cdot)$ we have

$$\begin{aligned} |\bar{u}^N(t) - \bar{u}^N(s)|_{V^*} &\le 2\sum_{i=s'}^{t'} h\left\{|\bar{K}(t_i^N, \bar{\theta}_i^N)|_{V^*} + |\bar{G}(t_i^N, \bar{u}_i^N)|_{V^*}\right\} + \sum_{i=s'}^{t'} h|\bar{f}_i^{*N}|_{V^*} \\ &\le 2\sqrt{(t'-s'+1)h}\left\{\left(\sum_{i=0}^{N} h|\bar{K}(t_i^N, \bar{\theta}_i^N)|^2_{V^*}\right)^{\frac{1}{2}}\right. \\ &\quad \left. + \left(\sum_{i=0}^{N} h|\bar{G}(t_i^N, \bar{u}_i^N)|^2_{V^*}\right)^{\frac{1}{2}} + \left(\sum_{i=0}^{N} h|\bar{f}_i^{*N}|^2_{V^*}\right)^{\frac{1}{2}}\right\}. \end{aligned}$$

Using Lemmas 3.1 and 4.2, we see that

$$\begin{aligned} &|\bar{u}^N(t) - \bar{u}^N(s)|_{V^*} \\ &\le 2\sqrt{|t-s| + 2h}\left\{C_2|\bar{\bar{\theta}}^N|_{L^2(0,T;V)} + C_5|\bar{u}^N|_{C([0,T];L^2(\Omega))} + |f^*|_{L^2(0,T;V^*)}\right\} \\ &\le 2\sqrt{|t-s| + 2h}\left\{C_2C_7 + C_5C_8T^{\frac{1}{2}} + |f^*|_{L^2(0,T;V^*)}\right\}, \end{aligned}$$

which shows the desired conclusion. □

By virtue of Lemmas 4.2, 4.3, and the Ascoli–Arzela's theorem it is easily seen that $\{\bar{u}^N\}_{N=1}^{\infty}$ is relatively compact in $C([0,T]; V^*)$, so that there exists a subsequence $\{N_m\} \subset \{N\}$ with $N_m \to \infty$ as $m \to \infty$ and there exists a function $\bar{u} \in C([0,T]; V^*)$ such that

$$\bar{u}^{N_m} \to \bar{u} \quad \text{in } C([0,T]; V^*), \quad \text{and hence} \quad \bar{\bar{u}}^{N_m} \to \bar{u} \quad \text{in } L^\infty(0,T; V^*). \tag{4.10}$$

Making use of the convergence (4.19), we get the convergence of $\{\bar{\bar{\theta}}^{N_m}\}_{m=1}^{\infty}$ as follows.

Lemma 4.4. *There exists a function $\bar{\theta} \in L^2(0,T; L^2(\Omega))$ such that*

$$\bar{\bar{\theta}}^{N_m} \to \bar{\theta} \quad \textit{in } L^2(0,T; L^2(\Omega)) \textit{ as } m \to +\infty,$$

and the pair of $\{\bar{\theta}, \bar{u}\}$ satisfies that $\bar{u} \in \bar{A}(\bar{\theta})$.

Proof. Since $\{\bar{\bar{\theta}}^{N_m}\}_{m=1}^{\infty}$ is bounded in $L^2(0,T; L^2(\Omega))$ by (4.4), it follows from (4.10) and (3.4) that

$$\begin{aligned} 0 &= \lim_{m,m' \to +\infty} \int_0^T \langle \bar{\bar{u}}^{N_m}(t) - \bar{\bar{u}}^{N_{m'}}(t), \bar{\bar{\theta}}^{N_m}(t) - \bar{\bar{\theta}}^{N_{m'}}(t)\rangle_{V^*,V} dt \\ &\ge \lim_{m,m' \to +\infty} C_1 \int_0^T |\bar{\bar{\theta}}^{N_m}(t) - \bar{\bar{\theta}}^{N_{m'}}(t)|^2_{L^2(\Omega)} dt. \end{aligned}$$

This implies that $\{\bar{\bar{\theta}}^{N_m}\}_{m=1}^{\infty}$ is a Cauchy sequence in $L^2(0,T; L^2(\Omega))$, so that $\bar{\bar{\theta}}^{N_m}$ converges in $L^2(0,T; L^2(\Omega))$ to some function $\bar{\theta}$. Also, the estimate (4.5) and the convergence (4.10) imply that $\bar{\bar{u}}^{N_m}$ converges weakly in $L^2(0,T; L^2(\Omega))$ to

$\bar{u}$ as $m \to +\infty$. This fact, together with the demi-closedness of $\bar{A}$, shows that $\bar{u} \in \bar{A}(\bar{\theta})$. □

Proof of Theorem 2.1. We have seen above that $\bar{\theta} \in L^2(0,T;V)$, $\bar{u} \in C([0,T];V^*) \cap L^\infty(0,T;L^2(\Omega))$ and

$$\bar{\bar{\theta}}^{N_m} \to \bar{\theta} \quad \text{in } L^2(0,T;L^2(\Omega)) \text{ and weakly in } L^2(0,T;V), \tag{4.11}$$

$$\bar{\bar{u}}^{N_m} \to \bar{u} \quad \text{in } L^\infty(0,T;V^*) \text{ and weakly-}* \text{ in } L^\infty(0,T;L^2(\Omega)), \tag{4.12}$$

as $m \to +\infty$. Now (4.2) implies that

$$\frac{d}{dt}\bar{u}^{N_m}(t) + \bar{\bar{K}}^{N_m}(t) + \bar{\bar{G}}^{N_m}(t) = \bar{\bar{f}}^{*N_m}(t) \quad \text{in } V^* \text{ a.e. } t \in (0,T), \tag{4.13}$$

where $\bar{\bar{f}}^{*N_m}(t) := \bar{f}_i^{*N_m}$ for each $t \in [t_{i-1}^{N_m}, t_i^{N_m})$ and $i = 1,2,\ldots,N_m$. Since $\{\bar{\bar{K}}^{N_m}\}$ is bounded in $L^2(0,T;V^*)$, we may assume that

$$\bar{\bar{K}}^{N_m} \to \tilde{K} \quad \text{weakly in } L^2(0,T;V^*) \quad \text{as } m \to +\infty \tag{4.14}$$

for some function $\tilde{K} \in L^2(0,T;V^*)$. By the expression (3.4) of K it is easy to see that

$$|\bar{\bar{K}}^{N_m}(t) - \bar{K}(t, \bar{\theta}^{N_m}(t))|_{V^*} \le C_{11} h |\bar{\theta}^{N_m}(t)|_V \quad \text{for all } t \in [0,T],$$

where C_{11} is a positive constant independent of N. Noting this together with (4.14), we see from (ii) of Lemma 3.1 that $\tilde{K} = \bar{K}(\cdot,\bar{\theta})$, since $K(\cdot,\bar{\theta}^{N_m}) \to \bar{K}(\cdot,\bar{\theta})$ weakly in $L^2(0,T;V^*)$. Similarly, by (4.10), (4.12) and (iii) of Lemma 3.1, we have

$$\bar{\bar{G}}^{N_m} \to \bar{G}(\cdot,\bar{u}) \quad \text{weakly in } L^2(0,T;V^*) \quad \text{as } m \to +\infty. \tag{4.15}$$

Now, it follows from (4.13) with (4.14) and (4.15) that

$$\frac{d}{dt}\bar{u}^{N_m} \to \frac{d}{dt}\bar{u} \quad \text{weakly in } L^2(0,T;V^*) \quad \text{as } m \to +\infty.$$

Furthermore, letting $m \to +\infty$ in (4.13), we see that

$$\frac{d}{dt}\bar{u}(t) + \bar{K}(t,\bar{\theta}(t)) + \bar{G}(t,\bar{u}(t)) = \bar{f}^*(t) \quad \text{in } V^* \quad \text{for a.e. } t \in [0,T],$$

that is, $\{\bar{\theta},\bar{u}\}$ satisfies (3.2) as well as (3.3). Therefore, since $\bar{u} \in \bar{A}(\bar{\theta})$, $\{\bar{\theta},\bar{u}\}$ is a solution of ($\bar{\text{P}}$). Finally we write this solution as the function of (t,x) by putting

$$\theta(t,x) := \bar{\theta}(t,\mathbf{y}(t,x)), \text{ and } u(t,x) := \bar{u}(t,\mathbf{y}(t,x)).$$

Then we see that $\{\theta,u\}$ satisfies (D1) and (D2). □

Acknowledgment
This work has been carried out also with partial support offered by the IMATI of CNR, Pavia, Italy. The first author is grateful to professor Pierluigi Colli and his research group in the Department of Mathematics of Pavia University for their warm hospitality.

References

[1] H. Brézis, *'Equations et inéquations non linéaires dans les espaces vectoriels en dualité*, Ann. Inst. Fourier (Grenoble), **18** (1968), 115–175.

[2] A. B. Crowley, *Mathematical modelling of heat flow in Czochralski crystal pulling*, IMA J. Appl. Math., **30** (1983), 173–189.

[3] A. Damlamian, *Some results on the multi-phase Stefan problem. Comm. Partial Differential Equations*, **2** (1977), 1017–1044.

[4] E. DiBenedetto and M. O'Leary, *Three-dimensional conduction-convection problems with change of phase*, Arch. Ration. Mech. Anal., **123** (1993), 99–116.

[5] J. Francu, *Modelling of the Czochralski flow*, Abstr. Appl. Anal., **3** (1998), 1–40.

[6] T. Fukao, *Transmission problems for degenerate parabolic equations*, pp.103–112 in *Elliptic and Parabolic Problems*, World Scientific, New Jersey-London-Singapore-Hong Kong, 2002.

[7] T. Fukao, N. Kenmochi and I. Pawłow, *Transmission problems arising in Czochralski process of crystal growth*, pp.228–243 in *Mathematical Aspects of Modeling Structure Formation Phenomena*, GAKUTO Internat. Ser. Math. Sci. Appl., Vol. **17**, Gakkōtosho, Tokyo, 2002.

[8] T. Fukao, N. Kenmochi and I. Pawłow, *Stefan problems in non-cylindrical domains arising in Czochralski process of crystal growth*, to appear.

[9] N. Kenmochi, *Nonlinear operators of monotone type in reflexive Banach spaces and nonlinear perturbations*, Hiroshima Math. J., **4** (1974), 229–263.

[10] N. Kenmochi and I. Pawłow, *A class of nonlinear elliptic-parabolic equations with time-dependent constraints*, Nonlinear Anal., **10** (1986), 1181–1202.

[11] N. Kenmochi and I. Pawłow, *Controlled Czochralski crystal growth*, J. Soci. Inst. Cont. Engi., **35** (1996), 944–950.

[12] I. Pawłow, *Three-phase boundary Czochralski model*, pp.203–227 in *Mathematical Aspects of Modeling Structure Formation Phenomena*, GAKUTO Internat. Ser. Math. Sci. Appl., Vol. **17**, Gakkōtosho, Tokyo, 2002.

(T. Fukao) Department of Mathematics, Graduate School of Science and Technology, Chiba University, 1-33 Yayoi-cho, Inage-ku, Chiba, 263-8522, Japan
E-mail address: `t-fukao@math.e.chiba-u.ac.jp`

(N. Kenmochi) Department of Mathematics, Faculty of Education, Chiba University, 1-33 Yayoi-cho, Inage-ku, Chiba, 263-8522, Japan
E-mail address: `kenmochi@math.e.chiba-u.ac.jp`

(I. Pawłow) Systems Research Institute of the Polish Academy of Sciences, ul. Newelska 6, 01-447 Warszawa, Poland
E-mail address: `pawlow@ibspan.waw.pl`

International Series of Numerical Mathematics, Vol. 147, 167–175

Quasi-static Melting of Crystals: Experiments and Analysis

Martin E. Glicksman, Afina Lupulescu, and Matthew B. Koss

Abstract. The kinetics of melting pivalic acid (PVA) dendrites was observed under convection-free conditions on STS-87 as part of the United States Microgravity Payload Mission (USMP-4) flown on *Columbia* in 1997 during STS-87. The analysis of video data show that PVA dendrites melt without relative motion with respect to the quiescent melt phase. Dendritic fragments display shrinking to extinction, with fragmentation occurring at higher initial supercoolings. Individual fragments follow a characteristic time-dependence derived elsewhere. The microgravity melting kinetics against which the experimental observations are compared is based on conduction-limited quasi-static melting under shape-preserving conditions. Agreement between the analytic theory and our experiments is found when the melting process occurs under shape-preserving conditions as measured using the $\mathcal{C}/\mathcal{A}$ ratio of individual needle-like crystal fragments.

1. Introduction

1.1. Isothermal Dendritic Growth Experiment (IDGE)

The thermodynamic driving force for crystallization in the case of a pure substance is the melt supercooling, $\Delta T \equiv T_m - T_\infty$, where T_m is the melting point of the material, and T_∞ is the temperature of the melt established far from the heat emitting dendritic solid-liquid interface. The crystallization process for free growth requires that the latent heat released be transported away from the solid-liquid interface by thermal conduction to the cooler (supercooled) melt. Theories [5]–[14] of dendritic growth in pure materials are best tested under conditions of conduction-controlled heat transfer, a situation possible only where gravity is reduced to nearly zero, where buoyancy-induced natural convection is virtually eliminated [3]–[9]. The Isothermal Dendritic Growth Experiment, or IDGE, provides basic terrestrial and microgravity data on the kinetics of dendritic solidification.

The IDGE flew in near-Earth orbit aboard *Columbia* three times, and proved that Ivantsov's conduction solution [9] describes steady-state dendritic growth for the test materials succinonitrile (SCN) and pivalic acid (PVA). These cubic organic crystals are 1) transparent, 2) easily purified, and 3) have conveniently low melting points. IDGE instruments, on board the STS-87, provided CCD images

FIGURE 1. Video sequence (frame height ~ 1.5 cm) showing the progressive melting of PVA dendrites in microgravity. As the dendrites fragment, individual crystallites remain motionless within the melt. Video data were analyzed frame-by-frame for the melting of the needle crystal shown in the second panel and reported in section 5.

(telemetered to Earth during each flight), NASA-processed 35-mm film negatives (available postflight), and, for the first time, real-time streaming of 30 frame per second (fps) full gray-scale video data telemetered via the K-band (high-frequency) antenna.

1.2. Mushy zones

A mushy zone is defined as a two-phase region, consisting of the dendritic solid and its melt phase. Mushy zones occur in ingots, castings, and welds, [16], [18]. We studied 4-9's purity PVA under nearly gravity-free conditions (mean quasi-static acceleration of $10^{-7}g_0$, where $g_0 = 9.8$ m/s^2 is the normal terrestrial gravitational acceleration). The melting of a mushy zone in a pure substance under convection-free conditions should be kinetically controlled by conduction-limited heat transfer, provided that convection is absent. Convection, if present, alters the thermal fields during melting, because sedimentation occurs between the solid and the melt, inducing relative motion between the phases.

2. Video data

The freezing and melting observations analyzed and reported here were recorded as 30 fps video downlinked from microgravity experiments conducted in low-earth orbit during November–December 1997. On-board video signals generated within the IDGE were telemetered to Earth via *Columbia's* K-band antenna, and recorded by NASA using a high-resolution VTR on 8-mm tapes. Two hours of 30 fps video frames were transferred onto a single image buffer file as 640×480, 255 gray-scale, digitized images. For each dendritic freezing/melting cycle, the video data were exported as individual .tif files for every tenth frame.

For the last minute of melting, where the thermal fields in the IDGE thermostat remain at a stable temperature, approximately 1.8K above T_m, individual .tif files were exported for every video frame. When melting begins, the dendritic

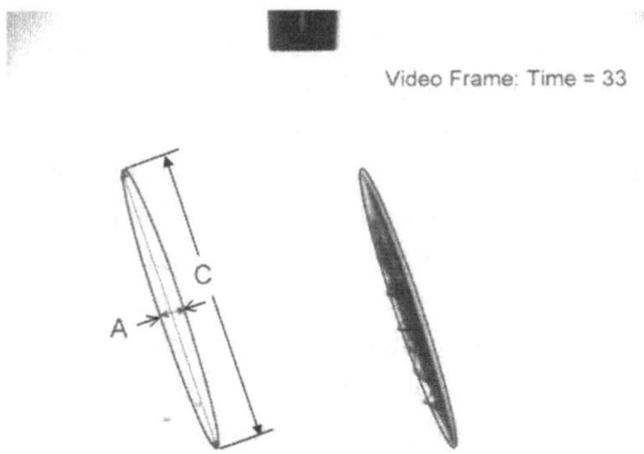

FIGURE 2. Sample digitized video frame for image analysis of a melting PVA dendritic fragment. The image analyzer calculates the equivalent ellipse for the fragment, allowing determination at each time of its major axis length, $\mathcal{C}(t)$, and its $\mathcal{C}/\mathcal{A}$ ratio.

side-arms first shorten, and then larger secondary branches detach from the primary stem. Figure 1 shows several video frames representative of a few stages of the total melting sequence. In microgravity, it is important to note that the individual crystalline elements of the mushy zone remain motionless.

3. Data Analysis

3.1. Image processing

The shapes of the melting dendritic fragments were determined using commercial image analysis software. Figure 2 shows the shape of a melting dendrite stem outlined using this software tool. The equivalent ellipse of the two-dimensional image of this fragment is calculated on the basis of its major axis, $\mathcal{C}$, and minor axis, $\mathcal{A}$. The particles are of course three-dimensional objects approximated here as smooth prolate spheroids of decreasing size. Image analysis conducted on sequential video frames allowed the measurement of the major and minor axes of the equivalent ellipsoidal fragment as a function of time.

4. Theory

Quasi-static and moving boundary solutions are well known [19] [2] for the growth of spherical crystals for which ($\mathcal{C}/\mathcal{A} = 1$). For the more general cases of freezing ellipsoids and hyperboloids, moving boundary solutions were developed by Ham [7]. Ham's similarity solutions for ellipsoids apply to the case of crystal growth, not melting. We developed and published elsewhere a quasi-static model for the conduction melting of prolate spheroids [4] .

The quasi-static melting or freezing of a crystal is based on determining the heat current entering or leaving its *static* solid-melt interface, and then formulating kinematic equations for the crystal's shrinkage or growth. The rate of change of the particle's volume is assumed to occur slowly enough so that the surrounding transport field has sufficient time to adjust to the changing particle. The thermal transport field used only approximates the time-dependent solution to the conduction equation as a quasi-static solution based on Laplace's equation. The key condition determining the applicability of such approximations is that the Stefan number — the intensity measure for melting or freezing — is small, i.e., $St \ll 1$.

The axis-symmetric temperature field developed around a melting particle, $T(\xi, \eta)$, is non-dimensionalized as the potential, ϑ, defined as

$$\vartheta(\xi, \eta) \equiv \frac{T(\xi, \eta) - T_m}{T_\infty - T_m}, \tag{1}$$

where T_∞ is the temperature of the melt infinitely far away from the particle. Here the three-dimensional coordinate system chosen is the separable prolate spheroidal system (ξ, η, φ). These coordinates correspond to isotherms as ($\xi = const.$), and to heat flow surfaces as ($\eta = const$). The crystal-melt interface is defined as the isotherm $T(\Xi, \eta = T_m)$, which defines the prolate spheroidal interface $\xi = \Xi$.

The rate of melting (or freezing) is determined by the sign and magnitude of the denominator in Eq.(1). A dimensionless parameter based on that temperature difference is the Stefan number, defined here as

$$St \equiv \frac{T_\infty - T_m}{\Delta H_f / C_\ell^p}, \tag{2}$$

where ΔH_f is the molar enthalpy of fusion, and C_ℓ^p is the molar specific heat of the melt at constant pressure. Specifically, the Laplacian potential surrounding a prolate spheroid with the fixed interfacial potential $\vartheta(\Xi) = 0$, embedded in a melt where the far-field potential $\vartheta(\infty) = 1$, is given in these curvilinear-coordinates as

$$\vartheta(\xi, \eta) = 1 - \frac{\log \frac{\xi+1}{\xi-1}}{\log \frac{\Xi+1}{\Xi-1}}. \tag{3}$$

A kinetic expression has been found in [4] for quasi-static melting or freezing of prolate spheroids, namely

$$\mathcal{C}(t) = \sqrt{\mathcal{C}_0^2 - \frac{16\alpha_\ell St}{\Xi \log \frac{\Xi+1}{\Xi-1}} \left(\frac{\mathcal{C}}{\mathcal{A}}\right)^2 t}, \tag{4}$$

where $\mathcal{C}/\mathcal{A} = \Xi/\sqrt{\Xi^2 - 1}$. Equation (4) may be non-dimensionalized as

$$\frac{\mathcal{C}(t)}{\mathcal{C}_0} = \sqrt{1 - K_{prol} \cdot St \cdot Fo}, \tag{5}$$

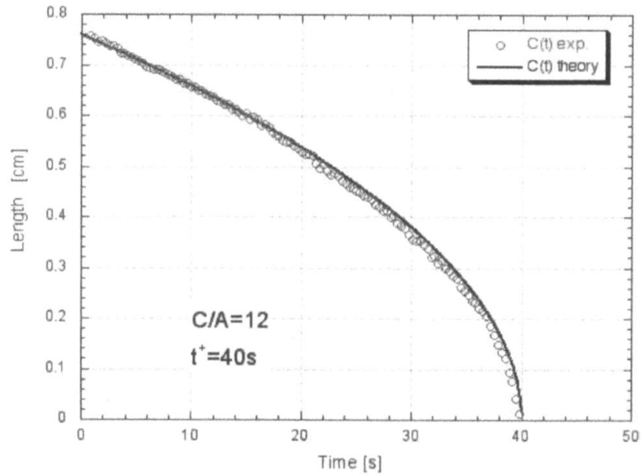

FIGURE 3. Melting kinetics of a dendritic pivalic acid (PVA) crystal fragment ($\mathcal{C}/\mathcal{A} = 12$) in microgravity. Experimental melting data (open symbols) obtained through digital analysis of 230 digitized video frames measure $\mathcal{C}(t)$ versus time. The solid curve is the theoretical prediction from Eq.(4) fitted with a Stefan number equal to 0.057. The fit with quasi-static theory is based on the observed extinction time $t^{\dagger} = 40$s.

where the kinetic coefficient, K_{prol}, is defined as

$$K_{prol} \equiv \frac{16}{\Xi \log \frac{\Xi+1}{\Xi-1}} \left(\frac{\mathcal{C}}{\mathcal{A}}\right)^2, \tag{6}$$

and the Fourier number (dimensionless transformation time) is defined as

$$Fo \equiv \frac{\alpha_\ell}{\mathcal{C}_0^2} t. \tag{7}$$

5. Results

5.1. Melting rates

We analyzed two PVA crystals melting in microgravity, each formed originally as a dendrite created in separate IDGE growth cycles. These dendrites grew from pure molten pivalic acid that was supercooled 0.400K and 0.421K, respectively. The mushy zones consisted of approximately 4% solid when melting was initiated. The time needed to melt this mushy zone completely was approximately 12 minutes. The last large fragments to melt were selected for quantitative analysis. These fragments had initial lengths of $\mathcal{C}_0 = 0.760$cm and 0.695cm, respectively.

To calculate the melting kinetics of these fragments one requires the $\mathcal{C}/\mathcal{A}$ ratio, the thermal diffusivity of the melt, and the Stefan number imposed by the

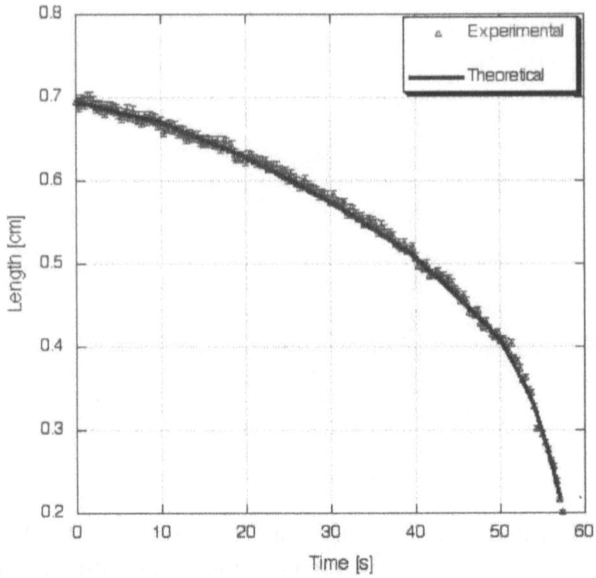

FIGURE 4. Melting kinetics of a dendritic pivalic acid (PVA) crystal fragment ($\Delta T = 0.421K$) in microgravity.

melting cycle. The thermistors arranged within the IDGE thermostat were optimized for measuring the supercooling *prior* to dendritic growth quite precisely ($\pm$0.002K). The subsequent temperature history within the growth chamber, especially during the melting portion of an experimental growth cycle, is, unfortunately, less certain, because temperature uncertainties arose from the lack of synchronization between the video clock and the mission clock. In view of this situation, one may instead adopt a self-consistent method for estimating the Stefan number in the melt during each melting sequence. The basis of the value of the self-consistent Stefan number is found from the $\mathcal{C}/\mathcal{A}$ ratio (which was equal to 12 for the entire melt cycle), the initial length, $C_0 = 0.760$, and the length $C(t)$ at a later time t, at which the fragment has decreased due to melting. In this instance the process of melting was followed to extinction, so $\mathcal{C}(t) = 0$, and $t = 40$s.

The crystal melting data reported in Figure 3 were fit in their entirety using the self-consistent Stefan number, which corresponds to the melt being superheated 0.63K above its equilibrium melting point, $T_m = 35.97K$. The melting kinetics predicted over the entire the lifetime of this dendrite fragment using eq.(4) is in good agreement with experiment. The video data show only small disparities from the theoretical predictions.

The second crystal analyzed, initially formed at a supercooling of 0.421K, did not melt with a constant $\mathcal{C}/\mathcal{A}$ ratio. These melting data were "sectorized" into 5 sequential melting periods of 10 seconds each, except for the last sector, which was 4 seconds long. Sectorization allowed calculation of the $\mathcal{C}/\mathcal{A}$ ratio, which was changing in this case. The same process of selecting a Stefan number was used, except each sector of melting was independently calculated using the initial and final lengths for that sector. Figure 4 shows the observed data and the theoretical predictions using the sectorized approach described here.

6. Conclusions

1. Dendritic mushy zones were formed in supercooled high-purity PVA melts in microgravity. At the end of each experimental cycle, the mushy zone was progressively melted by raising the temperature of the thermostat. Video data at 30 fps of the melting process for one cycle (0.4K supercooling) are reported here. Upon raising the temperature, the mushy zone melts into numerous fragments, many of which may be characterized as prolate spheroids.
2. Individual dendritic fragments remain motionless in the low-gravity environment, progressively melting toward extinction at nearly constant $\mathcal{C}/\mathcal{A}$ ratios.
3. Moving boundary solutions, such as those proposed by Ham for ellipsoidal crystals, are not applicable to the melting process. The theory describes the kinetics of conduction melting of prolate spheroids of arbitrary $\mathcal{C}/\mathcal{A}$ ratios.
4. The quasi-static analytic results are limited to small Stefan numbers ($St \ll 1$).
5. The observed melting kinetics in microgravity of two prolate spheroidal crystals with differing aspect ratios were found to be well described by the analysis.
6. The Stefan number required to compare theory and experiment was estimated using the time of melting and the observed lengths at the start and end of the melting.

7. Acknowledgements

We dedicate this paper to the memory of the entire crew of astronauts who tragically perished aboard the Shuttle *Columbia* on February 1, 2003. Dr. Kalpana Chawla, the Mission Specialist lost on flight STS-112 assisted us in 1997 on STS-87 in obtaining the video data reported herein. We are deeply grateful for her dedication and assistance. The authors also acknowledge the financial support provided for this research by the National Aeronautics and Space Administration, Division of Biological and Physical Sciences Research, Code U, Washington, D.C., under Grant NAG 8-1696.

References

[1] Glicksman, M.E., *Fundamentals of Dendritic Crystal Growth.* Crystal Growth of Electronic Materials. Chapter 5. Ed. E. Kaldis, Elsevier Science Publishers, (1985), 57–69.

[2] Glicksman, M.E., *Diffusion in Solids: Field Theory, Solid-state Principles, and Applications.* John Wiley and Sons, New York, 2000.

[3] Glicksman, M.E., Koss, M.B., Bushnell, L.T., Lacombe, J.C., Winsa, E.A., *The Isothermal Dendritic Growth Experiment.* Materials Science Forum **215-216** (1996), 179–190.

[4] M.E. Glicksman, A. Lupulescu, and M.B. Koss, *Melting in Microgravity*, J. Thermophysics and Heat Transfer **17**, no. 1 (2003), 69–76.

[5] Glicksman, M.E., Marsh, S.P., *The Dendrite.* Handbook of Crystal Growth. Ed. D.T.J. Hurle, Elsevier Science Publishers, Amsterdam, (1993), 1075–1122.

[6] Glicksman, M.E., Schaefer, R.J., Ayers, J. D., *Dendritic Growth-A Test of Theory,* Metall. Trans. A **7A** (1976), 1747–1759.

[7] Ham, F.S., *Shape-Preserving solutions of the time-dependent diffusion equation.* Quart. J. Appl. Math. **17** (1959), 137–145.

[8] Huang, S.C., Glicksman, M.E., *Fundamentals of dendritic solidification-I. Steady-state tip growth.* Acta Metallurgica **29** (1981), 701–715.

[9] Ivantsov, G.P., *The Temperature Field Around Spherical , Cylindrical, and Needle-Shaped Crystals which Grow in Supercooled Melts.* Dokl. Akad. Nauk. USSR **58**, No. 4 (1947), 567–569.

[10] LaCombe, J.C., Koss, M.B., Corrigan, D.C., Lupulescu, A., Tennenhouse, L.A., *Implications of the interface shape on steady-state dendritic crystal growth.* Journal of Crystal Growth **206** (1999), 331–344.

[11] LaCombe, J.C., Koss, M.B., Fradkov, V.E., Glicksman, M.E., *Three-dimensional dendrite-tip morphology.* Physical Review, E, **52**, No. 3 (1995), 2778–2786.

[12] Langer,J.S., *Nonequilibrium physics and the origins of complexity in nature.* Princeton Series in Physics. Critical Problems in Physics: Chapter 2, Princeton University Press, 1996.

[13] Langer, J.S., Muller-Krumbhaar, H., *Theory of dendritic growth. I. Elements of a stability analysis.* Acta Metallurgica **26** (1978), 1681–1687.

[14] Lipton, J., Glicksman, M.E., Kurz, W., *Equiaxed Dendrite Growth in Alloys at Small Supercooling.* Metall. Trans. A **18A** (1987), 341–345.

[15] Lupulescu, A., LaCombe, J.C., Frei, J.E., Glicksman, M.E., Koss, M.B., *USMP-4 Microgravity PVA Dendritic Growth Data -Preliminary Analysis-.* Microgravity Symposium: Proceedings TMS Meeting, Nashville, 2000, CD-ROM.

[16] Marsh, S.P., Glicksman, M.E., *Overview of geometric effects on coarsening of mushy zones.* Metallurgical and Materials Transactions A **27A** (1996), 557–567.

[17] Mullins,W.W., Sekerka S.F., *Morphological stability of a particle growing by diffusion of heat flow.* J. Appl. Phys. **34** (1961), 323–329.

[18] Paradies, C.J., Smith, R.N., Glicksman, M.E., *The influence of convection during solidification on fragmentation of the mushy zone of a model alloy.* Metallurgical and Materials Transactions A **28A** (1997), 875–883.

[19] Sekerka, R.F., Jeanfils, C.L., Heckel, R.W., *The Moving Boundary Problem.* Lectures on the Theory of Phase Transformations, Ed. H.I. Aaronson, The Metallurgical Society, AIME, New York, (1982), 117–169.

[20] Trivedi, R., Mason, J.T., *The effects of interface attachment kinetics on solidification interface morphologies.* Metall. Trans. A **22A** (1991), 235–249.

(M.E. Glicksman, A. Lupulescu) Materials Science and Engineering Department,
Rensselaer Polytechnic Institute,
Troy, NY 12180-3590,
USA

E-mail address: glickm@rpi.edu, lupula@rpi.edu

(Matthew B. Koss) Physics Department,
College of the Holy Cross,
P.O. Box 143A,
Worcester, MA 01610-2395,
USA
E-mail address: mkoss@holycross.edu

International Series of Numerical Mathematics, Vol. 147, 177–187

A Reduced Model for Simulating Grain Growth

Reiner Henseler, Barbara Niethammer and Felix Otto

Abstract. We introduce a reduced model to study normal grain growth in two dimensions, which is based on the gradient flow structure of the mean curvature flow. In the simplified model, we restrict this gradient flow to grain boundaries which are straight lines. We present simulations for a large number of grains and we also study the effect of anisotropies.

1. Introduction

Many materials have a polycrystalline structure which consists of a large number of single crystals, or grains, with different crystallographic orientations. In the last stage of the evolution, the grain boundaries move under capillary forces to reduce their surface energy and, as a result, the grain structure coarsens.

It is of fundamental importance for technological purposes to predict the time dependence of the characteristic scale of inhomogeneity, given for example by the average grain radius. What is often observed in experiments is the so called normal grain growth, which means that materials with different histories develop a certain universal large-time behavior. In particular, it implies power growth laws for the average grain size and self-similarity of the distribution of grain sizes and topological class (the number of neighbors).

In modelling grain growth, attention has mainly focused on two-dimensional grain growth, as it appears for example in thin films and foils, due to the simpler geometry and easier access to experimental data. The mean curvature flow coupled with equilibrium of forces at the triple junctions is widely accepted as an appropriate model for the evolution of the grain boundaries. To explore normal grain growth by simulations, one would need to simulate this complex motion for thousands of grains over a large time window. Due to this complexity, there is a large interest in deriving reduced models which allow for efficient numerical simulations.

The so called mean-field models are mostly based on the von Neumann–Mullins law, which gives an exact relation between the growth rate of the area of a grain and its topological class. Based on this relation, mean-field models for the distribution of grain sizes and topological class have been suggested and analyzed (see e.g. [Fra88, FU94, FUK88] and the references therein). However, a significant disadvantage of this approach is that the von Neumann–Mullins relation only holds in two dimensions and if all possible anisotropies are neglected.

In this paper we investigate a different approach, which is based on the fact that the mean curvature flow coupled with equilibrium of forces at the triple junctions has an interpretation as a gradient flow. It is the gradient flow of the surface energy with respect to the L^2 norm for the normal velocities of the grain boundaries. Our strategy is to restrict this gradient flow to grain boundaries which are just straight lines. Thus, we can parametrize the grain network by the vertices and their neighbor relations, which significantly reduces the computational effort. On the other hand, it is conceptually simple to introduce anisotropic boundary energies and mobilities. A similar approach has been explored by a Japanese group and is known in the applied literature as "vertex models" [KNN89, NFK94].

In Section 2 we explain in detail our reduced model, while in Section 3 we present the first results of simulations for an isotropic surface energy for Voronoi data with 20000 vertices. We also consider the case of anisotropic surface energies, where the energy depends on the relative orientation of the neighboring grains.

For more details on the physical background and references to the applied literature we refer to the excellent review article [FU94].

2. The model

2.1. The gradient flow structure of the basic model

In the sequel, we consider models for grain growth in two space dimensions. Thus, our objects are networks of curves which meet in triple junctions. Since we are not interested in the influence of boundary conditions, we consider one-periodic networks. In the sequel, we denote by $\gamma_k, k = 1, ..., M$, the boundaries, or edges, of the grains, i.e. the smooth curves with endpoints in triple junctions, which will be denoted by $X_i, i = 1, ..., L$. The system of grain boundaries is denoted by $\Gamma := \bigcup_{k=1}^{M} \gamma_k$.

We first describe how the mean curvature flow coupled with equilibrium of forces at the triple junctions has a natural interpretation as a gradient flow (see also [TC94] for the case without triple junctions). In the abstract framework we have a differentiable manifold $\mathcal{M}$, and a vector field f, which attaches a tangent vector $f(x) \in T_x\mathcal{M}$ to every point $x \in \mathcal{M}$. The vector field f defines a dynamical system $\dot{x} = f(x)$. A gradient flow is a dynamical system where f is the negative gradient $-\text{grad}E$ of a function E on $\mathcal{M}$. The notion of a gradient requires a Riemannian structure, that is, a metric tensor g on $\mathcal{M}$. Then, the extended version of $\dot{x} = -\text{grad}E_x$ is

$$g_{x(t)}(\dot{x}(t), y) + \langle \text{diff} E_{x(t)}, y \rangle = 0 \quad \text{for all } y \in T_{x(t)}\mathcal{M} \text{ and } t$$

and we find for $y = \dot{x}(t)$ that the value of E decreases along trajectories.

In the present setting, the manifold $\mathcal{M}$ is just given by all the one-periodic networks Γ consisting of smooth curves which meet in triple junctions. The tangent space $T_\Gamma \mathcal{M}$ of a point $\Gamma \in \mathcal{M}$ can be conveniently characterized by all admissible normal velocities v of the curves. Notice that we require the network to remain connected. Thus the normal velocities of the edges meeting in a vertex are not independent. The velocity $\dot{X}_i$ of a vertex can then be expressed by the normal velocities of the incoming edges.

For the metric tensor we take in the following

$$g_\Gamma(v, \tilde{v}) := \int_{\Gamma \cap (0,1)^2} v\tilde{v}\, ds = \sum_{k=1}^{M} \int_{\gamma_k \cap (0,1)^2} v\tilde{v}\, ds \qquad \forall v, \tilde{v} \in T_\Gamma \mathcal{M}.$$

Here, we assume that the mobilities of the grain boundaries are all equal and normalized to one. Otherwise, the inverse mobility would show up as a weight function in the scalar product. With this choice of g, we have also implicitly assumed that the mobility of the vertices is much larger than the mobility of the boundaries, such that the motion is only limited by boundary kinetics. The vertices are then just geometrical objects and their motion is fully defined by the motions of the adjoining boundaries.

The energy of the system is just the surface energy. In this paper we restrict our attention to the case in which the surface tension is constant on each edge and may depend on the relative orientation of the neighboring grains. Thus, the energy is given by

$$E(\Gamma) := \int_\Gamma \sigma(\alpha)\, ds = \sum_{k=1}^{M} \sigma(\alpha_k)\, |\gamma_k|$$

where σ denotes the surface tension, which may depend on the "misorientation" angle α.

The first variation of the energy in the direction of an element $\tilde{v}$ of the tangent space can be easily computed (see e.g. [KL01]):

$$\langle \mathrm{diff} E, \tilde{v} \rangle = \sum_{k=1}^{M} \sigma(\alpha_k) \int_{\gamma_k \cap (0,1)^2} \kappa \tilde{v} \,+\, \dot{X}_k \cdot \sigma(\alpha_k) b_k \big|_{X_k^1}^{X_k^2}.$$

Here, $X_k^i, i = 1, 2$ denote the endpoints of γ_k and b_k its tangent.

Hence the gradient flow is given by

$$v = -\sigma(\alpha_k)\kappa \qquad \text{on each } \gamma_k \tag{1}$$

where κ is the mean curvature of the edge, and at the triple point there holds mechanical equilibrium

$$\sum_{j_i} \sigma(\alpha_{j_i}) b_{j_i} = 0,$$

where the sum extends over the three edges meeting in X_i and the b_{j_i} point in the direction of X_i. If σ is independent of α, this just means that the curves meet in a 120 degree angle, otherwise it implies Young's law that the ratios of surface energies and the sin of the opposite angle are equal.

This evolution is well-defined [BR93, KL01] until two or more vertices collide, after which topological rearrangements may take place. This happens when either an edge or a whole grain vanishes. In the first case an unstable fourfold vertex is produced, which immediately splits up again, usually in such a way that two new vertices are connected by a new edge. In this case, two neighboring grains decrease their topological class, whereas the two other grains increase it. When a three-sided grain vanishes, all neighboring grains loose an edge and thus decrease their topological class. In principle, grains with more than three edges can vanish. For the numerical simulations, we assume that this unlikely situation does not occur and a neighbor switching takes place before a grain with more than three edges becomes small.

We recall that (1) implies for dimensional reasons that, if the typical length scale shows a scaling behavior, it must grow proportional to $t^{1/2}$ and consequently the energy density decreases as $t^{-1/2}$.

2.2. Restricted gradient flow

Our aim is now to simplify the dynamical system described in the previous section in such a way that simulations of large systems become feasible while the essential properties of the dynamics are still kept. In grain growth, one usually observes that the grain boundaries do not have large curvatures. This motivates a simplification which restricts the dynamics to networks whose curves are just straight lines. Now the advantage of the gradient flow perspective becomes apparent. A gradient flow can be naturally restricted to a lower dimensional submanifold $\mathcal{N} \subset \mathcal{M}$. Both the energy and the metric tensor restrict to $\mathcal{N}$ in a canonical way. Consequently, we restrict the gradient flow to the submanifold of networks where vertices are connected by straight lines. Then, this submanifold is completely characterized by the vertices, together with the information on which of these vertices are connected:

$$\mathcal{N} = \{X = \{X_i\}_i \ \& \text{ neighbor relations}\}.$$

The tangent space $T_X\mathcal{N}$ in a point $X \in \mathcal{N}$ can then be described by the velocities of the vertices

$$T_X\mathcal{N} = \{V = \{V_i\}_i\}.$$

In the sequel, we will denote by γ_{ij} the edge, which connects the vertices X_i and X_j. It is parametrized by

$$\gamma_{ij} = sX_i + (1-s)\,X_j, \ \ s \in [0,1],$$

and its normal velocity is given by

$$v_{ij} = (sV_i + (1-s)\,V_j) \cdot \vec{n}_{ij},$$

where $\vec{n}_{ij}$ is the normal to γ_{ij}. With the definition $\sum_i \sum_{j_i} = \sum_i \sum_{j, X_j conn.to X_i}$ the metric tensor can be easily computed as

$$\begin{aligned} g_X(V,V) &= \frac{1}{2}\sum_i \sum_{j_i} |\gamma_{ij}| \int_0^1 |v_{ij}|^2 \, ds \\ &= \frac{1}{2}\sum_i \sum_{j_i} \frac{1}{3} |\gamma_{ij}| \left[(V_i \cdot \vec{n}_{ij})^2 + (V_i \cdot \vec{n}_{ij})(V_j \cdot \vec{n}_{ij}) + (V_j \cdot \vec{n}_{ij})^2 \right] \\ &\stackrel{\text{def}}{=} V \cdot AV \end{aligned}$$

where the elements of the matrix A are given by

$$a_{ij} = \begin{cases} \frac{1}{3} \sum\limits_{m_i} |\gamma_{im}| \, \vec{n}_{im} \otimes \vec{n}_{im} & : \quad i = j \\ \frac{1}{6} |\gamma_{ij}| \, \vec{n}_{ij} \otimes \vec{n}_{ij} & : \quad X_i \text{ connected to } X_j \\ 0 & : \quad \text{otherwise} \end{cases}$$

Now we turn to the energy, which is given by

$$E(X) = \frac{1}{2} \sum_i \sum_{j_i} \sigma_{ij} |\gamma_{ij}| ,$$

where σ_{ij} denotes the surface tension of the edge γ_{ij} and the gradient is just

$$(\nabla E(X))_i = \sum_{j_i} \sigma_{ij} \frac{X_i - X_j}{|\gamma_{ij}|} .$$

Thus, in each time t, we can determine the direction of steepest descent $\dot{X}(t)$. The restricted gradient flow is thus a solution of the linear system of equations

$$\begin{aligned} X(0) &= X_0 \\ A(X(t)) \dot{X}(t) &= -\nabla E(X(t)) \quad \forall t \end{aligned} \tag{2}$$

Notice that in this reduced setting there is no angle condition at the vertices anymore.

We now describe how we deal with topological rearrangements in the corresponding time discretized evolution. If an edge γ_{ij} becomes smaller than a parameter ε during the evolution we proceed as follows. If γ_{ij} belongs to a triangle we remove the triangle and replace it by a vertex with position in the center of mass of the triangle (cf. Fig. 1).

Otherwise we replace the small edge by a fourfold vertex X_0 and introduce a new infinitesimally small edge $(X_0^+, X_0^-) = (X_0, X_0)$ with new neighbor relations as indicated in Fig. 2.

We determine the velocities $\dot{X}_0^+, \dot{X}_0^-$, for which we assume $\dot{X}_0^+ = -\dot{X}_0^-$, as the direction of steepest descent in the local configuration with $X_1, \ldots, X_4$ fixed. Then the energy is given by

$$\begin{aligned} E(X_0^+, X_0^-) = \sigma_{10}|X_1 - X_0^+| + \sigma_{20}|X_2 - X_0^-| + \sigma_{30}|X_3 - X_0^-| \\ + \sigma_{40}|X_4 - X_0^+| + \sigma_{00}|X_0^+ - X_0^-| . \end{aligned}$$

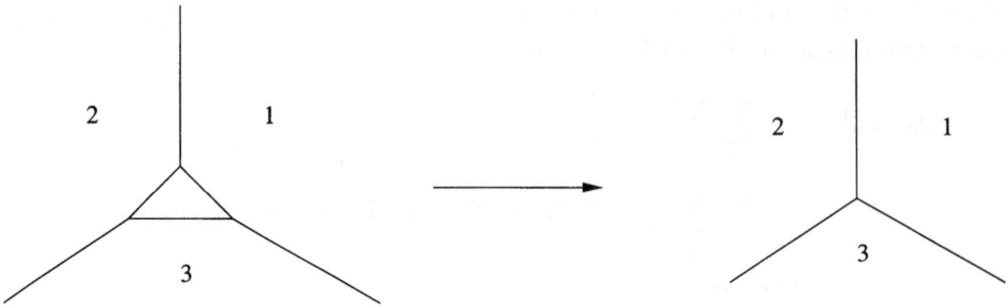

FIGURE 1. Annihilation of triangles

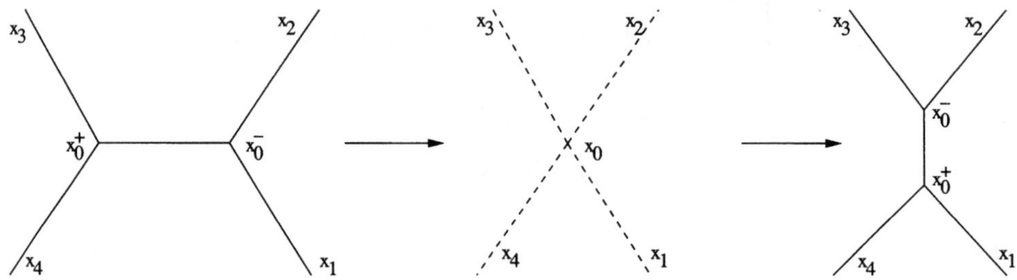

FIGURE 2. Change of neighbor relations

For general velocities in $\dot{X}_0^+, \dot{X}_0^-$, denoted by V^+, V^- with $V^+ = -V^-$, we find

$$\begin{aligned}\nabla E(X_0^+, X_0^-)(V^+, V^-) &= -\sigma_{10}t_1 \cdot V^+ - \sigma_{20}t_2 \cdot V^- - \sigma_{30}t_3 \cdot V^- \\ &\quad - \sigma_{40}t_4 \cdot V^+ + \sigma_{00}|V^+ - V^-| \\ &= (\sigma_{10}t_1 - \sigma_{20}t_2 - \sigma_{30}t_3 + \sigma_{40}t_4) \cdot (-V^+) + 2\sigma_{00}|V^+|\end{aligned}$$

where $t_i, i = 1, \ldots, 4$ denote the tangents of the edges (X_0, X_i) pointing towards X_i. The metric tensor is given by

$$g_{X_0}(V, V) = \frac{1}{3}\Big((V^+ \cdot \vec{n}_1)^2 + (V^+ \cdot \vec{n}_2)^2 + (V^+ \cdot \vec{n}_3)^2 + (V^+ \cdot \vec{n}_4)^2\Big).$$

Then, $\dot{X}_0^+ = -\dot{X}_0^-$ is determined as a solution V of

$$\begin{aligned}&\frac{1}{3}(\vec{n}_1 \otimes \vec{n}_1 + \vec{n}_1 \otimes \vec{n}_1 + \vec{n}_1 \otimes \vec{n}_1 + \vec{n}_1 \otimes \vec{n}_1) \cdot V \\ &\qquad + (\sigma_{10}t_1 - \sigma_{20}t_2 - \sigma_{30}t_3 + \sigma_{40}t_4) \cdot V + 2\sigma_{00}\frac{V}{|V|} = 0.\end{aligned} \tag{3}$$

It is easy to see that for (3) to be solvable it is necessary that

$$|\sigma_{10}t_1 - \sigma_{20}t_2 - \sigma_{30}t_3 + \sigma_{40}t_4| > 2|\sigma_{00}|.$$

If this condition is not satisfied, we simply set

$$\dot{X}_0^+ = -(\sigma_{10}t_1 - \sigma_{20}t_2 - \sigma_{30}t_3 + \sigma_{40}t_4),$$

i.e. $\dot{X}_0^+$ is the negative force. We also compute the analogous solution for the configuration without neighbor switching. We then obtain the configuration which minimizes the energy fastest. In most cases neighbor switching is favorable, as it can be expected, since otherwise the corresponding edge would not have become small in the first place. Finally, we update $X_0^+ = X_0^+ + \delta \dot{X}_0^+ / |\dot{X}_0^+|$ etc. for small δ such that the new edge is larger than ε.

We have already pointed out that one advantage of this model is that it allows — contrary to the mean field models presented in [FU94] — to include anisotropic surface energies and it can in principle be extended to three dimensional grain growth. The applications are by no means restricted to the cases considered here. It can also be easily extended to allow for surface energies which depend on the normal of the grain boundaries or for anisotropic boundary mobilities. These cases will be considered in forthcoming work.

3. Simulations

In the numerical algorithm we start with a periodic Voronoi polygon with N_0 vertices. We check whether we can find an edge of length smaller than $\varepsilon = 10^{-4}\overline{\Gamma}$, where $\overline{\Gamma}$ denotes the mean edge length. If we find a small edge, we proceed as described in Section 2.2, solving (3) by a Newton method. Then we solve (2) by a conjugate gradient algorithm and update the system according to the solution $\dot{X}$ of (2). The time step size Δt is chosen in each time step as $\Delta t = 0.1\overline{\Gamma} \min_k |\gamma_k|$.

3.1. Isotropic case

We first present results for $N_0 = 20000$ for the case of an isotropic surface energy, i.e. for $\sigma_{ij} = 1$. Fig. 3 shows the network at several times during the evolution.

We are mainly interested in the question whether the system evolves in the expected self-similar manner. To that purpose, we investigate the evolution of the moments of the edge length distribution which should grow proportional to $t^{1/2}$. In Fig. 4(a) we plot

$$\mu_i(|\gamma|) := \left(\frac{1}{M}\sum_{k=1}^{M} |\gamma_k|^i\right)^{1/i}, \qquad \text{for } i = \tfrac{1}{3}, \tfrac{1}{2}, 1, 2, 3$$

in a log-log plot. We see clearly that after a certain transition time, all the moments show the expected scaling behavior before finite size effects dominate in the end. We plot in Fig. 4(b) the energy which decreases proportional to $t^{-1/2}$ as expected

.

In Fig. 5 we plot the edge length and topological distribution respectively, where we averaged over the self-similar time window $\ln t = 10^{-3}$ to $\ln t = 5 \cdot 10^{-2}$.

3.2. Anisotropic versus isotropic case

We now want to investigate the influence of anisotropies. Therefore, we initially attach to each grain a random orientation $\phi \in [0, \pi)$, i.e. we assume that the lattice

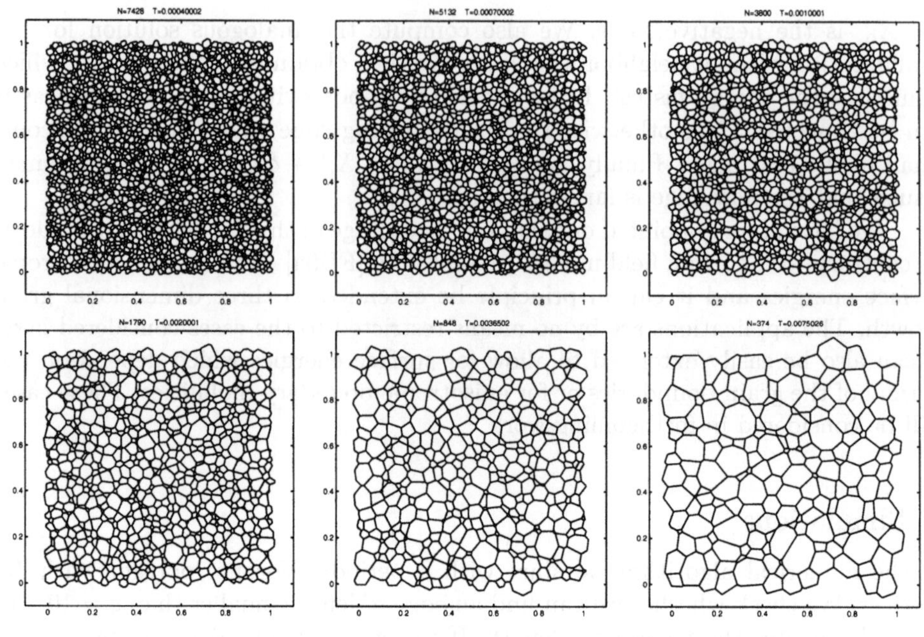

FIGURE 3. Evolution of network in the isotropic case

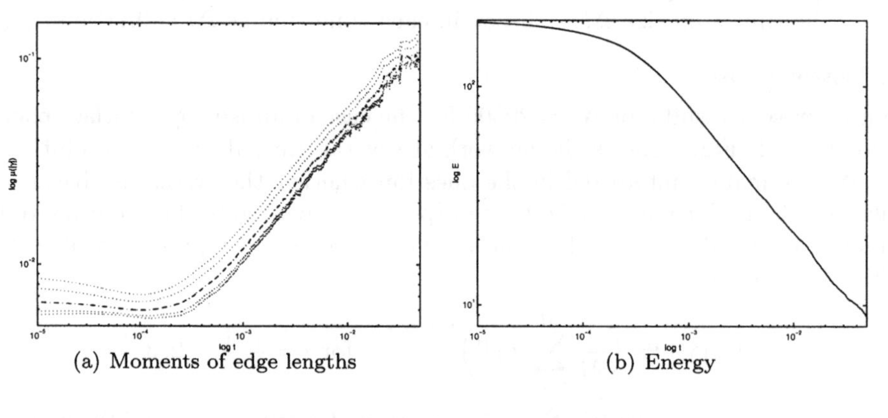

(a) Moments of edge lengths (b) Energy

FIGURE 4. Isotropic case, $N_0 = 20000$

structure is symmetric with respect to rotations of angle π. (We obtain similar results for different symmetries.) Then each edge is assigned a misorientation angle

$$\alpha_{ij} = |\phi_{k_i} - \phi_{k_j}|,$$

where k_i, k_j denote the grains which are separated by γ_{ij}. With this notation we choose

$$\sigma_{ij} = \theta + (1-\theta)\frac{6}{\pi^2}\alpha_{ij}(\pi - \alpha_{ij}),$$

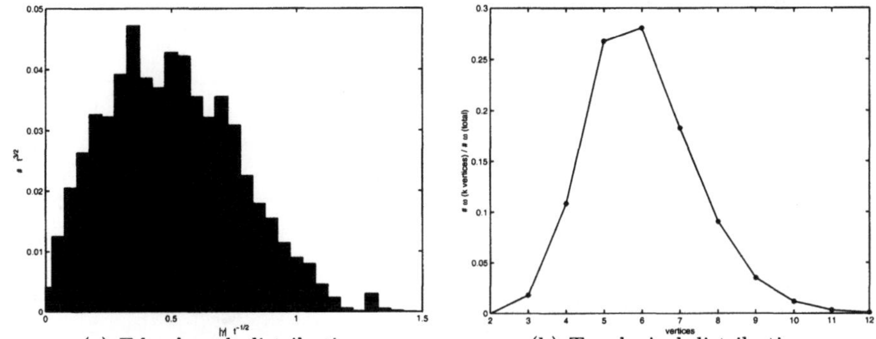

(a) Edge length distribution (b) Topological distribution

FIGURE 5. Isotropic case, $N_0 = 20000$

where $\theta \in [0,1]$ measures the strength of the anisotropy. The factor $\frac{6}{\pi^2}$ is normalization such that $\int_0^\pi \sigma(\alpha)\, d\alpha = \pi$.

In the sequel, we compare the results for initial data with $N_0 = 2000$ in the isotropic case $\theta = 1$ and in the anisotropic case with $\theta = 2/5$.

Fig. 6 shows the evolution of the moments of the edge length distribution. We see that also in the anisotropic case the system reaches a self-similar regime, the transition period being however slightly longer.

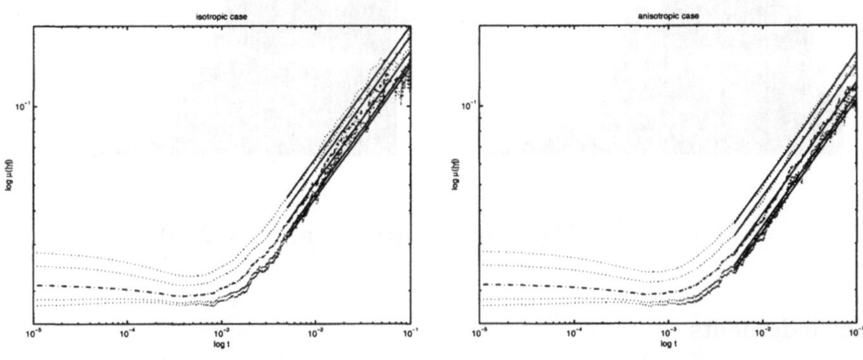

FIGURE 6. Moments of edge lengths, $N_0 = 2000$

That the evolution is slower is also seen in Fig. 7, which shows how the number of edges evolves, in the second picture in a log-log plot.

In Figures 8 and 9 we plot the edge length and topological distributions respectively. Even though the amount of data currently at hand is probably too small to draw any firm conclusion, we observe that the edge length distribution is narrower in the anisotropic case and has the peak at smaller edge lengths. The topological distribution seems to have a larger tail in the anisotropic case and is

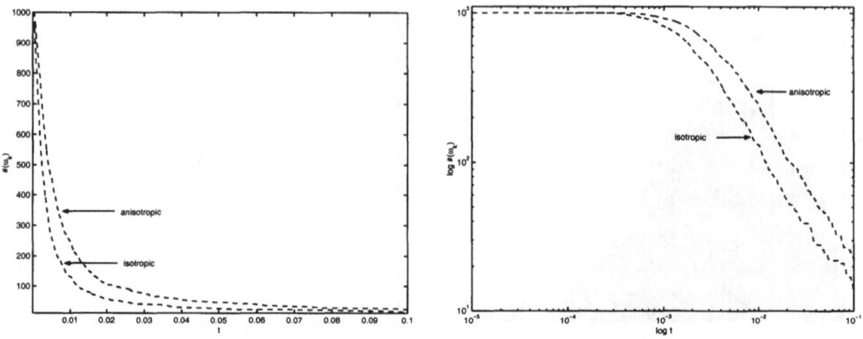

FIGURE 7. Edge numbers, $N_0 = 2000$

more symmetric around the peak. All these tendencies have also been observed in [NFK94] for a slightly different type of anisotropy, as well as in simulations of the full mean curvature flow [Kin].

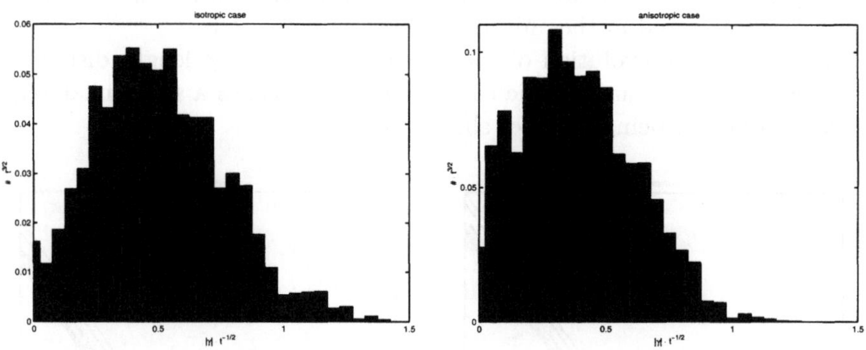

FIGURE 8. Edge length distribution, $N_0 = 2000$

Acknowledgments

The authors gratefully acknowledge support by the DFG through SFB611 at the University of Bonn.

References

[BR93] L. Bronsard and F. Reitich, *On three-phase boundary motion and the singular limit of a vector-valued Ginzburg–Landau equation*, Arch. Rat. Mech. Anal. **124, 4** (1993), 355–379.

[Fra88] V. E. Fradkov, *A theoretical investigation of two-dimensional grain growth in the 'gas' approximation*, Phil. Mag. Letters **58, 6** (1988), 271–275.

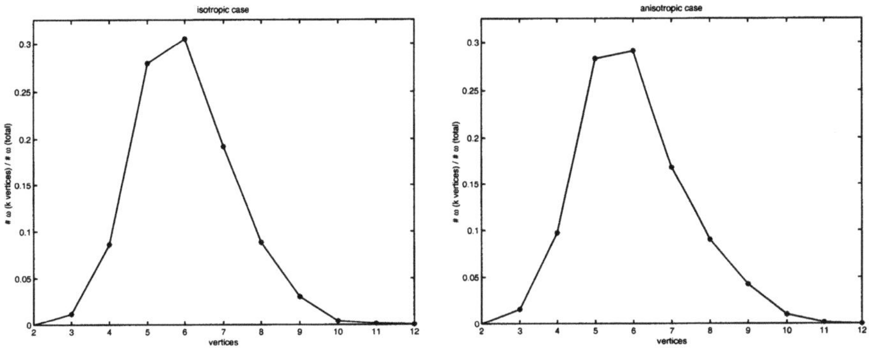

FIGURE 9. Topological distribution, $N_0 = 2000$

[FU94] V. E. Fradkov and D. Udler, *2d normal grain growth: topological aspects*, Advances in Physics **43** (1994), 739–789.

[FUK88] V. E. Fradkov, D. Udler, and R. E. Kris, *Computer simulation of two-dimensional normal grain growth (the 'gas' approximation)*, Phil. Mag. Letters **58, 6** (1988), 277–283.

[Kin] D. Kinderlehrer, private communication.

[KL01] D. Kinderlehrer and C. Liu, *Evolution of grain boundaries*, Math. Mod. Meth. Appl. Sc. **4, 11** (2001), 713–729.

[KNN89] K. Kawasaki, T. Nagai, and K. Nakashima, *Vertex models for two-dimensional grain growth*, Philosophical Magazine B **60** (1989), 399–421.

[NFK94] T. Nagai, K. Fuchizaki, and K. Kawasaki, *Orientation effect on grain growth*, Physica A **204** (1994), 450–463.

[TC94] J. E. Taylor and J. W. Cahn, *Linking anisotropic sharp and diffuse surface motion laws via gradient flows*, J. Stat. Phys. **77** (1994), 183–197.

Institut für Angewandte Mathematik
Wegelerstr. 6 and 10
53115 Bonn, Germany
E-mail address: `henseler@iam.uni-bonn.de`, `barbara@iam.uni-bonn.de`, `otto@iam.uni-bonn.de`

International Series of Numerical Mathematics, Vol. 147, 189–196

On a Reaction-Diffusion System for a Population of Hunters and Farmers

Danielle Hilhorst, Masayasu Mimura, and Rémi Weidenfeld

1. Introduction

Our starting point is an article by Aoki, Shida and Shigesada [1], who present travelling wave solutions of a reaction-diffusion system modelling the spread of farmers in a region occupied by hunter-gatherers. More precisely, they consider three different populations, namely the original farmer population, the converted farmers and the hunter-gatherers. We denote their densities at position x and at time t by $F(x,t)$, $C(x,t)$ and $H(x,t)$ respectively. The dynamics of the populations is governed by the following laws:

1. All the individuals move randomly with constant diffusion coefficient $D > 0$, which is assumed to be the same for all the three groups;
2. Hunter-gatherers who meet farmers become themselves farmers at the rate $e > 0$;
3. The growth of the three populations is logistic, that is to say that it is expressed by terms of the form $F(1 - \lambda F)$.

Aoki, Shida and Shigesada [1] derive the following reaction-diffusion system for the population densities

$$F_t = D\Delta F + r_F F\Big[1 - (F+C)/K\Big] \tag{1.1}$$

$$C_t = D\Delta C + r_C C\Big[1 - (F+C)/K\Big] + e(F+C)H \tag{1.2}$$

$$H_t = D\Delta H + r_H H\Big[1 - H/L\Big] - e(F+C)H, \tag{1.3}$$

where $r_F, r_C, r_H > 0$ are respectively the intrinsic growth rates of the original farmers, the converted farmers and the hunter-gatherers, and where $K > 0$ and $L > 0$ are the carrying capacities of the farmers and the hunter-gatherers respectively. They discuss the spreading of farmers into a region occupied by hunter-gatherers by using travelling wave solutions, which they compute numerically.

A special case of the system (1.1)–(1.3) is the case in which $C = H = 0$. Then F satisfies the classical well known Fisher equation :

$$F_t = D\Delta F + r_F F\Big[1 - F/K\Big], \tag{1.4}$$

which has been thoroughly studied by many authors (see [2], for instance).

A second special case is given by $H = 0$ and $C \neq 0$. The equations satisfied by F and C have the form

$$F_t = D\Delta F + r_F F\Big[1 - F/K\Big] - kFC \tag{1.5}$$

$$C_t = D\Delta C + r_C C\Big[1 - C/K\Big] - \alpha kFC, \tag{1.6}$$

where $\alpha > 0$ is a constant (see [3], for instance). The system (1.5)–(1.6), with in general different diffusion coefficients, is called a competition-diffusion system and the qualitative behavior of solutions has been studied by several authors (see for instance [4] and [5]).

A third special case is the case that $H \neq 0$ and $r_F = r_C$. Then, if we denote the density of the total population of farmers $F + C$ by T, we deduce from (1.1)–(1.3) the system

$$T_t = D\Delta T + r_T T\Big[1 - T/K\Big] + eTH \tag{1.7}$$

$$H_t = D\Delta H + r_H H\Big[1 - H/L\Big] - eTH, \tag{1.8}$$

where $r_T = r_F = r_C$. The system (1.7)–(1.8) seems to be similar to (1.5)–(1.6) except for the sign of the coupling term. This difference is very important from the ecological viewpoint. The system (1.7)–(1.8) is called a predator-prey system with logistic growth, i.e. each population can grow in the absence of the other.

Among the simplified systems, we consider (1.7)–(1.8). The problem which we study is

$$(P) \quad \begin{cases} u_t = d_u \Delta u + \gamma_u u(1 - u + a_u v) & \text{in } \Omega \times (0,\infty), \\ v_t = d_v \Delta v + \gamma_v v(1 - v - a_v u) & \text{in } \Omega \times (0,\infty), \\ \dfrac{\partial}{\partial \nu} u = \dfrac{\partial}{\partial \nu} v = 0 & \text{on } \partial\Omega \times (0,\infty), \\ u(x,0) = u_0(x), \quad v(x,0) = v_0(x) & \text{for all } x \in \Omega; \end{cases}$$

with the hypotheses

- H_d : $d_u, d_v > 0$,
- H_a : $\gamma_u, \gamma_v, a_u, a_v > 0$,
- H_Ω : Ω is a smooth bounded domain of $\mathbb{R}^N$,
- H_0 : The functions $u_0, v_0 \in C(\bar{\Omega})$ are non identically equal to zero and nonnegative. We suppose that $0 \leq u_0, v_0 \leq M$.

We prove that as $t \to \infty$, the solution (u, v) of Problem (P) converges to some constant pair (u^∞, v^∞), which depends on the parameters. More precisely, in the case that $1 - a_v \leq 0$, we have that $(u^\infty, v^\infty) = (1, 0)$, while if $1 - a_v > 0$, then $(u^\infty, v^\infty) = (u^*, v^*)$ where u^* and v^* are determined by the equalities

$$1 - u^* + a_u v^* = 1 - v^* - a_v u^* = 0.$$

This indicates that the large time behavior of solutions of (P) is quite simple, because it is described by the corresponding ODE system in absence of diffusion.

An important problem concerns the transient behavior of solutions. Unfortunately, in most cases, we have to resort to numerical methods. In order to perform an analytical study, we consider the case in which the conversion coefficients a_u and a_v between u and v are very large. In this case, the condition $1 - a_v \leq 0$ holds. We prove that the solution pair first becomes close to a pair $(u^0, 0)$ where u^0 is the solution of a parabolic problem, before converging to $(1,0)$ as $t \to \infty$.

In none of the cases, does a free boundary problem occur. However, if we relaxed some hypotheses on the coefficients or considered a slightly different scaling, we would obtain cases where the limit problem is indeed represented as a free boundary problem. This will be stressed in the concluding remarks.

2. The main results

We study the existence, uniqueness and large time behavior of solutions of Problem (P). We prove the following result :

Theorem 2.1. *Problem (P) has a unique classical solution pair (u, v), which satisfies*

$$0 < u(x,t) \leq M_1, \quad 0 < v(x,t) \leq M_2 \quad \textit{for all } (x,t) \in \Omega \times (0,\infty), \tag{2.1}$$

where $M_1 := \max(1 + a_u M_2, M)$ and $M_2 := \max(1, M)$. Moreover the solution pair $(u,v)(t)$ converges to

$$\begin{cases} (1,0) & \textit{if } 1 - a_v \leq 0, \\ (u^*, v^*) & \textit{if } 1 - a_v > 0, \end{cases}$$

in $\left[C(\bar{\Omega})\right]^2$ as $t \to \infty$, where (u^*, v^*) is uniquely determined by the equalities

$$1 - u^* + a_u v^* = 1 - v^* - a_v u^* = 0.$$

The next result concerns the singular limit when a_u and a_v tend to infinity. More precisely, we investigate the limit of the solutions $\{(u^\epsilon, v^\epsilon)\}$ of the sequence of problems (P^ϵ)

$$(P^\epsilon) \quad \begin{cases} u_t = d_u \Delta u + \gamma_u u(1-u) + \dfrac{1}{\epsilon} uv & \text{in } \Omega \times (0,\infty), \\ v_t = d_v \Delta v + \gamma_v v(1-v) - \lambda \dfrac{1}{\epsilon} uv & \text{in } \Omega \times (0,\infty), \\ \dfrac{\partial}{\partial \nu} u = \dfrac{\partial}{\partial \nu} v = 0 & \text{on } \partial\Omega \times (0,\infty), \\ u(x,0) = u_0(x), \quad v(x,0) = v_0(x) & \text{for all } x \in \Omega, \end{cases}$$

with $\lambda > 0$. Fig. 1 shows the time evolution of (u^ϵ, v^ϵ) when ϵ is very small. It appears that v^ϵ vanishes due to the strong competition between u^ϵ and v^ϵ. Then u^ϵ expands and occupies the whole domain. The dynamics of (u^ϵ, v^ϵ) is described by the following theorem :

Theorem 2.2. *Let u^0 be the solution of the problem*

$$(P^0)\quad \begin{cases} u^0_t = d_u \Delta u^0 + \gamma_u u^0(1-u^0) & \textit{in } \Omega \times (0,T), \\ \dfrac{\partial}{\partial \nu} u^0 = 0 & \textit{in } \partial\Omega \times (0,T), \\ u^0(x,0) = u_0(x) + \dfrac{v_0(x)}{\lambda} & \textit{for all } x \in \Omega. \end{cases}$$

Then for all $T > 0$

$$(u^\epsilon, v^\epsilon) \to (u^0, 0) \quad \textit{in } \left[L^2(\Omega \times (0,T))\right]^2 \quad \textit{as } \epsilon \to 0. \tag{2.2}$$

3. Main idea of the proofs

In the proof of the stabilization for large time (Theorem 2.1), a major role is played by the nonnegative solutions of the system

$$\begin{cases} \gamma_u r(1 - r + a_u s) = 0, \\ \gamma_v s(1 - s - a_v r) = 0, \end{cases}$$

which are given by $(0,0)$, $(1,0)$, $(0,1)$ and (u^*, v^*), where

$$u^* = \frac{1 + a_u}{1 + a_u a_v} \quad \text{and} \quad v^* = \frac{1 - a_v}{1 + a_u a_v}.$$

Note that if $v^* > 0$, that is $1 > a_v$, those four zeros are of interest to us since they lie in the set $\{r \geq 0, s \geq 0\}$ whereas the only zeros to be considered are $(0,0)$, $(1,0)$ and $(0,1)$ in the case in which $1 \leq a_v$. Note that in the first case (u^*, v^*) is the only stable zero of the corresponding ODE system whereas in the second case the only stable zero is given by $(1,0)$.

In the sequel we present the main ideas of the proofs of the results which we obtain. We refer the reader to [6] for complete proofs. We prove Theorem 2.1 by showing the lemmas 3.1 and 3.2 below.

Lemma 3.1. *Suppose that $1 - a_v \leq 0$. If (u,v) is the solution pair of Problem (P) then*

$$(u,v)(t) \to (1,0) \quad \textit{in } \left[C(\bar{\Omega})\right]^2 \quad \textit{as } t \to \infty.$$

Sketch of the proof. The idea is to construct suitable upper and lower solutions for u and for v. Let $\underline{u}$ be the solution of the problem

$$(F)\quad \begin{cases} \underline{u}_t = d_u \Delta \underline{u} + \gamma_u \underline{u}(1 - \underline{u}) & \text{in } \Omega \times (0,\infty), \\ \dfrac{\partial}{\partial \nu} \underline{u} = 0 & \text{on } \partial\Omega \times (0,\infty), \\ \underline{u}(x,0) = u_0(x) & \text{for all } x \in \Omega. \end{cases}$$

It follows from the comparison principle that

$$u \geq \underline{u} \quad \text{in } \Omega \times (0,\infty). \tag{3.1}$$

Since the function $\underline{u}(t)$ tends to the stable steady state 1 as $t \to \infty$, we deduce that for all $\mu > 0$ there exists $T = T(\mu) > 0$ such that

$$u(x,t) \geq 1 - \mu \quad \text{for all } (x,t) \in \Omega \times (T, \infty). \tag{3.2}$$

The inequality $1 - a_v \leq 0$ together with the comparison principle imply that $v \leq \bar{v}$ in $\Omega \times (T, \infty)$, where $\bar{v}$ is the unique solution of the problem

$$(F') \quad \begin{cases} \bar{v}_t = d_v \Delta \bar{v} + \gamma_v \bar{v}(a_v \mu - \bar{v}) & \text{in } \Omega \times (T, \infty), \\ \dfrac{\partial}{\partial \nu} \bar{v} = 0 & \text{on } \partial\Omega \times (T, \infty), \\ \bar{v}(x,T) = v(x,T) & \text{for all } x \in \Omega. \end{cases}$$

Since $\bar{v}(t) \to a_v \mu$ as $t \to \infty$ with μ arbitrary, we deduce that $v(t) \to 0$ as $t \to \infty$. In turn the comparison principle ensures that $u \leq \bar{u}$ in $\Omega \times (T, \infty)$ for sufficiently large T, where $\bar{u}$ satisfies

$$(F'') \quad \begin{cases} \bar{u}_t = d_u \Delta \bar{u} + \gamma_u \bar{u}(1 + \mu - \bar{u}) & \text{in } \Omega \times (T, \infty), \\ \dfrac{\partial}{\partial \nu} \bar{u} = 0 & \text{on } \partial\Omega \times (T, \infty), \\ \bar{u}(x,T) = u(x,T) & \text{for all } x \in \Omega. \end{cases}$$

Therefore $u(t) \leq 1 + \mu$ in Ω for t large enough, which, together with (3.2), implies that $u(t) \to 1$ as $t \to \infty$. □

Lemma 3.2. *Suppose that* $1 - a_v > 0$. *If* (u, v) *is the solution pair of Problem* (P), *then*

$$(u,v)(t) \to (u^*, v^*) \quad \text{in } \left[C(\bar{\Omega})\right]^2 \quad \text{as } t \to \infty.$$

Sketch of the proof. The starting point is the equality for all $0 < \tau < T$

$$\begin{aligned} \mathcal{F}(T) = \mathcal{F}(\tau) &+ \int_\tau^T \int_\Omega \left(d_u \gamma_v a_v \frac{u^*}{u^2} |\nabla u|^2 + d_v \gamma_u a_u \frac{v^*}{v^2} |\nabla v|^2 \right) \\ &+ \gamma_u \gamma_v \int_\tau^T \int_\Omega \left(a_v (u^* - u)^2 + a_u (v^* - v)^2 \right), \end{aligned} \tag{3.3}$$

where $\mathcal{F}$ is given by

$$\mathcal{F}(t) := \int_\Omega \Big(\gamma_v a_v (u^* \log u - u) + \gamma_u a_u (v^* \log v - v) \Big)(t),$$

so that $-\mathcal{F}$ is a Lyapunov functional for Problem (P). We set $\tau = 1$ and let $T \to \infty$ to deduce, also using the boundedness of u and v, that

$$\int_1^\infty \int_\Omega \left(\gamma_u \gamma_v a_v (u^* - u)^2 + \gamma_v \gamma_u a_u (v^* - v)^2 \right) \leq c_1 + |\mathcal{F}(1)|,$$

for a certain positive constant c_1, which in turn implies that

$$(u,v)(t) \to (u^*, v^*) \quad \text{in } \left[L^2(\Omega)\right]^2 \quad \text{as } t \to \infty.$$

□

Finally, we consider the case in which $1/\epsilon$ multiplies the interaction terms and prove Theorem 2.2.

Sketch of the proof of Theorem 2.2. We first remark that the upper bound for v^ϵ, namely M_2, does not depend on ϵ, but the upper bound for u^ϵ does. Multiplying the equation for v^ϵ by v^ϵ and integrating over $\Omega \times (\tau, T)$ give

$$\begin{aligned}\frac{1}{2}\int_\Omega (v^\epsilon)^2(T) + d_v \int_\tau^T \int_\Omega |\nabla v^\epsilon|^2 + \frac{\lambda}{\epsilon}\int_\tau^T \int_\Omega u^\epsilon (v^\epsilon)^2 \\ = \frac{1}{2}\int_\Omega (v^\epsilon)^2(t) + \int_\tau^T \int_\Omega \gamma_v (v^\epsilon)^2 (1 - v^\epsilon),\end{aligned} \tag{3.4}$$

which in turn implies that

$$\frac{\lambda}{\epsilon}\int_\tau^T \int_\Omega u^\epsilon (v^\epsilon)^2 \leq \frac{1}{2}|\Omega|(M_2)^2 + (T - \tau)|\Omega|\gamma_v (M_2)^2. \tag{3.5}$$

We deduce from the fact that $u^\epsilon \geq \underline{u}$, where $\underline{u}$ is the solution of Problem (F), that for all positive times $\tau > 0$ there exists a constant $\mu = \mu(\tau) > 0$ such that

$$u^\epsilon(x,t) \geq \mu \quad \text{for all } (x,t) \in \Omega \times (\tau, \infty) \text{ and } \epsilon > 0.$$

Therefore, the inequality (3.5) implies that

$$\int_\tau^T \int_\Omega (v^\epsilon)^2 \leq \frac{c_2 \epsilon}{\mu \lambda},$$

for some positive constant c_2. Letting $\epsilon \to 0$ and $\tau \to 0$, we deduce that

$$v^\epsilon \to 0 \quad \text{in } L^2(\Omega \times (0,T) \text{ as } \epsilon \to 0.$$

It follows from a result due to Simon [7] that the sequence $\{u^\epsilon\}$ is precompact in $L^2(\Omega \times (0,T))$. Therefore, there exists a subsequence of $\{(u^\epsilon, v^\epsilon)\}$, which we denote again by $\{(u^\epsilon, v^\epsilon)\}$, and a function $u \in L^2(\Omega \times (0,T))$ such that $(u^\epsilon, v^\epsilon) \to (u, 0)$ in $\left[L^2(\Omega \times (0,T))\right]^2$ as $\epsilon \to 0$. Let $\varphi \in C^{2,1}(\bar{\Omega} \times [0,T])$ be an arbitrary function such that $\dfrac{\partial \varphi}{\partial \nu} = 0$ on $\partial\Omega \times [0,T]$ and $\varphi(T) = 0$. Then

$$\begin{aligned}\int_0^T \int_\Omega \left[(u^\epsilon + \frac{1}{\lambda}v^\epsilon)\varphi_t + (d_u u^\epsilon + \frac{d_v}{\lambda}v^\epsilon)\Delta\varphi\right] + \int_\Omega (u_0 + \frac{1}{\lambda}v_0)\varphi(0) \\ + \int_0^T \int_\Omega \Big(\gamma_u u^\epsilon (1 - u^\epsilon) + \frac{\gamma_v}{\lambda} v^\epsilon (1 - v^\epsilon)\Big)\varphi = 0.\end{aligned} \tag{3.6}$$

Letting $\epsilon \to 0$ in equation (3.6) yields

$$\int_0^T \int_\Omega \left[u\varphi_t + d_u u \Delta\varphi\right] + \int_\Omega (u_0 + \frac{1}{\lambda}v_0)\varphi(0) + \int_0^T \int_\Omega \gamma_u u(1-u)\varphi = 0,$$

which implies that u is the unique solution of Problem (P^0) and concludes the proof of Theorem 2.2. □

4. Concluding remarks

As we stated in the introduction, for our scaling limits the limiting problem as $\epsilon \to 0$ of Problem (P^ϵ) is not a free boundary problem. However, when d_u and d_v are small and γ_u and γ_v are large, the large time behavior is the same as in the previous case, that is $(u, v)(t) \to (1, 0)$ as $t \to \infty$, but the transient behavior is totally different. There appears a free boundary between u and v. Fig. 2 shows one example. Coming back to the original three component system (1.1)–(1.3) with $r_F \neq r_C$, a formal computation of the limiting problem as $e \to \infty$ in (1.2) and (1.3) gives a free boundary problem. However, these cases are still out of reach from the mathematical viewpoint and will be the purpose of future research.

References

[1] K. Aoki, M. Shida and N. Shigesada, *Travelling wave solutions for the spread of farmers into a region occupied by hunter-gatherers.* Theor. Pop. Biol. **50** (1996), 1–17.

[2] P.C. Fife, *Asymptotic states for equations of reaction and diffusion.* Bull. Amer. Math. Soc. **84** (1978), 693–726.

[3] F. Rothe, *Convergence to the equilibrium state in the Volterra–Lotka diffusion equations.* J. Math. Biol. **3** (1974), 319–324.

[4] N. Dancer, D. Hilhorst, M. Mimura and L.A. Peletier, *Spatial segregation limit of a competition-diffusion system.* Europ. J. of Applied Mathematics **10** (1999), 97–115.

[5] S.-I. Ei and E. Yanagida, *Dynamics of interfaces in competition-diffusion systems.* SIAM J.Appl. Math. **54** (1994), 1355–1373.

[6] R. Weidenfeld, *Large time behavior and singular limits for a reaction-diffusion system arising in population dynamics.* in preparation.

[7] J. Simon, *Compact sets in the space $L^p(0,T;B)$.* Ann. Mat. Pura Appl. **146** (1987), 65–96.

(Danielle Hilhorst) Laboratoire de Mathématique (UMR 8628), Université de Paris-Sud,
91405 Orsay Cedex, France
E-mail address: `Danielle.Hilhorst@math.u-psud.fr`

(Masayasu Mimura) Department of Mathematical and Life Sciences, Graduate School of Science,
Hiroshima University, 1-3-1, Kagamiyama, Higashi-Hiroshima 739-8526, Japan
E-mail address: `mimura@math.sci.hiroshima-u.ac.jp`

(Rémi Weidenfeld) Laboratoire de Mathématique (UMR 8628), Université de Paris-Sud,
91405 Orsay Cedex, France
E-mail address: `Remi.Weidenfeld@math.u-psud.fr`

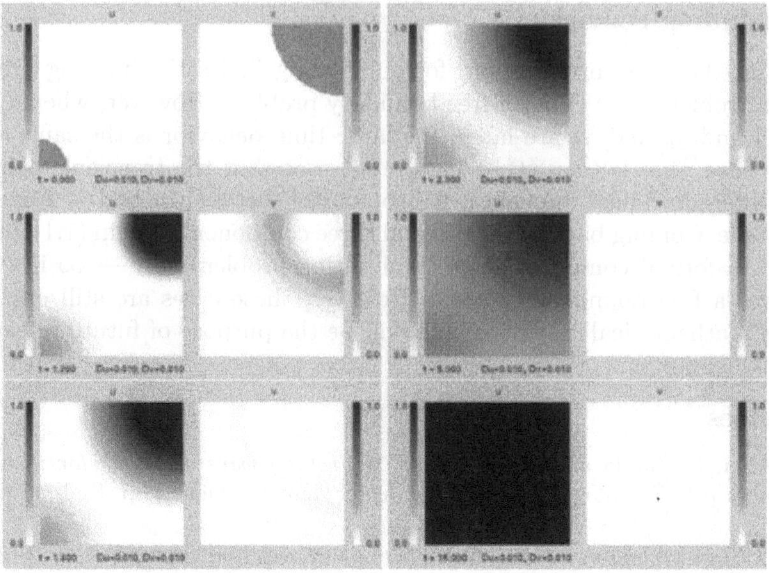

FIGURE 1. $d_u = d_v = 0.01$, $\gamma_u = \gamma_v = 0.5$, $a_u = 1000.0$, $a_v = 500.0$, $\Omega = [0,1] \times [0,1]$

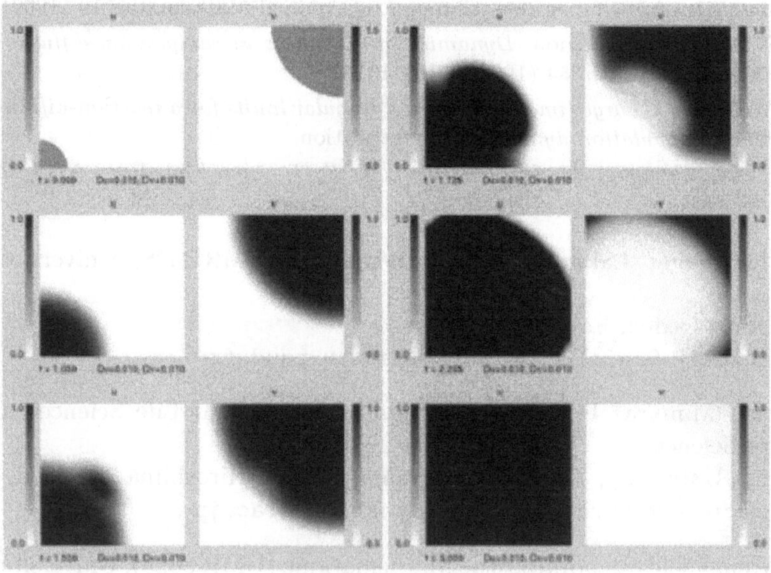

FIGURE 2. $d_u = d_v = 0.01$, $\gamma_u = \gamma_v = 10.0$, $a_u = 2.0$, $a_v = 1.0$, $\Omega = [0,1] \times [0,1]$

International Series of Numerical Mathematics, Vol. 147, 197–205

The Stochastic Geometry of the Crystallization Process of Polymers

Alessandra Micheletti and Vincenzo Capasso

Abstract. A Johnson–Mehl tessellation arises as a random division of a given bounded region in a d-dimensional Euclidean space, generated by a stochastic birth-and-growth process, also known as a germ-grain process in stochastic geometry. A typical example is the crystallization of a polymer from an amorphous liquid phase by cooling; in this case a grain (crystal) is formed by growth of a germ (nucleus) born at a random time and space location.

The kinetic parameters , i.e. the nucleation rate and the growth rate, may both depend upon space and time. By assuming that at contact points of the growth fronts grains stop growing (impingement), the spatial region will be divided into cells, and interfaces (n-facets) at different Hausdorff dimensions (cells,faces, edges, vertices) appear. Here we extend previous results regarding the evolution of the morphology of the resulting tessellation to the general case of space and time heterogeneous birth-and-growth processes. In particular evolution equations for the n-facet densities are provided and a technique to estimate these densities from digitized images is proposed.

1. Introduction

Johnson–Mehl tessellations are generated by birth-and-growth processes, which may model many real world phenomena, like crystallization processes, spread of fires in the woods, spread of a pollutant in the environment, etc. [8, 11, 14].

Such a process may be represented as a spatially marked point process $\phi = \{(T_i, X_i)\}_{i\in\mathbf{N}}$, where T_i represents the random time of birth of the i-th germ and X_i its random spatial location. Once born, each germ generates a grain growing at the surface (growth front) with a speed $G(x,t) > 0$ which may, in general, be assumed space-time dependent.

When two growing cells meet, they stop growing at the contact points, thus forming interfaces (n-facets) at various Hausdorff dimensions; this phenomenon is called *impingement* (see figures 1, 2).

In many applications the characterization of the final morphology of the tessellation is of interest, in particular the geometric measures of n-facets. We denote by Θ^t the closed set in $\mathbf{R}^d$ representing the region occupied by all grains by time t, and by $\Theta^t(X_0, T_0)$ a grain born at point X_0 and time T_0 and freely grown

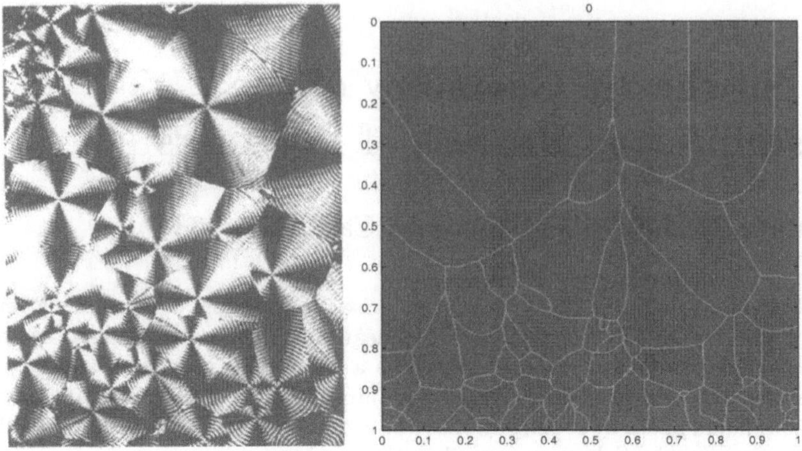

FIGURE 1. A photograph from a real experiment showing a Johnson–Mehl tessellation of polymeric crystals and a simulation of a spatially non homogeneous crystallization process

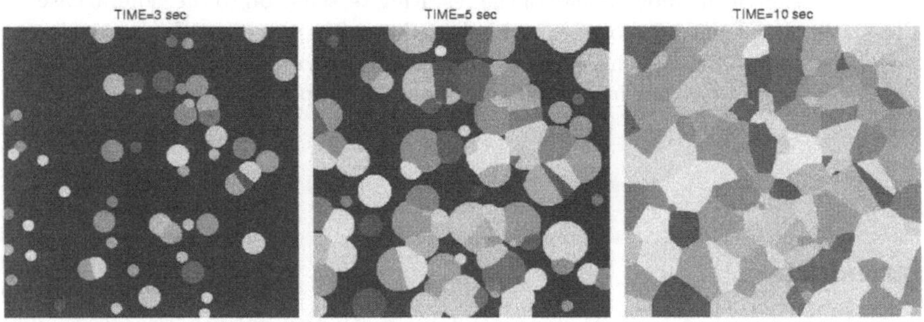

FIGURE 2. A simulation of a 2D birth-and-growth process leading to a Johnson–Mehl tessellation

up to time t. In a birth-and-growth process with nucleation events $\{(X_j, T_j)\ 0 \leq T_1 \leq T_2 \leq \ldots\}$, the occupied region at time t is then given by

$$\Theta^t = \bigcup_{T_j < t} \Theta^t(X_j, T_j). \tag{1}$$

The set Θ^t in (1) is also known as *germ-grain model* or *generalised Boolean model* (see [2] and references therein). We will assume that the birth-and-growth process takes place in a bounded region $E \subset \mathbf{R}^d$, $d = 1, 2, 3$, and that $G(x,t) > 0$ for all $x \in E, t \in \mathbf{R}_+$, so that the time of completion of the tessellation (i.e. the time at which all the available space is filled by the growing grains) is a.s. finite.

We also assume that the marked point process (MPP) ϕ that models the birth process of the germs has a stochastic intensity ν given by

$$\nu(dx \times dt) = \alpha(x,t)(1 - I_{\Theta^{t-}}(x))dt\ dx$$

where α, also known as the *free space nucleation rate* in some applications, is a real valued measurable function on $E \times \mathbf{R}_+$ such that $\alpha(t,\cdot) \in \mathcal{L}^1(E)$, for all $t > 0$ and such that $0 < \int_0^T dt \int_E \alpha(x,t)dx < \infty$ for any $0 < T < \infty$. The term $(1 - I_{\Theta^{t-}}(x))$ is responsible for the fact that no new nuclei can be born at time t in a zone already occupied by a growing cell. We will denote by

$$\nu_0(dx \times dt) = \alpha(x,t)dtdx$$

the *free space* intensity.

We will assume throughout the paper that the kinetic parameters of birth $(\alpha(x,t))$ and of growth $(G(x,t))$ are given deterministic fields in $E \times \mathbf{R}_+$. If both fields are spatially homogeneous, i.e. they are independent from the spatial coordinate x, then the birth-and-growth process is spatially *stationary*. Explicit expressions for the densities of n-facets in the stationary case have been found in [12] (see also [13]); particular cases have also been considered in [9]. A discussion of the general case, in which the crystallization process is coupled with temperature, so that in particular α and G may depend upon an underlying field (temperature, in this case), can be found in [3], and more extensively in [7], Chapter 5.

2. The Random Tessellation of Space

At any time t during the crystallization process, the available space (and in particular the crystalline phase) is randomly divided into cells due to impingement, forming a so called **(incomplete) Johnson–Mehl tessellation** [8, 11] (the tessellation is **incomplete** at time t if a region with amorphous material is still present). Here we will consider random tessellations in $\mathbb{R}^d$ generated by birth-and-growth processes.

2.1. Basic Definitions

Let us denote by $\tau_i(y)$ the time at which a point $y \in E$ is reached by a crystal freely grown (disregarding impingement) from a nucleus $a_i = (X_i, T_i)$, where X_i is the random space location and T_i the random time of birth. We admit $\tau_i(y) = \infty$, which corresponds to the case in which the point y is never covered by the i-th crystal.

Definition 2.1. *The* **crystal** $C_i(t)$ *of the (incomplete) tessellation, generated by the nucleus* $a_i = (X_i, T_i)$ *at the time of observation* t *is the non-empty set*

$$C_i(t) = \{y \in E |\ \tau_i(y) \leq t \ \text{ and } \ \tau_i(y) \leq \tau_j(y)\ \forall j \neq i\}.$$

We will denote by $C_e(t)$ the **amorphous region** at time t, i.e.

$$C_e(t) \colon = \{y \in E |\ \tau_i(y) > t,\ \forall i \in \mathbb{N}\}.$$

	$d=1$	$d=2$	$d=3$
$n=0$	**vertex**=intersection of 2 crystals	**vertex**=intersection of 3 crystals	**vertex**=intersection of 4 crystals
$n=1$	**crystal**	**edge**=intersection of 2 crystals	**edge**=intersection of 3 crystals
$n=2$		**crystal**	**face**=intersection of 2 crystals
$n=3$			**crystal**

TABLE 1. Specific meaning of "n-facet" in spaces of dimension d

Let us now introduce a rigorous concept of "interface":

Definition 2.2. *An* **n-facet** *at time t ($0 \leq n \leq d$) is the non-empty intersection between $m+1$ crystals or between m crystals and the amorphous region, i.e.*

$$F_n(t, k_0, \ldots, k_m) = C_{k_0}(t) \cap C_{k_1}(t) \cap \cdots \cap C_{k_m}(t),$$

with $m = d - n$ and $k_0, \ldots, k_m \in \mathbb{N} \cup \{e\}$.

Note that in the previous definition

- $d =$ dimension of the space in which the tessellation takes place;
- $n =$ dimension of the interface under consideration;
- $m+1 =$ number of crystals (including C_e) necessary for the formation of such an interface,

(see also Table 1).

Consider now the set of all n-facets at time t

$$\Xi_n(t) = \bigcup_{k_0,\ldots,k_m \in \mathbb{N}\cup\{e\}}^{\neq} F_n(t, k_0, \ldots, k_m),$$

for $n < d$.

For any Borel set B one can define the *n-facet mean content* of B at time t as the measure

$$\mathcal{M}_{d,n}(t, B) = \frac{1}{(m+1)!} E\left[\sum_{k_0,\ldots,k_m \in \mathbb{N}\cup\{e\}}^{\neq} \lambda_n(B \cap F_n(t, k_0, \ldots, k_m))\right], \quad \text{for } n < d$$

$$\mathcal{M}_{d,d}(t, B) = E\left[\sum_{k_0 \in \mathbb{N}} \lambda_d(B \cap F_d(t, k_0))\right] \tag{2}$$

where λ_n is the n-dimensional Hausdorff measure and the symbol $\neq$ denotes that the indices $k_0, \ldots, k_m$ must be distinct (and thus also the corresponding crystals). Note that, with the previous definition, $\mathcal{M}_{d,d}(t, B)$ is the d-dimensional volume of the portion of the set B occupied by crystals at time t.

If $\mathcal{M}_{d,n}(t,\cdot) \ll \nu_d$, where ν_d is the d-dimensional Lebesgue measure [4], there exists a density $\mu_{d,n}(x,t)$ such that for all Borel sets B in $\mathbb{R}^d$

$$\mathcal{M}_{d,n}(t,B) = \int_B \mu_{d,n}(x,t)dx. \tag{3}$$

Definition 2.3. *The function $\mu_{d,n}(x,t)$ defined by (3) is called local mean n-facet density of the (incomplete) tessellation at time t .*

In particular $\mu_{d,d}(x,t)$ and $\mu_{d,d-1}(x,t)$ are the mean local volume density and the mean local surface density, respectively, of the crystallized region at time t (elsewhere denoted by $V_V(x,t)$ and $S_V(x,t)$, respectively; see e.g. [14] for the nomenclature). The mean local volume density is also the crystallinity. A complete characterization of a crystallization pattern at time t is given in terms of the family of densities $\mu_{d,n}(x,t)$ for $n = 0, 1, \ldots, d$.

Evolution equation for the whole set of spatial densities $\mu_{d,n}(x,t)$ of n-facets have been recently obtained in [6] in term of the hazard function h (see [3] for the definition of h).

Theorem 2.4. *Under sufficient regularity conditions on the kinetic parameters $\alpha(x,t)$ and $G(x,t)$ (given deterministic fields), we have*

$$\frac{\partial}{\partial t}\mu_{d,n}(x,t) = c_{d,n}\frac{h_{m+1}(x,t)}{(m+1)!}(1-\xi(x,t))[G(x,t)]^{-m} \tag{4}$$

where for $d = 2, 3$

$$h_k(x,t) = [h(x,t)]^k, \qquad k = 2, 3,$$

and $c_{d,n}$ are numerical constants depending only upon the dimensions d and n

$$c_{d,n} = \frac{\Gamma(\frac{dm+n+1}{2})\Gamma(\frac{d}{2})^{m+1}}{(m+1)!\Gamma(\frac{dm+n}{2})\Gamma(\frac{d+1}{2})^m\Gamma(\frac{d-m+1}{2})}.$$

System (4) is closed by the evolution equation for the hazard function, which is given by (see [5])

$$h(x,t) = G(x,t)S_{ex}(x,t).$$

The evolution equation for $S_{ex}(x,t)$ is provided in [3].

In the spatially homogeneous case the measure $\mathcal{M}_{d,n}(t,\cdot)$ is translation-invariant, so that

$$\mathcal{M}_{d,n}(t,B) = \mu_{d,n}(t)\nu_d(B),$$

where $\nu_d(B)$ is the d-dimensional volume of B. Then, by the use of the hazard function $h(t)$, which in this case has an explicit expression in terms of the kinetic parameters $\alpha(t)$ and $G(t)$, the following expression for the densities $\mu_{d,n}$ of an (incomplete) Johnson–Mehl tessellation can be derived [9, 12, 11]:

$$\mu_{d,n}(t) = c_{d,n}\int_0^t \left[s_d \int_0^\tau \left(\int_s^\tau G(u)du \right)^{d-1} \alpha(s)ds \right]^{m+1} G(\tau)p(\tau)d\tau \tag{5}$$

where $d = 1, 2, 3$, $n = 0, \ldots, d$, $m = d - n$ and s_d is the surface of the unit ball in $\mathbb{R}^d$.

3. Estimating the Local Density of Interfaces

Here the problem of estimating the measure of the boundaries of the cells of a 2D or a 3D tessellation on digitized images is tackled. Note that, since e.g. a line in a 2D digitized image is represented by pixels, which are 2-dimensional objects, the task of measuring its length can be difficult. Here a method is proposed to estimate the measure of $(d-1)$-dimensional objects in a d-dimensional space.

3.1. Estimate of the Density of $d-1$-Facets via the Spherical Contact Distribution Function

For the sake of simplicity, let us denote by

$$\Xi(t) := \Xi_{d-1}(t) = \bigcup_i \partial C_i(t)$$

the random set of the boundaries of the cells of a Johnson–Mehl tessellation at time t in $\mathbf{R}^d$ (here we will only deal with the cases relevant in the applications, i.e. $d = 2, 3$). Under sufficient regularity conditions on the growth and nucleation rates, this set may be assumed to have Hausdorff dimension $d-1$.

Definition 3.1. *The local spherical contact distribution function H_s associated with a random set Σ [4] is*

$$H_s(r, x) := P(x \in \Sigma \oplus b(0, r) |\ x \notin \Sigma),$$

where $\oplus$ is the Minkowski addition ($A \oplus B = \{a + b |\ a \in A, b \in B\}, \forall A, B \subseteq \mathbf{R}^d$), and $b(0, r)$ is a d-dimensional ball of radius r centered at the origin.

Since our random set $\Xi(t)$ has Hausdorff dimension $d-1$, for any fixed $x \in \mathbf{R}^d$ we have

$$P(x \notin \Xi(t)) = 1,$$

so that the spherical contact distribution function associated with $\Xi(t)$, which will be denoted by $H_s(r, x, t)$, is

$$H_s(r, x, t) := P(x \in \Xi(t) \oplus b(0, r)). \tag{6}$$

The estimate of $\mu_{d,d-1}$ is then possible thanks to the following theorem, which has been proven in [10].

Theorem 3.2. *Let $\Xi(t) = \bigcup_i \partial C_i(t)$ be the set of the boundaries of the cells of an (incomplete) Johnson–Mehl tessellation at time t having Hausdorff dimension $d-1$. Let $\Xi_r(t)$ be its parallel set of radius r, i.e.*

$$\Xi_r(t) = \Xi(t) \oplus b(0, r).$$

If the condition

$$\lim_{r \to 0} \frac{\nu_d[\Xi_r(t) \cap b(x, \varepsilon)]}{r} = 2\nu_{d-1}(\partial \Xi(t) \cap b(x, \varepsilon)) \tag{7}$$

is satisfied for all $x \in \mathbf{R}^d$ and for all $\varepsilon \in \mathbf{R}_+$ then the function $H_s(r,x,t)$ is differentiable at $r=0$ for ν_d-almost all x, and its derivative satisfies

$$\frac{d}{dr} H_s(r,x,t)_{|r=0} = 2\mu_{d,d-1}(x,t). \tag{8}$$

It can be proved via the Coarea Formula that Condition (7) is fulfilled by a set $\Xi \subset \mathbf{R}^d$ if the function

$$\begin{aligned} f\colon \quad \mathbf{R}^d &\to \mathbf{R}_+ \\ x &\mapsto d(x,\Xi) \end{aligned}$$

(where d is the euclidean distance) is Lipschitz [1].

Formula (8) may be used to estimate $\mu_{d,d-1}$ from an estimator of H_s. Suppose to recover a black-and-white image of the boundaries of the simulated crystals, as shown in Figure 3. Note from (6) that $H_s(r,x,t)$ equals the local volume density of the parallel set $\Xi_r(t)$ of $\Xi(t)$, which is in general nonzero, being $\Xi_r(t)$ a set of Hausdorff dimension d.

An unbiased estimator of $H_s(r,x,t)$ is then

$$\widehat{H}_s(r,x,t) = \frac{1}{n}\sum_{i=1}^{n} \mathbf{1}_{\Xi_r(t)}(x_i),$$

where x_i are points of a grid overlapped to a suitable window of observation. An estimator of $\frac{d}{dr}H_s(r,x,t)_{|r=0}$ may be obtained by numerical approximation, thus an estimator of the interface density $\mu_{d,d-1}(x,t)$ can be recovered.

The result of an estimation performed on simulated data is shown in Figure 3.

Acknowledgments

Financial support is acknowledged to the EU under the TMR-Network *Differential Equations in Industry and Commerce*, grant N.FMRXCT97-0117, to the Italian MURST/cofin. Program "Stochastic Processes with Spatial Structure" and to the Italian CNR contract n. 98.03635.ST74

References

[1] L. Ambrosio, N. Fusco, D. Pallara, *Functions of Bounded Variation and Free Discontinuity Problems*, Oxford Clarendon Press, 2000.

[2] A.J. Baddeley, *A crash course in stochastic geometry*, in Stochastic Geometry. Likelihood and Computation (O.E. Barndorff-Nielsen et al. Eds.), Chapman&Hall/CRC, Boca Raton, 1999.

[3] V. Capasso, R. Escobedo, C. Salani, *Moving bands and moving boundaries for an hybrid model of crystallization of polymers.* In this volume (2003).

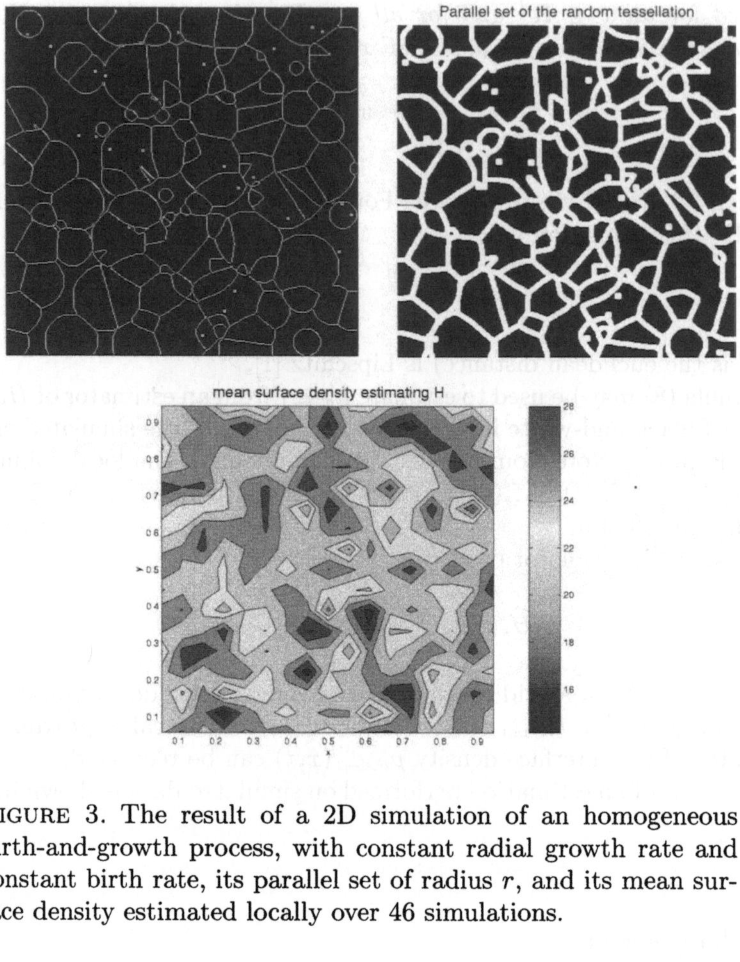

FIGURE 3. The result of a 2D simulation of an homogeneous birth-and-growth process, with constant radial growth rate and constant birth rate, its parallel set of radius r, and its mean surface density estimated locally over 46 simulations.

[4] V. Capasso, A. Micheletti, *Local spherical contact distribution function and local mean densities for inhomogeneous random sets.* Stochastics and Stoch. Rep. **71** (2000), 51–67.

[5] V. Capasso, A. Micheletti, *The hazard function of an inhomogeneous birth-and-growth process.* Preprint n.39/2001, Dip. di Matematica, Universita' di Milano, 2001.

[6] V.Capasso, A.Micheletti, M.Burger, *Densities of n-facets of incomplete Johnson–Mehl tessellations generated by inhomogeneous birth-and-growth processes.* Quaderno n. 38/2001, Dip. di Matematica, Università di Milano.

[7] V.Capasso, Ed. *Mathematical Modelling for Polymer Processing. Polymerization, Crystallization, Manufacturing*, Springer, 2002.

[8] Johnson, W.A., Mehl, R.F. *Reaction Kinetics in processes of nucleation and growth.* Trans. A.I.M.M.E. **135** (1939), 416–458.

[9] J.L.Meijering, *Interface area, edge length, and number of vertices in crystal aggregates with random nucleation.* Philips Res. Rep. **8** (1953), 270–290.

[10] A. Micheletti, *The surface density of a random Johnson–Mehl tessellation*, Quaderno n. 17/2001, Dip. di Matematica, Università di Milano, 2001.

[11] A. Micheletti, V. Capasso, G. Eder, *The density of the n-facets of an incomplete Johnson–Mehl tessellation.* Preprint of the Institute for Industrial Mathematics, Johannes-Kepler University of Linz (Austria), 1997.

[12] J. Møller, *Random Johnson–Mehl tessellations.* Adv. Appl. Prob. **24** (1992), 814–844.

[13] J. Møller, *Topics in Voronoi and Johnson–Mehl tessellations, in Stochastic Geometry.* Likelihood and Computation (O.E. Barndorff-Nielsen et al. Eds.), Chapman&Hall/CRC, Boca Raton, 1999.

[14] Stoyan, D., Kendall, W.S., Mecke, J., *Stochastic Geometry and its Application*, John Wiley & Sons, New York, 1995.

(A. Micheletti) MIRIAM and Department of Mathematics,
University of Milan, via Saldini 50, 20133 Milano, Italy
E-mail address: `Alessandra.Micheletti@mat.unimi.it`

(V. Capasso) MIRIAM and Department of Mathematics,
University of Milan, via Saldini 50, 20133 Milano, Italy
E-mail address: `Vincenzo.Capasso@mat.unimi.it`

International Series of Numerical Mathematics, Vol. 147, 207–223

A Posteriori Error Control of Free Boundary Problems

Ricardo H. Nochetto*

Abstract. A posteriori error estimators are computable quantities depending on the discrete solution(s) and the data, which provide upper (and lower) bounds for the error; they are thus instrumental for adaptivity. This paper assesses the derivation of a posteriori error estimators in the context of free boundary problems (FBPs). The discussion is based on simple examples, from contact problems to degenerate parabolic equations and curvature driven flows, and addresses both space and time discretization as well as both energy and maximum norm estimates (including interface error estimates).
Keywords: a posteriori error analysis, contact problems, degenerate parabolic, interface error estimates, curvature.

AMS subject classification: 35K55, 65M12, 65M15, 65M60.

1. Residuals and A Posteriori Error Estimators

We consider estimating the error committed in approximating the PDE

$$u_t + \mathfrak{F}(u) = f \tag{1}$$

solely relying on the data and the computed solution U. This yields the so-called *a posteriori* error estimates and is instrumental in modifying meshes and time steps (adaptivity). Adaptive error control is indeed a popular and exciting area of current research in science and engineering. The purpose of this paper is to survey recent results for (1) within the framework of *free boundary problems* (FBPs).

Error and Residual. Let U be the approximate solution, and let $\mathcal{R}$ be the *residual*

$$\mathcal{R} := -U_t - \mathfrak{F}(U) + f, \tag{2}$$

that is the amount by which U misses to be a solution of (1). If $e := u - U$ denotes the error, then it satisfies the error equation

$$e_t + \mathfrak{F}(u) - \mathfrak{F}(U) = \mathcal{R} \tag{3}$$

Ideally, we would like to derive estimates of the form

$$C_1 \|\mathcal{R}\|_{\mathcal{Y}} \le \|e\|_{\mathcal{X}} \le C_2 \|\mathcal{R}\|_{\mathcal{Y}} \tag{4}$$

* Partially supported by NSF Grants DMS-9971450 and DMS-0204670, and NSF-DAAD Grant INT-0129243.

in suitable functional spaces $\mathcal{X}$ and $\mathcal{Y}$. The upper bound shows that the quantity $\|\mathcal{R}\|_\mathcal{Y}$ is reliable, whereas the lower bound indicates that it is efficient; we will focus on *reliability* in this paper. Key issues are to identify suitable pairs $(\mathcal{X}, \mathcal{Y})$ and evaluate the norm $\|\cdot\|_\mathcal{Y}$ in practice, which usually is a negative Sobolev norm.

Failure of Duality for FBPs. A popular technique from linear PDE theory to establish (4) is duality [10, 11, 12]. If $\mathfrak{F}$ is Lipschitz, we can rewrite (3) as

$$e_t + \mathcal{A}e = \mathcal{R},$$

where

$$\mathcal{A} := \int_0^1 \nabla\mathfrak{F}(su + (1-s)U)\, ds.$$

If we formally multiply by $\varphi \in C^1(0,T)$ and integrate by parts over $(0,T)$, we get

$$\langle e(T), \varphi(T)\rangle = \langle e(0), \varphi(0)\rangle + \int_0^T \langle e, \varphi_t - \mathcal{A}^*\varphi\rangle + \int_0^T \langle \mathcal{R}, \varphi\rangle, \tag{5}$$

where $\mathcal{A}^*$ is the adjoint of $\mathcal{A}$. An error estimate follows by selecting φ in (5) as the solution of the *backward dual* problem

$$\varphi_t - \mathcal{A}^*\varphi = 0 \quad \text{in } (0,T), \qquad \varphi(T) = \rho, \tag{6}$$

and using *strong* stability properties of φ, such as a bound for φ_t in terms of ρ, for evaluating $\varphi(0)$ and $\langle \mathcal{R}, \varphi\rangle$. This program may fail to apply to the present low regularity setting because

- $\mathfrak{F}$ may not be Lipschitz, and so $\mathcal{A}$ and $\mathcal{A}^*$ may be singular or not even defined;
- (6) may make no sense or φ_t may be unbounded;
- φ is not computable because it depends on the unknown u;
- Replacing $\mathcal{A}$ by $\nabla\mathfrak{F}(U)$ to determine φ would lead to additional terms involving further derivatives of the possibly singular $\nabla\mathfrak{F}$.

Therefore, except in special circumstances such as [5, 24], duality does not quite lead to rigorous error control for FBPs.

Plan of the Paper. We show below several novel techniques for error control applicable to FBPs. In §2 we present pointwise error estimates in space via contact problems and apply them for error control of free boundaries. In §3 we discuss error control for time discretizations using the implicit Euler method for (1) with $\mathfrak{F}$ being either a subdifferential or an angle-bounded operator (Hilbert theory), or accretive (Banach theory). We conclude in §4 with some extensions.

2. Error Control in Space: Contact Problems

2.1. Setting and Challenges

Let Ω be a bounded, polyhedral, not necessarily convex domain in $\mathbb{R}^d$ with $d \in \{1, 2, 3\}$. Let $f \in L^\infty(\Omega)$ be a load function, $\chi \in H^1(\Omega) \cap C^{0,\alpha}(\bar{\Omega})$ be a lower obstacle, with $0 < \alpha \leq 1$, which satisfies $\chi \leq 0$ on $\partial\Omega$. Let $\mathcal{K}$ be the following non-empty, closed and convex subset of $\mathring{H}^1(\Omega)$:

$$\mathcal{K} := \{v \in \mathring{H}^1(\Omega) \mid v \geq \chi \text{ a.e. in } \Omega\}.$$

The variational formulation of the continuous *obstacle* problem reads as follows:

$$u \in \mathcal{K}: \qquad \langle \nabla u, \nabla(u-v)\rangle \leq \langle f, u-v\rangle \quad \forall v \in \mathcal{K}. \tag{7}$$

Hereafter, $\langle \varphi, \psi\rangle$ denotes the scalar product in $L^2(\Omega)$ as well as the duality pairing between $\mathring{H}^1(\Omega)$ and $H^{-1}(\Omega) = \mathring{H}^1(\Omega)^*$; we refer to [15, 16, 28]. We are interested in the approximation of both the solution u in $\mathcal{X} := L^\infty(\Omega)$ and the *contact set*

$$\Lambda := \{u = \chi\} := \{x \in \Omega \mid u(x) = \chi(x)\},$$

or equivalently of the *free boundary* or *interface*

$$\mathcal{F} := \partial\{u > \chi\} \cap \Omega.$$

The location of Λ is encoded in the non-positive functional $\sigma \in H^{-1}(\Omega)$

$$\langle \sigma, \varphi\rangle := \langle f, \varphi\rangle - \langle \nabla u, \nabla\varphi\rangle \quad \forall \varphi \in \mathring{H}^1(\Omega), \tag{8}$$

which plays the role of a *multiplier* for the unilateral constraint. In fact, we have $\sigma = f + \Delta\chi$ in the interior of the contact set $\Lambda = \{u = \chi\}$, where σ is typically < 0, and $\sigma = 0$ in the open *non-contact* set $\Omega \setminus \Lambda = \{u > \chi\}$.

Given a shape-regular partition $\mathcal{T}_h$ of Ω, the set of nodes of $\mathcal{T}_h$ is denoted by $\mathcal{N}_h$, and the subset of interior nodes by $\mathring{\mathcal{N}}_h$. Let $\mathring{\mathbb{V}}_h$ indicate the space of continuous piecewise affine finite element functions over $\mathcal{T}_h$ and $\mathring{\mathbb{V}}_h := \mathbb{V}_h \cap \mathring{H}^1(\Omega)$. The nodal basis functions of $\mathbb{V}_h$ are given by $(\phi_z)_{z\in\mathcal{N}_h}$, and they form a *partition of unity* of Ω, that is $\sum_{z\in\mathcal{N}_h} \phi_z = 1$. Let I_h be the Lagrange interpolation operator onto $\mathbb{V}_h$. Let $\chi_h := I_h\chi$ be the discrete obstacle. The discrete counterpart $\mathcal{K}_h$ of $\mathcal{K}$ is then

$$\mathcal{K}_h := \{v_h \in \mathring{\mathbb{V}}_h \mid v_h \geq \chi_h \text{ in } \Omega\}.$$

Note that it is sufficient to check the unilateral constraint of $\mathcal{K}_h$ only at the nodes. The set $\mathcal{K}_h$ is non-empty, convex, closed but in general not a subset of $\mathcal{K}$ (non-conforming approximation). The *discrete* obstacle problem reads as follows:

$$u_h \in \mathcal{K}_h: \quad \langle \nabla u_h, \nabla(u_h - v_h)\rangle \leq \langle f, u_h - v_h\rangle \quad \forall v_h \in \mathcal{K}_h. \tag{9}$$

Let $\omega_z = \text{supp}(\phi_z)$ be a generic finite element star and γ_z be the union of all interior interelement boundaries (sides) in ω_z. If J_h denotes the jump of the normal derivative of u_h across the interior sides, we define the set of *full-contact nodes* to be

$$\mathcal{C}_h = \{z \in \mathcal{N}_h \mid u_h = \chi_h \text{ and } f \leq 0 \text{ in } \omega_z,\ J_h \leq 0 \text{ on } \gamma_z\}$$

and denote the discrete *full-contact set* and its complement by

$$\Omega_h^0 = \Big\{x \in \Omega \mid \sum_{z\in\mathcal{C}_h} \phi_z(x) = 1\Big\}, \qquad \Omega_h^+ = \Omega\backslash\Omega_h^0.$$

Furthermore we set $\Gamma_h^0 = \Gamma \cap \Omega_h^0$ and $\Gamma_h^+ = \Gamma \cap \Omega_h^+$, along with

$$\omega_z^+ = \omega_z \cap \Omega_h^+, \qquad \gamma_z^+ = \gamma_z \cap \Omega_h^+. \tag{10}$$

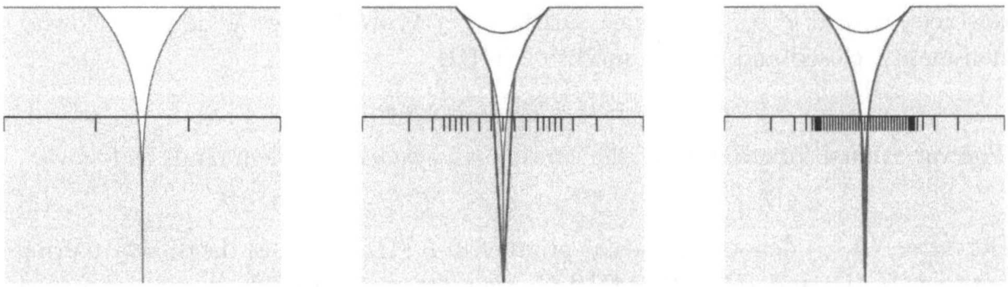

FIGURE 1. Localization for an obstacle with downward cusp and $f = -1$.

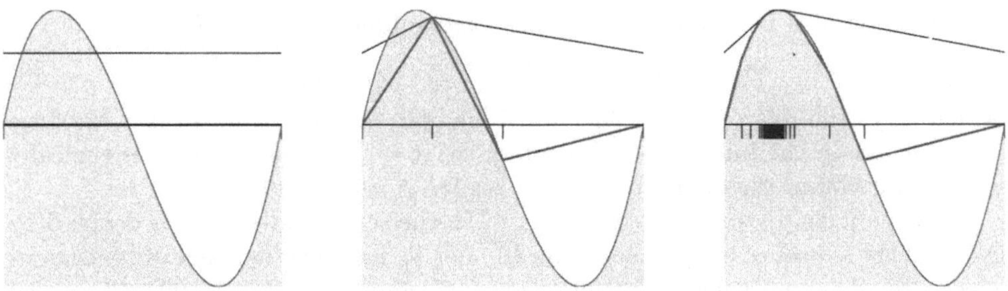

FIGURE 2. Localization for an obstacle not resolved on coarse grids.

To shed some light on the inherent difficulties in estimating $\|e\|_{L^\infty(\Omega)}$ we consider the following 1d examples. The estimator should possess the properties:

- Be insensitive to f within Ω_h^0, because f should not dictate the mesh quality in Ω_h^0 (*localization*). Figure 1 shows this since meshes are rather coarse in Ω_h^0.
- Be insensitive to the obstacle within Ω_h^+ and thus reduce to the usual estimator for the (unconstrained) Laplacian. This effect is shown in Figure 1: the refinement is due to the curvature generated by $f = -1$ but not to the cusp of χ pointing downwards. Overall there is an excellent accuracy for u in the non-contact set even though the approximation of χ is rather rough.
- Be able to detect the non-conforming situation $\chi > u_h$, and refine accordingly. This is depicted in Figure 2, where the concave part of χ above u_h is detected early on and thereby the solution is lifted up.

Let $\Pi_h : L_1(\Omega) \to \mathring{\mathbb{V}}_h$ be the interpolation operator of [4]; see also [27]. Such a Π_h is both *positivity preserving* and *second order* accurate, and plays a crucial role in defining the discrete counterpart σ_h of σ. Namely, we let $\sigma_h \in H^{-1}(\Omega)$ be $\langle \sigma_h, \varphi \rangle := \sum_{z \in \mathcal{N}_h} \langle \sigma_h, \varphi \phi_z \rangle$, where

$$\begin{aligned} \langle \sigma_h, \varphi \phi_z \rangle := & \int_{\Omega_h^0} f \varphi \phi_z + \int_{\Gamma_h^0} J_h \varphi \phi_z \\ & + \int_{\Omega_h^+} f \, (\Pi_h \varphi)(z) \, \phi_z + \int_{\Gamma_h^+} J_h \, (\Pi_h \varphi)(z) \, \phi_z, \end{aligned} \tag{11}$$

for all $z \in \mathcal{N}_h$ and $\varphi \in \mathring{H}^1(\Omega)$. Note that $\Pi_h \varphi$ is evaluated at the node z, and is thus a *constant* for each $z \in \mathcal{N}_h$. It turns out that

$$z \in \mathcal{N}_h \setminus \mathcal{C}_h \implies \langle \sigma_h, \varphi \phi_z \rangle = (\Pi_h \varphi)(z) \, s_z, \tag{12}$$

where s_z is a *nodal* multiplier:

$$s_z := \int_\Omega f \phi_z + \int_\Gamma J_h \phi_z \quad \forall z \in \mathcal{N}_h. \tag{13}$$

Moreover, s_z satisfies $s_z \le 0$ whenever $z \in \mathring{\mathcal{N}}_h \cup \mathcal{C}_h$ and $\sigma_h \le 0$. We finally let

$$\Lambda_h := \bigcup \{ \omega_z : z \in \mathring{\mathcal{N}}_h \cup (\mathcal{C}_h \cap \partial\Omega) \text{ and } s_z < 0 \} \tag{14}$$

The discrete multiplier σ_h satisfies $\sigma_h \le 0$, cancels with σ inside the discrete full contact set Ω_h^0, and localizes to the non-contact set Ω_h^+. Even though σ_h is non-computable, we only need the nodal multiplier s_z in (14) and Theorem 2.1.

2.2. Pointwise A Posteriori Error Estimates for Solutions

Let η_z be the star-based residual indicator

$$\eta_z := h_z^2 \| (f - \hat{f}_z) \, \phi_z \|_{L_\infty(\omega_z^+)} + h_z \, \| J_h \, \phi_z \|_{L_\infty(\gamma_z^+)}, \tag{15}$$

where ω_z^+, γ_z^+ are defined in (10) and $\hat{f}_z$ is any constant provided $\int_{\Omega_h^+} f \phi_z + \int_{\Gamma_h^+} J_h \phi_z = 0$ and 0 otherwise. The interior residual reduces to *oscillation* of f.

Theorem 2.1 (Reliability). *Let u be the continuous solution of* (7) *and let u_h be the discrete solution of* (9), *respectively. Then the following global a posteriori upper bound holds:*

$$\| u - u_h \|_{L_\infty(\Omega)} \le \mathcal{E}_h, \tag{16}$$

where the error estimator $\mathcal{E}_h$ is given by

$$\begin{aligned} \mathcal{E}_h := & \; C_* |\log h_{\min}|^2 \max\nolimits_{z \in \mathcal{N}_h} \eta_z && \textit{(localized residual)} \\ & + \| (\chi - u_h)^+ \|_{L_\infty(\omega_z)} + \| (u_h - \chi)^+ \|_{L_\infty(\Lambda_h)}, && \textit{(localized obstacle approx.)} \end{aligned}$$

C_* is a constant solely depending on the mesh regularity, and Λ_h is defined in (14).

The derivation of this optimal bound is done as follows. We first note that both σ and σ_h are designed to restore the equality in (7) and (9). We then form the so-called *Galerkin functional* $\mathcal{G}_h \in H^{-1}(\Omega)$:

$$\langle \mathcal{G}_h, \varphi \rangle = \langle \nabla(u - u_h), \nabla\varphi \rangle + \langle \sigma - \sigma_h, \varphi \rangle = - \langle \nabla u_h, \nabla\varphi \rangle + \langle f - \sigma_h, \varphi \rangle .$$

This concept, introduced in [32], plays the role of residual in the linear setting. A simple manipulation reveals the *localization* effect built in (11), namely,

$$\langle \mathcal{G}_h, \varphi \rangle = \int_{\Omega_h^+} f[\varphi - \Pi_h\varphi] + \int_{\Gamma_h^+} J_h\, [\varphi - \Pi_h\varphi]. \tag{17}$$

We next let $w \in \mathring{H}^1(\Omega)$ be the the Riesz representation of $\mathcal{G}_h$, namely $-\Delta w = \mathcal{G}_h$, and use (17) to derive the pointwise estimate

$$\|w\|_{L_\infty(\Omega)} \le C|\log h_{\min}|^2 \max_{z\in\mathcal{N}_h} \eta_z .$$

Such w is employed to correct u_h and thereby construct barrier functions for (7) which are of the form $u_h + w \pm \|w\|_{0,\infty;\Omega}$ plus additional terms involving $u_h - \chi$ (last two terms of $\mathcal{E}_h$). The *continuous* maximum principle allows us to compare these barriers with u and derive (16). A full proof is given in [26]. We also refer to [25, 26] for lower bounds, and to [14] for total variation problems.

Example: Lipschitz Obstacle. Let $\Omega := \{x \in \mathbb{R}^2 : |x|_1 < 1\}$, $f := -5$, and the obstacle be $\chi(x) := \text{dist}(x, \partial\Omega) - \frac{1}{5}$. The corresponding multiplier σ is a measure singular with respect to the Lebesgue measure since it has a mass supported along the edges of the pyramid within the contact set.

FIGURE 3. Discrete solution and obstacle on grids aligned with the edges of the obstacle (left) and non-aligned with them (right).

Figure 3 displays u_h and χ_h over two meshes, one aligned and the other non-aligned with the edges of χ: in the latter case, the representation χ_h of χ is rather crude. In contrast, the aligned meshes provide excellent approximability of χ, and the full localization property of $\mathcal{E}_h$ is clearly reflected in the meshes on the left column of Figure 4: the refinement along the diagonals is avoided. This is a crucial improvement of the estimator of [26] with respect to the *partially* localized estimator of [25]. The refinement for the non-aligned meshes is instead due to the lack of resolution of χ and thus of u (see right column of Figure 4). Note that the non-aligned grids do not satisfy the discrete maximum principle, which is not required for Theorem 2.1 to hold. These results were obtained with ALBERT [31].

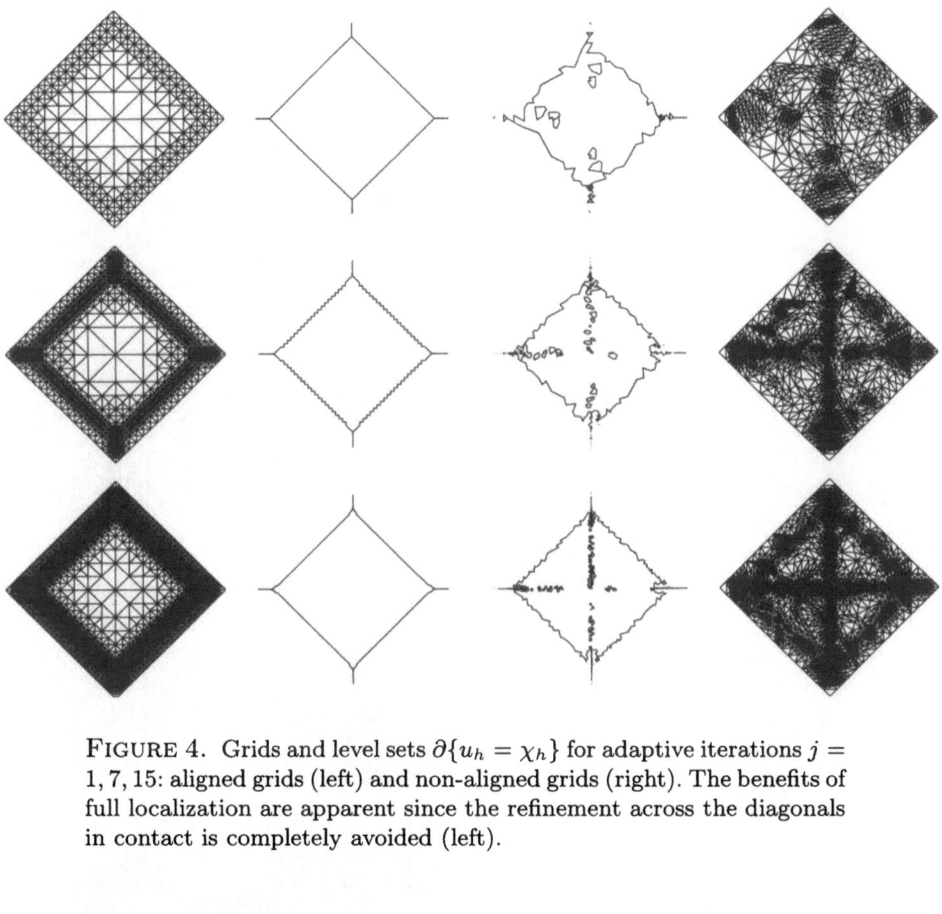

FIGURE 4. Grids and level sets $\partial\{u_h = \chi_h\}$ for adaptive iterations $j = 1, 7, 15$: aligned grids (left) and non-aligned grids (right). The benefits of full localization are apparent since the refinement across the diagonals in contact is completely avoided (left).

2.3. A Posteriori Error Estimates for Interfaces: Barrier Sets

Since the estimator $\mathcal{E}_h$ controls the pointwise error $\|u - u_h\|_{L_\infty(\Omega)}$, we can obtain *a posteriori* information on Λ and also on the exact free boundary (or interface) $\mathcal{F}$. We need the additional *nondegeneracy* condition: there exists $\lambda > 0$ such that

$$\langle f, \varphi \rangle - \langle \nabla\chi, \nabla\varphi \rangle \leq -\lambda \int_\Omega \varphi \qquad \forall \varphi \in \mathring{H}^1(\Omega), \varphi \geq 0. \tag{18}$$

Condition (18) guarantees *stability* of the exact free boundary $\mathcal{F}$ and is due to CAFFARELLI [2]; see also [15, §2.10]. This condition has been crucial in the a priori error analysis of free boundaries [1, 20] as well. If

$$r_h^2 := \frac{2d}{\lambda}\Big(2\mathcal{E}_h + \|(\chi_h - \chi)^+\|_{L_\infty(\{u_h \le \chi_h + \mathcal{E}_h\})}\Big), \tag{19}$$

we then define $K := \{\operatorname{dist}(\cdot, \partial\Omega) \ge r_h\}$ and the *barrier sets*

$$\Lambda^* := \{u_h \le \chi + \mathcal{E}_h\}, \qquad \Lambda_* := \Big\{\operatorname{dist}\big(\cdot, \{u_h \ge \chi_h + \mathcal{E}_h\}\big) \ge r_h\Big\}. \tag{20}$$

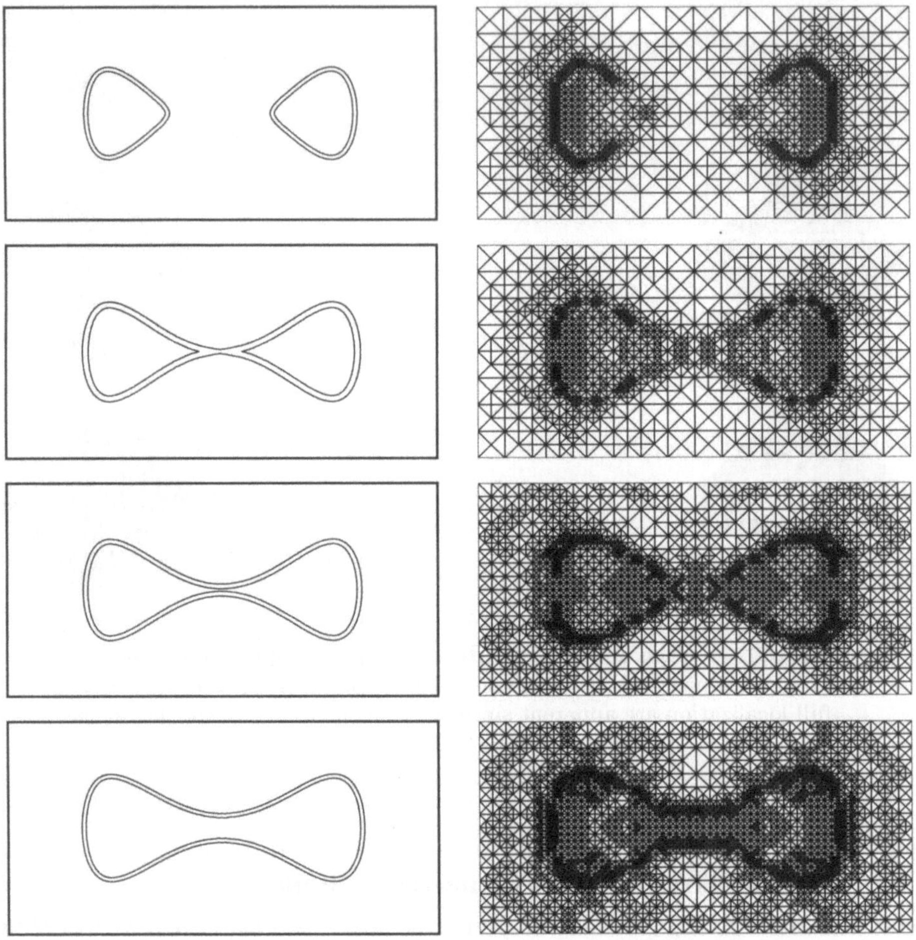

FIGURE 5. From balls to bones: Interface barriers for tolerance ≈ 0.01 (left) and adaptive grids for tolerance ≈ 0.15 (right) in 2d for forcing term $f = 0, -5.9, -8.1, -15$ (from top to bottom). The distance of the barriers is ≈ 0.05 for all the four forces.

Theorem 2.2 (A posteriori control of contact set and interface). *The set Λ^* is an upper barrier set for the exact contact set $\Lambda = \{u = \chi\}$, i.e. $\Lambda \subset \Lambda^*$.*

Moreover, if the stability condition (18) *holds, then the set Λ_* is a lower barrier set for Λ in the sense that $\Lambda_* \cap K \subset \Lambda \cap K$, whence*

$$\mathcal{F} \cap K \subset \left(\Lambda^* \cap K\right) \setminus \operatorname{int}\left(\Lambda_* \cap K\right).$$

A proof is given in [26]. Note that λ dictates the thickness of the strips $\Omega \setminus K$ and $\left(\Lambda^* \cap K\right) \setminus \operatorname{int}\left(\Lambda_* \cap K\right)$, and thus acts as a condition number for the free boundary. Also, if $\chi \in H^2(\Omega)$, one can adaptively compute the condition number λ by $\lambda = -\sup_{\{u_h \le \chi_h + \mathcal{E}_h\}}(f + \Delta\chi)$ [26]. This is used in the following example.

Example: From Balls to Bones. Let $\Omega = (-2,2) \times (-1,1)$ and the obstacle χ be $\chi(x) = 10 - 6(x_1^2 - 1)^2 - 20(|x|^2 - x_1^2)$, which consists of two hills connected by a saddle. Results, reported in Figure 5, were obtained with ALBERT [31]. The left column in Figure 5 depicts the interface barriers for four constant loads $f = 0, -5.9, -8.1, -15$ for about the same tolerance $\tau \approx 0.01$; the exterior curves correspond to $\partial\Lambda^*$ whereas the interior curves display $\partial\Lambda_*$. There is a topological change, with a double-cusp singularity at the origin, between the extreme values $f = 0, -15$ corresponding to f_{crit}. The barrier sets Λ_*, Λ^* give a reliable range for f_{crit}; see [26]. Although $\mathcal{F}$ develops a singularity, it is worth noticing that $u \in W^2_\infty(\Omega)$ and thus no special refinement is needed to approximate u. Moreover, $\lambda = 16$ in Ω for the 4 loads, which shows that the double-cusp is not due to lack of stability. The interface estimate of Theorem 2.2 thus applies and provides a posteriori error control of the entire free boundary including the double-cusp.

3. Error Control in Time

Given the abstract PDE in either a Hilbert or a Banach space

$$u_t + \mathfrak{F}(u) = 0, \tag{21}$$

we now derive error estimates without regularity assumptions on $\mathfrak{F}$ for

$$\frac{1}{\tau_n}\big(U_n - U_{n-1}\big) + \mathfrak{F}(U_n) = 0. \tag{22}$$

Let $U(t)$ be the piecewise linear interpolant of $\{U_n\}_{n=0}^N$ and set $e := u - U$.

3.1. Hilbert Theory: Subdifferential Operators

Let $\mathcal{H}$ be a Hilbert space, with scalar product $\langle\cdot,\cdot\rangle$ and norm $|\cdot|$, and let $\phi : \mathcal{H} \to \mathbb{R} \cup \{\infty\}$ be a proper lower semicontinuous convex function, with domain $D(\phi)$. Let $\mathfrak{F} = \partial\phi : \mathcal{H} \to 2^{\mathcal{H}}$ be the *subdifferential* of ϕ, with domain $D(\mathfrak{F}) \subset D(\phi)$. Then (21) is equivalent to the variational inequality

$$\langle u_t, u - v\rangle + \phi(u) - \phi(v) \le 0 \quad \forall v \in D(\phi), \tag{23}$$

and it satisfies the energy identity

$$|u_t(t)|^2 + \tfrac{d}{dt}\phi(u(t)) = 0 \quad \text{a.e. } t \in (0,T). \tag{24}$$

Energy Dissipation Estimator. On the other hand, (22) is equivalent to

$$\frac{1}{\tau_n}\langle U_n - U_{n-1}, U_n - v\rangle + \phi(U_n) - \phi(v) \le 0 \quad \forall v \in D(\phi), \tag{25}$$

which, upon taking $v = U_{n-1}$, yields the discrete energy inequality

$$-\mathcal{E}_n := \left|\frac{U_n - U_{n-1}}{\tau_n}\right|^2 + \frac{\phi(U_n) - \phi(U_{n-1})}{\tau_n} \le 0 \quad \forall 1 \le n \le N. \tag{26}$$

The *energy dissipation estimator* $\mathcal{E}_n \ge 0$ in (26) measures the deviation of $\{U_n\}_{n=0}^N$ from satisfaction of the conservation form (24).

Theorem 3.1. *If* $U_0 \in D(\phi)$, *then* $\max_{t\in[0,T]} |e(t)|^2 \le |e(0)|^2 + \sum_{n=1}^N \tau_n^2 \mathcal{E}_n$.

Proof. We first write a *continuous* evolution variational inequality for U

$$\langle U_t(t), U(t) - v\rangle + \phi(U(t)) - \phi(v) \le \mathcal{R}(t), \quad \forall v \in D(\phi), \tag{27}$$

where $\mathcal{R}(t)$ is a residual term, independent of the test function v, given by

$$\mathcal{R}(t) := \langle U'(t), U(t) - U_n\rangle + \phi(U(t)) - \phi(U_n), \quad \forall t \in (t_{n-1}, t_n).$$

We next estimate $\mathcal{R}(t)$. In view of the elementary identity

$$U(t) - U_n = \frac{t - t_n}{\tau_n}(U_n - U_{n-1}) = (t - t_n)U_t(t),$$

and the convexity of ϕ

$$\phi(U(t)) \le \frac{t_n - t}{\tau_n}\phi(U_{n-1}) + \frac{t - t_{n-1}}{\tau_n}\phi(U_n),$$

(26) implies $\mathcal{R}(t) \le (t_n - t)\mathcal{E}_n$. Finally, choosing $v = U(t)$ in (23), $v = u(t)$ in (27), and adding the two inequalities, we see that the terms involving ϕ cancel out and we end up with the error inequality

$$\tfrac{1}{2}\tfrac{d}{dt}|e(t)|^2 \le (t_n - t)\mathcal{E}_n \quad \text{a.e. } t \in (t_{n-1}, t_n).$$

Upon integration in time, this upper bound readily implies the assertion. □

Monotone Residual Estimator. The discrete counterpart of the dissipation energy inequality $\frac{d}{dt}|u_t(t)|^2 \le 0$, namely,

$$\tau_n^{-2}\left|U_n - U_{n-1}\right|^2 \le \tau_{n-1}^{-2}\left|U_{n-1} - U_{n-2}\right|^2 \quad \forall 1 \le n \le N, \tag{28}$$

can be derived as follows. Taking $v = U_n$ in (25) at the time-step $n - 1$, with $U_{-1} := U_0 - \tau_0\mathfrak{F}(U_0)$, we find

$$\tau_{n-1}^{-1}\langle U_{n-1} - U_{n-2}, U_{n-1} - U_n\rangle + \phi(U_{n-1}) - \phi(U_n) \le 0. \tag{29}$$

Dividing (29) by τ_n and replacing it into (26), we obtain (28) from the elementary inequality $2\langle v - w, v\rangle \ge |v|^2 - |w|^2$ for $v, w \in \mathcal{H}$. Recalling (22), we also deduce a crucial bound of $\mathcal{E}_n$ in terms of a *monotone residual* estimator $\mathcal{D}_n$:

$$\mathcal{E}_n \le \tau_n^{-1}\langle \mathfrak{F}(U_n) - \mathfrak{F}(U_{n-1}), U_n - U_{n-1}\rangle =: \mathcal{D}_n. \tag{30}$$

Combining (30) with Theorem 3.1 yields the following error estimate.

Theorem 3.2. $\max_{t\in[0,T]} |e(t)|^2 \le |e(0)|^2 + \sum_{n=1}^N \tau_n^2 \mathcal{D}_n$.

Rate of Convergence. We now prove that the estimates in Theorems 3.1 and 3.2 give rise to upper bounds of the error with respect to the largest time-step τ (*a priori* bounds). A simple first result is directly suggested by (26), which implies

$$0 \le \tau_n \mathcal{E}_n \le \phi(U_{n-1}) - \phi(U_n).$$

Therefore, we end up with a telescopic sum

$$\sum_{n=1}^N \tau_n^2 \mathcal{E}_n \le \tau \sum_{n=1}^N \big(\phi(U_{n-1}) - \phi(U_n)\big) = \tau\big(\phi(U_0) - \phi(U_N)\big).$$

Without loss of generality we may assume $\phi \ge 0$, for otherwise we may add an affine function. We then obtain a rate of convergence of order $O(\sqrt{\tau})$

Theorem 3.3. *If* $U_0 \in D(\phi)$, *then* $\max_{t\in[0,T]} |e(t)|^2 \le |e(0)|^2 + \phi(U_0)\,\tau$.

On the other hand, making use of (25) in conjunction with (30), we can rewrite the monotone residual estimator $\mathcal{D}_n$ as follows:

$$\mathcal{D}_n = \langle \mathfrak{F}(U_{n-1}) - \mathfrak{F}(U_n), \mathfrak{F}(U_n)\rangle.$$

By virtue of the property $\mathcal{D}_n \ge 0$ and the inequality $2\langle v-w, v\rangle \ge |v|^2 - |w|^2$, we then see that $\sum_{n=1}^N \mathcal{D}_n$ telescopes

$$2\sum_{n=1}^N \tau_n^2 \mathcal{D}_n \le \tau^2 \sum_{n=1}^N \big(|\mathfrak{F}(U_{n-1})|^2 - |\mathfrak{F}(U_n)|^2\big) \le \tau^2 |\mathfrak{F}(U_0)|^2.$$

We then obtain the following *linear* rate of convergence.

Theorem 3.4. *If* $U_0 \in D(\mathfrak{F})$, *then* $\max_{t\in[0,T]} |e(t)|^2 \le |e(0)|^2 + \frac{1}{2}|\mathfrak{F}(U_0)|^2\tau^2$.

We point out that no regularity of ϕ or $\mathfrak{F}$ has been used as well as no constraints between consecutive time-steps t_n. We stress that the *a priori* rate of convergence depends on the regularity of u_0, which dictates the size of $|e(0)|$. We also note that a linear rate of convergence cannot be expected in general for monotone operators without further structure [29]. We finally refer to [30] for extensions to the Wasserstein metric.

Examples. We consider two prototype parabolic FBPs for which duality does not give optimal results. The first one is the *parabolic obstacle problem* in $\mathcal{H} = L_2(\Omega)$:

$$u \ge 0: \quad \langle u_t, u - v\rangle + \langle \nabla u, \nabla(u-v)\rangle \le 0 \quad \forall\, v \ge 0.$$

If we denote by $Z_n := \Delta U_n - \frac{1}{\tau_n}(U_n - U_{n-1}) \le 0$ the discrete multiplier corresponding to the unilateral constraint $U_n \ge 0$, then Theorem 3.2 holds with

$$\mathcal{D}_n = \tau_n^{-1}\|\nabla(U_n - U_{n-1})\|^2_{L^2(\Omega)} + \tau_n^{-1}\langle Z_n - Z_{n-1}, U_n - U_{n-1}\rangle.$$

The second example is a *degenerate parabolic problem* in $\mathcal{H} = H^{-1}(\Omega)$ which, for β non-decreasing, reads

$$u_t - \Delta\beta(u) = 0.$$

Then Theorem 3.2 holds with

$$\mathcal{D}_n = \tau_n^{-1} \langle \beta(U_n) - \beta(U_{n-1}), U_n - U_{n-1} \rangle .$$

3.2. Hilbert Theory: Angle-Bounded Operators

Angled-bounded operators form a class of monotone operators larger than subdifferential for which optimal error control is possible [23]. They satisfy for $\gamma > 0$

$$\langle \mathfrak{F}(v) - \mathfrak{F}(w), w - z \rangle \leq \gamma \langle \mathfrak{F}(v) - \mathfrak{F}(z), v - z \rangle \quad \forall v, w, z \in D(\mathfrak{F}).$$

Theorem 3.5. *The results of Theorems* 3.2 *and* 3.4 *are valid for* $\mathfrak{F}$ *angle-bounded.*

Example. We consider the convection-diffusion equation in $\mathcal{H} = L^2(\Omega)$:

$$\mathfrak{F}(v) = -\nu^2 \Delta v + b \cdot \nabla v + cv.$$

If $|b(x)| \leq b_0$ and $d(x) = -\frac{1}{2}\mathrm{div} b(x) + c(x) \geq d_0^2$, then we prove in [23] that

$$\gamma^2 = \frac{1}{4}\Big(1 + \frac{b_0^2}{\nu^2(d_0^2 + \nu^2 p_0^2)}\Big),$$

where p_0 is the Poincaré constant for zero Dirichlet boundary values, along with the optimal a posteriori error estimate

$$\|u - U\|^2_{L_\infty(0,T;L_2(\Omega))}, \ \nu^2 \|\nabla(u - U)\|^2_{L_2(0,T;L_2(\Omega))} \leq \|u_0 - U_0\|^2_{L_2(\Omega)} + 2\gamma^2 \sum_{n=1}^{N} \tau_n^2 \Big(\int_\Omega \nu^2 \|\nabla(U_n - U_{n-1}\|^2_{L_2(\Omega)} + d(x) \|U_n - U_{n-1}\|^2_{L_2(\Omega)} \Big).$$

3.3. Banach Theory: Accretive Operators

We consider now a Banach space $\mathcal{B}$ with norm $\|\cdot\|$ and dual $\mathcal{B}^*$. Let $\langle \cdot, \cdot \rangle$ be the duality pairing between $\mathcal{B}$ and $\mathcal{B}^*$, and let $J : \mathcal{B} \to 2^{\mathcal{B}^*}$ be the subdifferential of the norm, which satisfies $\|v\| = \langle v, J(v) \rangle$ for all $v \in \mathcal{B}$. Let $\mathfrak{F} : \mathcal{B} \to 2^{\mathcal{B}}$ be a (multivalued) *accretive* operator, namely,

$$\big\langle \mathfrak{F}(u) - \mathfrak{F}(v), J(u - v) \big\rangle \geq 0 \quad \forall u, v \in D(\mathfrak{F}). \tag{31}$$

If $\bar{U}(t) = U_n$ for $t_{n-1} \leq t < t_n$ is a piecewise constant interpolant of $\{U_n\}_{n=0}^N$, then (22) also reads

$$U_t(t) + \mathfrak{F}(\bar{U}(t)) = 0 \quad \text{a.e. in } (0, T). \tag{32}$$

If, to derive an error bound, we naively take the difference of (21) and (32), and multiply it by $J(u(t) - \bar{U}(t))$ to exploit (31), we realize that $\langle u_t - U_t, J(u - \bar{U}) \rangle$ is no longer the derivative of $\|u - U\|$ as in §3.1. The difficulty of dealing with two strong nonlinearities, $\mathfrak{F}$ and J, arises and prevents us from deriving error bounds. We are thus forced to deal with the continuous and the discrete equations separately; below are the crucial steps of our argument. We refer to [21] for details.

Evolution Equation for the Continuous Solution. Since J is the subdifferential of the norm, (21) is in fact equivalent of the evolution equation

$$\frac{d}{dt}\|u(t)-v\| = -\big\langle \mathfrak{F}(u(t)), J(u(t)-v)\big\rangle \quad \forall v \in D(\mathfrak{F}), \tag{33}$$

which we will try to reproduce in the discrete setting, starting from (32).

Evolution Inequalities for the Discrete Solutions. The map $s \mapsto \|U(s)-V\|$ is convex on each interval $(t_{n-1}, t_n]$ and therefore its derivative is not decreasing and is bounded by its value at $s = t_n$. Then, for all $V \in \mathcal{B}$ and a.e. $t \in (0,T)$

$$\frac{d}{ds}\|U(s)-V\| = \langle U_s(s), J(U(s)-V)\rangle \le \langle U_s(s), J(\bar{U}(s)-V)\rangle. \tag{34}$$

Combining (32) with (34), for all $V \in \mathcal{B}$ and a.e. $s \in (0,T)$, we obtain

$$\frac{d}{ds}\|U(s)-V\| \le -\langle \mathfrak{F}(\bar{U}(s)), J(\bar{U}(s)-V)\rangle. \tag{35}$$

Doubling Variables. To combine (33) with (35) via (31), we would like to choose $v := U$ and $V := u$ and add the equations. However, it is *impossible* to write (33) and (35) at the same time "$t = s$", since we have derived the equations with respect to *time-independent* test "elements" v, V. Therefore we keep two distinct "time" variables t, s and choose $v := \bar{U}(s)$, $V := u(t)$, thus obtaining from (31)

$$\partial_s\|U(s)-u(t)\| + \partial_t\|\bar{U}(s)-u(t)\| \le 0. \tag{36}$$

Penalization. To prevent t and s from getting too far apart, we penalize their distance with a parameter $\varepsilon \in (0,T)$, and thus integrate (36) in the 2d strip

$$S^{\varepsilon}_{0,T} := \big\{(s,t) \in \mathbb{R}^2 : 0 \le t \le T, \quad t-\varepsilon \le s \le t\big\}. \tag{37}$$

In order to deal with negative values of s, we extend $U, \bar{U}$ in $(-\varepsilon, 0)$ by

$$\bar{U}(s) := U_0, \quad U(s) := U_0 - s\mathfrak{F}(U_0) \tag{38}$$

so that (32) still holds. Applying the Gauss–Green formula in $S^{\varepsilon}_{0,T}$ gives

$$\begin{aligned}\int_{T-\varepsilon}^{T} \|\bar{U}(s)-u(T)\|\, ds \le{}& \varepsilon\|u_0 - U_0\| \\ &+ \int_0^T \Big(\|\bar{U}(t)-u(t)\| - \|U(t)-u(t)\|\Big)\, dt \\ &+ \int_0^T \Big(\|U(t-\varepsilon)-u(t)\| - \|\bar{U}(t-\varepsilon)-u(t)\|\Big)\, dt.\end{aligned} \tag{39}$$

Stability. We take the difference between two consecutive equations (22), and multiply by $J(U_n - U_{n-1})$, to arrive at

$$\begin{aligned}\tau_n^{-1}\big\langle U_n - U_{n-1}, J(U_n - U_{n-1})\big\rangle + \big\langle \mathfrak{F}(U_n) - \mathfrak{F}(U_{n-1}), J(U_n - U_{n-1})\big\rangle \\ = \tau_{n-1}^{-1}\big\langle U_{n-1} - U_{n-2}, J(U_n - U_{n-1})\big\rangle.\end{aligned}$$

Making use of (31), together with property $\langle v, j(w)\rangle \le \|v\|$ for all $v \in \mathcal{B}$ with equality for $w = v$, we get

$$\tau_n^{-1}\|U_n - U_{n-1}\| \le \tau_{n-1}^{-1}\|U_{n-1} - U_{n-2}\| \quad \forall\, 1 \le n \le N.$$

If we now choose $U_{-1} = U_0 - \tau_0\mathfrak{F}(U_0)$ with $\tau_0 \le \varepsilon$, which is consistent with (38), recursion yields the following discrete stability estimate:

$$\sup_{t\in(0,T)} \|U_t(t)\| = \sup_{1\le n\le N} \frac{1}{\tau_n}\|U_n - U_{n-1}\| \le \|\mathfrak{F}(U_0)\|. \tag{40}$$

A Posteriori Error Estimate. Since

$$\|\bar U(t) - u(t)\| - \|U(t) - u(t)\| \le \|\bar U(t) - U(t)\|,$$
$$\|U(t-\varepsilon) - u(t)\| - \|\bar U(t-\varepsilon) - u(t)\| \le \|\bar U(t-\varepsilon) - U(t-\varepsilon)\|,$$

and $\int_{-\varepsilon}^0 \|\bar U(t) - U(t)\|dt = \frac{\varepsilon}{2}\|\mathfrak{F}(U_0)\|$, a simple manipulation of (39) implies

$$\begin{aligned}\frac{1}{\varepsilon}\int_{T-\varepsilon}^T \|U(s) - u(T)\|\,ds &\le \frac{1}{\varepsilon}\int_{T-\varepsilon}^T \big(\|\bar U(s) - u(T)\| + \|\bar U(s) - U(s)\|\big)ds \\ &\le \|u_0 - U_0\| + \frac{\varepsilon}{2}\|\mathfrak{F}(U_0)\| + \frac{2}{\varepsilon}\int_0^T \|\bar U(t) - U(t)\|\,dt.\end{aligned}$$

Since $\int_0^T \|U(t) - \bar U(t)\|\,dt = \frac12\sum_{n=1}^N \tau_n\|U_n - U_{n-1}\|$, and (40) leads to $\frac{1}{\varepsilon}\int_{T-\varepsilon}^T \|U(s) - U(T)\|\,ds \le \frac{\varepsilon}{2}\|\mathfrak{F}(U_0)\|$, we readily infer that

$$\|u(T) - U(T)\| \le \|u_0 - U_0\| + \varepsilon\|\mathfrak{F}(U_0)\| + \frac{1}{\varepsilon}\sum_{n=1}^N \tau_n\|U_n - U_{n-1}\|.$$

Upon *optimizing* ε, we conclude the desired a posteriori error bound of [21].

Theorem 3.6. *If $U_0 \in D(\mathfrak{F})$, then*

$$\|e(T)\| \le \|e(0)\| + 2\|\mathfrak{F}(U_0)\|^{1/2}\Big(\sum_{n=1}^N \tau_n\|U_n - U_{n-1}\|\Big)^{1/2}.$$

The idea of doubling the variables and penalizing goes back to Kružkov for first order conservation laws [17].

A Priori Error Estimate. To show that the above error bound exhibits the correct asymptotic order of convergence, we resort to (40) to deduce

$$\sum_{n=1}^N \tau_n\|U_n - U_{n-1}\| \le \tau T\|\mathfrak{F}(U_0)\|.$$

This yields the following celebrated a priori estimate of Crandall and Liggett [7], originally derived for constant time steps by a different technique:

$$\|e(T)\| \le \|e(0)\| + 2\|\mathfrak{F}(U_0)\|\sqrt{\tau T}. \tag{41}$$

Example. Consider a *degenerate* parabolic PDE $u_t - \Delta\beta(u) = 0$ in $\mathcal{B} = L_1(\Omega)$.

4. Extensions

We finally discuss briefly two extensions. We refer to [18, 19] for details.

4.1. Mean Curvature Flow of Graphs

Let a surface $\Gamma(t)$ be described by the graph $u\ : \Omega \times [0,T] \to \mathbb{R}$. Let

$$Qu := \left(1 + |\nabla u|^2\right)^{1/2}, \quad Nu := (\nabla u, -1)/Qu, \quad Vu := \partial_t u/Qu,$$

be the elementary area, the unit normal vector and the scalar normal velocity of $\Gamma(t)$. We assume that $\Gamma(t)$ evolves according to the geometric law

$$\frac{\partial_t u}{Qu} - \operatorname{div}\Big(\frac{\nabla u}{Qu}\Big) = f(x,t).$$

Note the presence of Qu in the first term, which makes the study of this equation much harder than (21). Let $U : [0,T] \to \mathbb{V}_h$ denote the semidiscrete finite element approximation

$$U(t) \in \mathbb{V}_h : \quad \int_\Omega \frac{\partial_t U}{QU} V + \frac{\nabla U \nabla V}{QU} = \int_\Omega f V \quad \forall V \in \mathbb{V}_h.$$

A key step in the analysis is to identify the correct geometric quantities to estimate. This was done by Deckelnick and Dziuk for the a priori analysis [8], and boils down to examining

$$E(t) := \int_\Omega |Nu - NU|^2 Qu + \int_0^t \int_\Omega |Vu - VU|^2 Qu.$$

Theorem 4.1 (Conditional Estimate). *There exist two computable estimators $\mathcal{E}_\infty(t)$ and $\mathcal{E}_2(t)$, and a constant C_0 such that if $\mathcal{E}_\infty(t) \le \frac{1}{8}$ then $E(t) \le C_0 \mathcal{E}_2(t)$.*

4.2. Allen–Cahn Equation: Circumventing Gronwall's Inequality

We finally consider the *singularly perturbed* parabolic Cauchy problem

$$\varepsilon \langle \partial_t u, v\rangle + \varepsilon \langle \nabla u, \nabla v\rangle + \frac{1}{\varepsilon} \langle f(u), v\rangle = 0, \quad u(0) = u_0,$$

where $f(u) := u(u^2-1)$, and its finite element discretization: find $U_n \in \mathbb{V}_n$

$$\frac{\varepsilon}{\tau_n} \langle U_n - I_n U_{n-1}, V\rangle + \varepsilon \langle \nabla U_n, \nabla V\rangle + \frac{1}{\varepsilon} \langle f(I_n U_{n-1}), V\rangle$$
$$+ \frac{1}{\varepsilon} \langle f'(I_n U_{n-1})(U_n - I_n U_{n-1}), V\rangle = 0 \quad \forall\, V \in \mathbb{V}_n.$$

Let $U(t)$ be the piecewise linear interpolant of $\{U_n\}_{n=0}^N$ and $\mathcal{R}(t)$ be the *residual*

$$\langle \mathcal{R}(t), v\rangle := \varepsilon \langle \partial_t U, v\rangle + \varepsilon \langle \nabla U, \nabla v\rangle + \frac{1}{\varepsilon} \langle f(U), v\rangle \quad \forall v \in H^1(\Omega).$$

We assume that the initial condition u_0 is such that interfaces are already developed. Then we have the following *spectral estimate* [3, 9]:

$$\varepsilon \|\nabla v\|^2_{L_2(\Omega)} + \frac{1}{\varepsilon} \langle f'(u) v, v\rangle \ge -\lambda_0 \varepsilon \|v\|^2_{L_2(\Omega)}, \quad \forall\, v \in H^1(\Omega). \qquad (42)$$

Theorem 4.2 (Polynomial Error Control). *Let $\theta > 0$ be a tolerance, and let $\mathcal{E}_0, \mathcal{E}_1$ be parabolic error estimators based on $\mathcal{R}(t)$. If (42) is valid, then there exist constants $(\Lambda_i)_{i=0}^4$, depending exponentially on $\lambda_0 T$ but independent of ε, such that if*

$$\theta \leq \Lambda_0 \varepsilon^4, \quad \|e(0)\|_{L_2(\Omega)} \leq \Lambda_1 \theta, \quad \mathcal{E}_0 \leq \Lambda_2 \varepsilon \theta, \quad \mathcal{E}_1 \leq \lambda_2 \varepsilon^2 \theta,$$

then

$$\|e\|_{L_\infty(0,T;L_2(\Omega))} \leq \theta.$$

We observe that the dependence on ε^{-1} is polynomial and not exponential, as it could be expected from a naive application of Gronwall's inequality. This result is inspired in the a priori analysis of FENG and PROHL [13].

References

[1] F. BREZZI AND L. A. CAFFARELLI, *Convergence of the discrete free boundaries for finite element approximations*, RAIRO Numer. Anal. **17** (1983), 385–395.

[2] L. A. CAFFARELLI, *A remark on the Hausdorff measure of a free boundary, and the convergence of coincidence set*, Boll. Unione Mat. Ital., V. Ser. A **18** (1981), 109–113.

[3] X. CHEN, *Spectrum for the Allen–Cahn, Cahn–Hilliard, and phase-field equations for generic interfaces*, Comm. Partial Differential Eqs. **19** (1994), 1371–1395.

[4] Z. CHEN AND R.H. NOCHETTO, *Residual type a posteriori error estimates for elliptic obstacle problems*, Numer. Math. **84** (2000), 527–548.

[5] Z. CHEN, R.H. NOCHETTO, A. SCHMIDT, *A characteristic Galerkin method with adaptive error control for the continuous casting problem*, Comput. Methods Appl. Mech. Engrg. **189** (2000), 249–276.

[6] Z. CHEN, R.H. NOCHETTO, AND A. SCHMIDT, *Adaptive finite element methods for diffuse interface models*, (to appear).

[7] M. G. CRANDALL AND T. M. LIGGETT, *Generation of semi-groups of nonlinear transformations on general Banach spaces*, Amer. J. Math. **93** (1971), 265–298.

[8] K. DECKELNICK AND G. DZIUK, *Error estimates for a semi-implicit fully discrete finite element scheme for the mean curvature flow of graphs*, Interfaces and Free Bound. **2** (2000), 341–359.

[9] P. DE MOTTONI AND M. SCHATZMANN, *Geometrical evolution of developed interfaces*, Trans. Amer. math. Soc. **347** (1995), 1533–1589.

[10] K. ERIKSSON AND C. JOHNSON, *Adaptive finite element methods for parabolic problems I: a linear model problem*, SIAM J. Numer. Anal. **28** (1991), 43–77.

[11] K. ERIKSSON AND C. JOHNSON, *Adaptive finite element methods for parabolic problems IV: nonlinear problems.*, SIAM J. Numer. Anal. **32** (1995), 1729–1749.

[12] K. ERIKSSON, C. JOHNSON, AND S. LARSSON, *Adaptive finite element methods for parabolic problems VI: Analytic semigroups*, SIAM J. Numer. Anal. **35** (198), 1315–1325.

[13] X. FENG AND A. PROHL, *Numerical analysis of the Allen-Cahn equation and approximation of the mean curvature flows*, Numer. Math. (to appear).

[14] F. Fierro and A. Veeser, *A posteriori error estimates and regularized total variation of characteristic functions*, SIAM J. Numer. Anal. (to appear).

[15] A. Friedman, *Variational Principles and Free-Boundary Problems*, Pure Appl. Math., John Wiley, New York, 1982.

[16] D. Kinderlehrer and G. Stampacchia, *An Introduction to Variational Inequalities and their Applications*, vol. 88 of Pure Appl. Math., Academic Press, New York, 1980.

[17] S.N. Kružkov, *First order quasilinear equations with several independent variables*, Mat. Sb. (N.S.) **81** (1970), 228–255.

[18] O. Lakkis and R.H. Nochetto, *A posteriori error analysis for the mean curvature flow of graphs*, (to appear).

[19] D. Kessler, R.H. Nochetto, and A. Schmidt, *A posteriori error control for the Allen-Cahn problem: Circumventing Gronwall's inequality*, (to appear).

[20] R. H. Nochetto, *A note on the approximation of free boundaries by finite elements*, RAIRO Model. Math. Anal. Numer. **20** (1986), 355–368.

[21] R.H. Nochetto and G. Savaré, *Nonlinear evolution equations governed by accretive operators: error control and applications*, (to appear).

[22] R.H. Nochetto, G. Savaré, and C. Verdi, *Error control for nonlinear evolution equations*, C.R. Acad. Sci. Paris, Serie I **326** (1998), 1437–1442.

[23] R.H. Nochetto, G. Savaré, and C. Verdi, *A posteriori error estimates for variable time-step discretizations of nonlinear evolution equations*, Comm. Pure Appl. Math. **53** (2000), 525–589.

[24] R.H. Nochetto, A. Schmidt, and C. Verdi, *A posteriori error estimation and adaptivity for degenerate parabolic problems*, Math. Comp. **69** (2000), 1–24.

[25] R.H. Nochetto, K.G. Siebert, and A. Veeser, *Pointwise a posteriori error control for elliptic obstacle problems*, Numer. Math. (to appear).

[26] R.H. Nochetto, K.G. Siebert, and A. Veeser, *Fully localized a posteriori error estimators and barrier sets for contact problems*, (to appear).

[27] R.H. Nochetto and L.B. Wahlbin, *Positivity preserving finite element approximation*, Math. Comp. **71** (2001), 1405–1419.

[28] J.-F. Rodrigues, *Obstacle Problems in Mathematical Physics*, vol. 134 of North-Holland Math. Stud., North-Holland, Amsterdam, 1987.

[29] J. Rulla, *Error analysis for implicit approximations to solutions to Cauchy problems*, SIAM J. Numer. Anal. **33** (1996), 68–87.

[30] G. Savaré, *Optimal error estimates for dissipative evolution problems*, this volume (2003).

[31] A.Schmidt and K.G. Siebert, *ALBERT: An adaptive hierarchical finite element toolbox*, Documentation, Preprint 06/2000 Universität Freiburg, 244 p.

[32] A. Veeser, *Efficient and reliable a posteriori error estimators for elliptic obstacle problems*, SIAM J. Numer. Anal. **39** (2001), 146–167.

Department of Mathematics and Institute for Physical Science and Technology, University of Maryland, College Park, MD 20742, USA
E-mail address: rhn@math.umd.edu

International Series of Numerical Mathematics, Vol. 147, 225–236

The Total Variation Flow

Matteo Novaga

Abstract. We consider the gradient flow of the total variation functional, stating general existence and uniqueness results. Particular attention is paid to self-similar solutions, which are partially classified. Finally, as an application, some explicit solutions of the denoising problem are given.

1. Introduction

The aim of this paper is to analyze the following variational parabolic equation

$$u_t - \operatorname{div}\left(\frac{Du}{|Du|}\right) = 0, \qquad u \in L^1_{\mathrm{loc}}(]0, +\infty[\,\times \mathbb{R}^N), \tag{1}$$

coupled with the initial condition

$$u(0, x) = u_0(x) \in L^1_{\mathrm{loc}}(\mathbb{R}^N), \tag{2}$$

some results on the problem (1)–(2) have been proved in [5], [6].

This equation corresponds to the L^2-gradient flow for the total variation functional

$$u \longrightarrow \int_{\mathbb{R}^N} |Du|,$$

starting from u_0. Such a flow has many applications to image denoising and reconstruction, and numerical simulations have been performed by several authors [13], [10], [9], [16], [15]. Existence and uniqueness for solutions of (1) has first been established in the case of bounded domains (see [4] and references therein) and then extended to $\mathbb{R}^N$ in [5] (see Theorem 2.4 below). The notion of solution employed in these papers is the so-called *entropy solution*, introduced by Kruzhkov [12] for scalar conservation laws, and first applied to parabolic equations in divergence form by Andreu *et al.* [2], in order to prove uniqueness with initial data in L^1_{loc}.

It has been proved in [3] that the solution of (1) reaches its asymptotic state in finite time, with an extinction profile which solves (up to a rescaling) the following eigenvalue problem

$$-\operatorname{div}\left(\frac{Du}{|Du|}\right) = u, \qquad u \in L^1_{\mathrm{loc}}(\mathbb{R}^N). \tag{3}$$

In [6] (see also Sections 3 and 4) the solutions of (3) are partially characterized in the case $N = 2$. Particular relevance is given to solutions which are in $W^{1,1}_{\mathrm{loc}}(\mathbb{R}^2)$

(see Proposition 3.4), or which are a finite sum of characteristic functions (see Theorems 4.1, 4.6 and 4.7).

The solutions u which have no jumps, i.e. $(-M) \vee u \wedge M \in W^{1,1}_{\rm loc}(\mathbb{R}^N)$ for any $M > 0$, can be constructed starting from the "fundamental solutions" $u(x) = \pm\frac{1}{|x|}$ (Proposition 3.4), and they can be obtained as limits of "tower solutions".

If a solution u of (3) is positive, then all the connected components of the sets $\{u \geq t\}$ are convex (see Proposition 3.1), and one can construct solutions which look like a tower, i.e. $u = \sum_{i=1}^{p} \alpha_i \chi_{B_i}$, for any choice of balls B_i such that $B_1 \supset B_2 \supset \ldots \supset B_p$ and for suitable constants $\alpha_i > 0$ (see Section 5). More generally, at least in two dimensions, one can construct more complicated towers (which can also oscillate) starting from sets which are not necessarily circles, but which must satisfy a suitable condition on the curvature of the boundary and on the mutual distance (see Theorem 4.6 and Section 5). These towers correspond to solutions of (3) which change sign, and in turn provide new solutions of problem (4) below.

Indeed, a related question is to find solutions of the so-called denoising problem, i.e.

$$\min_{u \in BV(\mathbb{R}^2)} \int_{\mathbb{R}^2} |Du| + \frac{1}{2\lambda} \int_{\mathbb{R}^2} (u-f)^2 \, dx, \tag{4}$$

and, in particular, to determine functions $f \in L^2(\mathbb{R}^2)$ for which the solution u of (4) is constructed by the soft-thresholding rule. In Propositions 6.1 and 6.2 below, it is shown how to construct solutions of problem (4), starting from solutions of (3).

2. Existence and uniqueness of solutions

In the sequel, we will denote by $L^1_w(]0,T[; BV(\mathbb{R}^N))$ the space of functions $w : [0,T] \to BV(\mathbb{R}^N)$ such that $w \in L^1(]0,T[\times\mathbb{R}^N)$, the maps $t \to \int_{\mathbb{R}^N} \phi dDw(t)$ are measurable for any $\phi \in C^1_0(\mathbb{R}^N;\mathbb{R}^N)$ and $\int_0^T |Dw(t)|(\mathbb{R}^N)dt < +\infty$. We will denote by $L^1_w(]0,T[; BV_{\rm loc}(\mathbb{R}^N))$ the space of functions $w : [0,T] \to BV_{\rm loc}(\mathbb{R}^N)$ such that $w\phi \in L^1_w(]0,T[; BV(\mathbb{R}^N))$ for any $\phi \in C^\infty_0(\mathbb{R}^N)$.

Definition 2.1. A function $u \in C([0,T]; L^2(\mathbb{R}^N))$ is called a *strong solution* of (1) if

$$u \in W^{1,2}_{\rm loc}(0,T; L^2(\mathbb{R}^N)) \cap L^1_w(]0,T[; BV(\mathbb{R}^N))$$

and there exists $z \in L^\infty\left(]0,T[\times\mathbb{R}^N;\mathbb{R}^N\right)$ with $\|z\|_\infty \leq 1$ such that

$$u_t = \operatorname{div} z \qquad \text{in } \mathcal{D}'\left(]0,T[\times\mathbb{R}^N\right)$$

and

$$\int_{\mathbb{R}^N} (u(t)-w)u_t(t) = \int_{\mathbb{R}^N} (z(t), Dw) - \int_{\mathbb{R}^N} |Du(t)| \quad \forall w \in L^2(\mathbb{R}^N) \cap BV(\mathbb{R}^N), \tag{5}$$

for a.e. $t \in [0,T]$.

Notice that a solution of (1) of class C^1 and such that $Du \neq 0$ is also a strong solution, setting $z := \frac{Du}{|Du|}$.

We recall from [5, Theorem 2] the following result.

Theorem 2.2. *Let $u_0 \in L^2(\mathbb{R}^N)$. Then there exists a unique strong solution u of (1) and (2) in $[0,T] \times \mathbb{R}^N$ for every $T > 0$. Moreover, if u and v are the strong solutions of (1) corresponding to the initial conditions $u_0, v_0 \in L^2(\mathbb{R}^N)$, then*

$$\|(u(t) - v(t)) \vee 0\|_2 \leq \|(u_0 - v_0) \vee 0\|_2 \quad \textit{for any } t > 0. \tag{6}$$

Since (1) is the gradient flow of a convex functional on $L^2(\mathbb{R}^N)$, which is an Hilbert space, Theorem 2.2 can be proved by means of classical techniques, up to minor modifications (see [7] for a general approach to these problems).

Notice that, putting $w = u(t)$ in (5), it follows

$$\int_{\mathbb{R}^N} (z, Du) = \int_{\mathbb{R}^N} |Du|. \tag{7}$$

Let us give a heuristic explanation of what the vector field z represents. Given a set $E \subseteq \mathbb{R}^N$ of finite perimeter, denote by $\partial^* E$ the reduced boundary of E in the sense of De Giorgi (see [1]). Condition (7) essentially means that z has unit norm and is orthogonal to the level sets of u. In some sense, z is invariant under local contrast changes. To be more precise, observe that if $u = \sum_{i=1}^p c_i \chi_{E_i}$ where E_i are sets of finite perimeter with disjoint closure, $c_i \in \mathbb{R}$ and

$$-\mathrm{div}\left(\frac{Du}{|Du|}\right) = f \in L^2(\mathbb{R}^N), \tag{8}$$

then also $-\mathrm{div}\,(\frac{Dv}{|Dv|}) = f$ for any $v = \sum_{i=1}^p d_i \chi_{E_i}$ where $d_i \in \mathbb{R}$ and $\mathrm{sign}(d_i) = \mathrm{sign}(c_i)$. In an informal way, this means that a local contrast change of a solution of (8) produces a new solution.

2.1. Entropy solutions

In order to generalize the previous result by allowing initial data which are only in $L^1_{\mathrm{loc}}(\mathbb{R}^N)$, let us introduce the set $\mathcal{P} \subset W^{1,\infty}(\mathbb{R})$ as follows

$$\mathcal{P} := \{p \in W^{1,\infty}(\mathbb{R}) : p' \geq 0, \ \mathrm{spt}(p') \text{ compact}\}.$$

Definition 2.3. A function $u \in C([0,T]; L^1_{\mathrm{loc}}(\mathbb{R}^N))$ is called an *entropy solution* of (1) and (2) if $u(t)$ converges to u_0 in $L^1_{\mathrm{loc}}(\mathbb{R}^N)$ as $t \to 0^+$,

$$p(u) \in L^1_w(]0,T[; BV_{\mathrm{loc}}(\mathbb{R}^N)) \qquad \forall p \in \mathcal{P},$$

and there exists $z \in L^\infty\left(]0,T[\times\mathbb{R}^N; \mathbb{R}^N\right)$ with $\|z\|_\infty \leq 1$ such that

$$u_t = \mathrm{div}\ z \qquad \text{in } \mathcal{D}'\left(]0,T[\times\mathbb{R}^N\right) \tag{9}$$

and

$$-\int_0^T \int_{\mathbb{R}^N} j(u-l)\eta_t + \int_0^T \int_{\mathbb{R}^N} \eta\, d|D(p(u-l))| + \int_0^T \int_{\mathbb{R}^N} z \cdot \nabla\eta\, p(u-l) \leq 0 \tag{10}$$

for all $l \in \mathbb{R}$, all $\eta \in C^\infty\left(]0,T[\times\mathbb{R}^N\right)$, with $\eta \geq 0$, $\eta(t,x) = \phi(t)\psi(x)$, being $\phi \in C_0^\infty\left(]0,T[\right)$, $\psi \in C_0^\infty(\mathbb{R}^N)$, and for all $p \in \mathcal{P}$, where $j(r) := \int_0^T p(s)\, ds$.

Inequality (10) is a weak form of equality (5); indeed, substituting u_t with $\operatorname{div} z$ in (5) and integrating by parts, using also $\|z\|_\infty \leq 1$ and the fact that η is nonnegative, one gets

$$\begin{aligned}\int_{\mathbb{R}^N} z \cdot \nabla\eta\, p(u-l) &= -\int_{\mathbb{R}^N} j(u-l)_t \eta - \int_{\mathbb{R}^N} \eta\, d(z, D\left(p(u-l)\right)) \\ &\geq -\int_{\mathbb{R}^N} j(u-l)_t \eta - \int_{\mathbb{R}^N} \eta\, d\left|(z, D\left(p(u-l)\right))\right|,\end{aligned}$$

which, after integration in time, gives the opposite inequality in (10).

From [5, Theorem 3, Proposition 4] one gets the following results.

Theorem 2.4. *Let $u_0 \in L^1_{\text{loc}}(\mathbb{R}^N)$. Then there exists a unique entropy solution of* (1) *and* (2) *in $[0,T] \times \mathbb{R}^N$ for all $T > 0$. Moreover, if $u_0, u_{0k} \in L^1_{\text{loc}}(\mathbb{R}^N)$ are such that $u_{0k} \to u_0$ in $L^1_{\text{loc}}(\mathbb{R}^N)$ and u, u_k denote the corresponding entropy solutions, then $u_k \to u$ in $C([0,T]; L^1_{\text{loc}}(\mathbb{R}^N))$ as $k \to +\infty$.*

Theorem 2.5. *Let $u_0 \in L^1_{\text{loc}}(\mathbb{R}^N)$ and u be the entropy solution of* (1) *and* (2). *Then*

$$p(u)_t \in L^2_{\text{loc}}(0,\infty; L^2_{\text{loc}}(\mathbb{R}^N)), \qquad t^{\frac{1}{2}} p(u)_t \in L^2_{\text{loc}}([0,\infty[; L^2_{\text{loc}}(\mathbb{R}^N)), \qquad \forall p \in \mathcal{P}.$$

Moreover, if $u_0 \geq -M$ for some $M > 0$, it follows

$$u'(t) \leq \frac{u(t)+M}{t} \qquad \text{for a.e. } t > 0.$$

Finally,

$$u_t \in L^1_{\text{loc}}\left(]0,T[; L^1_{\text{loc}}(\mathbb{R}^N)\right)$$

for any $T > 0$. A similar statement holds if $u_0 \leq M$ for some $M > 0$.

Remark 2.6. If $u_0 \in L^2(\mathbb{R}^N)$, then the strong solution of (1) and (2) coincides with the (unique) entropy solution.

3. Self-similar solutions

The aim of this section is to characterize the solutions of (3), when $N = 2$, i.e. to characterize the solutions of equation (1) which are self-similar. Therefore, we shall consider to the following eigenvalue problem

$$-\operatorname{div}\left(\frac{Du}{|Du|}\right) = u, \qquad u \in L^1_{\text{loc}}(\mathbb{R}^2). \tag{11}$$

An entropy solution of (11) is such that $p(u) \in BV_{\text{loc}}(\mathbb{R}^2)$, for any $p \in \mathcal{P}$. In particular, it follows that $u \in GBV_{\text{loc}}(\mathbb{R}^2)$.

We recall from [6] the following regularity result.

Proposition 3.1. *Let u be a solution of* (11). *Assume that $u \in L^p_{\mathrm{loc}}(\Omega)$ for some $p > 2$ and for some open set $\Omega \subseteq \mathbb{R}^2$. The following assertions hold.*

(a) *For any $t \in \mathbb{R}$ the sets $\{u > t\}$ and $\{u \geq t\}$ have a boundary of class $C^{1,\alpha}$ in Ω, for some $\alpha \in]0,1[$ ($C^{1,1}$ when $u \in L^\infty_{\mathrm{loc}}(\Omega)$).*

(b) *If $u \geq a$ in Ω (resp. $u \leq a$ in Ω) for some $a \in \mathbb{R}$ then, for any $t \in \mathbb{R}$, the curvatures κ of $\partial\{u > t\} \cap \Omega$ and h of $\partial\{u \geq t\} \cap \Omega$ satisfy $\kappa \geq a$ and $h \geq a$ (resp. $\kappa \leq a$ and $h \leq a$) in the sense of distributions.*

In the sequel, given a function u as in Proposition 3.1 and $t \in \mathbb{R}$, we shall always identify the set $\{u > t\}$ (resp. $\{u < t\}$) with its points of density one, which is an open set, and accordingly define $\{u \geq t\}$ as the complement of $\{u < t\}$.

Proposition 3.2. *Let u be a solution of* (11). *Assume that $u \in W^{1,1}_{\mathrm{loc}}(\Omega) \cap L^\infty_{\mathrm{loc}}(\Omega)$ for some open set $\Omega \subseteq \mathbb{R}^2$. Then for any $t \in \mathbb{R}$, $t \neq 0$, every connected component of $\partial\{u > t\} \cap \Omega$ is contained in the boundary of a ball of radius $1/|t|$.*

Proof. Let $t \in \mathbb{R}$, $t \neq 0$, let $\epsilon > 0$ and let Ω_ϵ be the open set defined as $\Omega_\epsilon := \{x \in \Omega : \ |u(x) - t| < \epsilon\}$. Let also $\gamma := \partial\{u > t\} \cap \Omega$. Since $u \in L^\infty_{\mathrm{loc}}(\Omega)$, by Proposition 3.1 the curve γ and the two curves $\gamma_\epsilon^- := \partial\{u > t - \epsilon\} \cap \Omega$, $\gamma_\epsilon^+ := \partial\{u < t + \epsilon\} \cap \Omega$ are of class $C^{1,1}$. Moreover, since $u \in W^{1,1}_{\mathrm{loc}}(\Omega)$, the two sets $\Sigma_\epsilon^- := \gamma_\epsilon^- \cap \gamma$ and $\Sigma_\epsilon^+ := \gamma_\epsilon^+ \cap \gamma$ are closed sets of zero $\mathcal{H}^1$-measure. Then the curve $\gamma \setminus (\Sigma_\epsilon^- \cup \Sigma_\epsilon^+)$ is contained in Ω_ϵ for any $\epsilon > 0$. Since γ is of class $C^{1,1}$, by (b) of Proposition 3.1 it follows that the curvature of γ belongs to $(t - \epsilon, t + \epsilon)$ for any $\epsilon > 0$. The conclusion follows by letting $\epsilon \to 0^+$. □

Note that if u is as in Proposition 3.2 then the set $\{u > t\}$ is a disjoint union of circles of radius $\frac{1}{|t|}$, for any $t \neq 0$ such that the boundary of $\{u > t\}$ is entirely contained in Ω.

The following result shows that there are no nontrivial solutions of (11) which are locally bounded and have no jumps.

Lemma 3.3. *Let $u \in W^{1,1}_{\mathrm{loc}}(\mathbb{R}^2) \cap L^\infty_{\mathrm{loc}}(\mathbb{R}^2)$ be a solution of* (11). *Then $u \equiv 0$.*

Proof. Assume by contradiction that $\sup u > 0$ (the case $\inf u < 0$ can be treated in a similar way). From Proposition 3.2 it follows that the set $\{u > t\}$ contains an open ball B_t of radius $\frac{1}{t}$, for any $t \in (0, \sup u)$. Fix $t \in (0, \sup u)$ and let $t^* := \sup_{B_t} u > t$. Then a connected component of the set $\{u = t^*\}$ is a closed ball $D_{t^*} \subset B_t$ of radius $\frac{1}{t^*}$. From (11) one gets

$$t^* = \frac{(t^*)^2}{\pi} \int_{D_{t^*}} u \, dx = \frac{(t^*)^2}{\pi} \int_{D_{t^*}} u \, dx = -\frac{(t^*)^2}{\pi} \int_{D_{t^*}} \operatorname{div} z \, dx = 2t^*,$$

which is a contradiction. □

Hereafter, we shall consider solutions u of (11) which satisfy the following property

$$\forall t \in \mathbb{R} \text{ there exists an open set } U_t \supset \partial\{u > t\} \text{ such that } u \in L^\infty_{\mathrm{loc}}(U_t). \qquad (12)$$

Observe that, if u satisfies (12), thanks to Proposition 3.1 all the level sets of u have boundary of class $C^{1,1}$.

The following proposition characterizes the solutions of (11) with have no jumps (see [6, Section 4]).

Proposition 3.4. *Assume that a solution u of (11) satisfies (12) and $(-M)\vee u\wedge M \in W^{1,1}_{\mathrm{loc}}(\mathbb{R}^2)$ for any $M > 0$. Then one of the following possibilities holds:*

(i) $u \equiv 0$;

(ii) *u is positive and the set $\{u > t\}$ is a ball of radius $\frac{1}{t}$, for any $t > 0$;*

(iii) *u is negative and the set $\{u < t\}$ is a ball of radius $-\frac{1}{t}$, for any $t < 0$;*

(iv) *u is nonnegative, $\{u > 0\}$ is a halfspace and the set $\{u > t\}$ is a ball of radius $\frac{1}{t}$, for any $t > 0$;*

(v) *u is nonpositive, $\{u < 0\}$ is a halfspace and the set $\{u < t\}$ is a ball of radius $-\frac{1}{t}$, for any $t < 0$;*

(vi) *both $\{u > 0\}$ and $\{u < 0\}$ are halfspaces, the set $\{u > t\}$ is a ball of radius $\frac{1}{t}$, for any $t > 0$, and the set $\{u < \tau\}$ is a ball of radius $-\frac{1}{\tau}$, for any $\tau < 0$.*

Proposition 3.4 essentially means that the solutions of (11) with no jumps can be obtained from the fundamental solutions $u(x) = \pm\frac{1}{|x|}$, by a suitable translation of the level-sets $\{u > t\}$ (or $\{u < t\}$).

4. Properties of the level-sets

In this section we shall consider "tower solutions" of (11), i.e. solutions of the type $u = \sum_{i=1}^{p} c_i \chi_{E_i}$ for a suitable choice of sets $E_i \subset \mathbb{R}^2$ and constants $c_i \in \mathbb{R}$.

Let us denote in the sequel by $C_0, C_1, \dots C_m$ $m+1$ open sets with boundary of class $C^{1,1}$ with the following properties:

$\overline{C_i} \subset C_0$ for any $i \in \{1, \dots, m\}$;

$\overline{C_i} \cap \overline{C_j} = \emptyset$ for any $i, j \in \{1, \dots, m\}$.

Choosing $k \in \{0, \dots, m\}$, let us define

$$F := C_0 \setminus \bigcup_{i=1}^{m} \overline{C_i}, \qquad \widehat{F} := F \cup \bigcup_{j=k+1}^{m} \overline{C_j}.$$

Let also ν^F be the exterior unit normal to ∂F.

We want to identify necessary and sufficient conditions on F in order to have $F = \{u = t\}$ and $\bigcup_{j=k+1}^{m} C_j = \{u > t\}$, for some $t \in \mathbb{R}$ and for some tower solution u of (11). Using (9), this implies that there exists a vector field $z \in L^\infty(F, \mathbb{R}^2)$ such that

$$\begin{cases} -\operatorname{div} z = t & \text{on } F, \\ \|z\|_\infty \le 1, \\ [z, \nu^F] = -1 & \mathcal{H}^1\text{-a.e. on } \partial C_i, i \in \{1, \dots k\}, \\ [z, \nu^F] = 1 & \mathcal{H}^1\text{-a.e. on } \partial C_j, j \in \{k+1, \dots m\}, \end{cases} \tag{13}$$

where $[z, \nu^F]$ denotes the trace of the scalar product between z and ν^F (see [5] for the precise definition and properties).

Notice that a problem closely related to (13) has already been considered in the literature related to capillarity problems, see [11], [8].

On the other hand, given a function $u = \sum_{i=1}^M t_i \chi_{F_i}$, where the sets $F_i \subset \mathbb{R}^2$ are disjoint, with boundary of class $C^{1,1}$ and such that $\bigcup_{i=1}^M \overline{F_i} = \mathbb{R}^2$, then u is a solution of (11) if and only if there exist vector fields $z_i \in L^\infty(F_i, \mathbb{R}^2)$ satisfying (13) for any $i \in \{1, \ldots, M\}$ and such that $[z, \nu_i^F] + [z, \nu_j^F] = 0$ on $\partial F_i \cap \partial F_j$.

It follows that the characterization of the tower solutions of (11) reduces to solve problem (13).

4.1. Bounded domains

Let us first consider the case in which the set F is bounded. Set

$$
\begin{aligned}
j_0 &:= \frac{\sum_{i=0}^k P(C_i) - \sum_{j=k+1}^m P(C_j)}{|F|}, \\
\mathcal{F}_F(E) &:= P(E, F) + \sum_{i=0}^k |\partial^* E \cap \partial C_i| - \sum_{j=k+1}^m |\partial^* E \cap \partial C_j| - j_0 |E|.
\end{aligned}
$$

By the Gauss–Green Formula, problem (13) reduces to

$$
\begin{cases}
-\operatorname{div} z &= j_0 \quad \text{on } F, \\
\|z\|_\infty &\leq 1, \\
[z, \nu^F] &= -1 \quad \text{on } \partial C_i, i \in \{0, \ldots k\}, \\
[z, \nu^F] &= 1 \quad \text{on } \partial C_j, j \in \{k+1, \ldots m\}.
\end{cases}
\tag{14}
$$

Finally, we denote by $\mathcal{A}$ the family of all sets $E \subseteq F$ such that $\partial E \cap \operatorname{int}(F)$ consists of pieces of circumferences of radius $1/j_0$, which meet ∂F tangentially and which span an angle less than or equal to π. We also require the contact angle to be π on $\bigcup_{i=0}^k \partial C_i$ and zero on $\bigcup_{j=k+1}^m \partial C_j$, and we impose that $\{\emptyset, F\} \in \mathcal{A}$.

The following result is proved in [6].

Theorem 4.1. *The following conditions are equivalent.*

(a) *There exists a vector field $z : F \to \mathbb{R}^2$ satisfying* (14).

(b) *There holds*

$$
j_0 \int_F w \leq \int_F |Dw| + \sum_{i=0}^k \int_{\partial C_i} w - \sum_{j=k+1}^m \int_{\partial C_j} w \quad \forall w \in BV(F). \tag{15}
$$

(c) *For any $E \subseteq F$ of finite perimeter, there holds*

$$
\mathcal{F}_F(E) \geq 0. \tag{16}
$$

(d) *There holds*

$$
\min_{E \in \mathcal{A}} \mathcal{F}_F(E) = 0. \tag{17}
$$

Remark 4.2. Notice that, when $k = 0$, j_0 tends to zero as C_0 tends to $\mathbb{R}^2$; in this case, the minimization problem (17) reduces to the problem considered in [5, Theorem 6].

Notice that, in most of the cases, the family $\mathcal{A}$ is a finite set, hence condition (d) is often easy to verify. However, Theorem 4.6 below will provide more explicit conditions, involving the curvature of ∂F and the distance between the sets C_i.

Let us start with the definition of the so-called *ball condition* [8].

Definition 4.3. Let $\Omega \subseteq \mathbb{R}^2$ be an open set with boundary of class $C^{1,1}$, and let $\rho > 0$. We say that Ω satisfies the *ρ-ball condition* if a ball of radius ρ can be rotated along $\partial\Omega$ in the interior of Ω such that no antipods of the ball lie on $\partial\Omega$.

We recall the following result [8, Theorem 4.1].

Lemma 4.4. *Let $\Omega \subseteq \mathbb{R}^2$ be an open set satisfying the ρ-ball condition, for some $\rho > 0$. Then $\sup_{\partial\Omega} \kappa \le \frac{1}{\rho}$. Moreover, given a ball $B \subset \Omega$ of radius ρ and tangent to $\partial\Omega$, the set $\Gamma \cap \partial B$ is connected for any connected component Γ of $\partial\Omega$, and it spans an angle less than π.*

Observe that, in general, the inequality $\sup_{\partial\Omega} \kappa \le \frac{1}{\rho}$ does not imply the ρ-ball condition for the set Ω. Let us also point out that if Ω is a convex set with boundary of class $C^{1,1}$ such that $\sup_{\partial\Omega} \kappa < \frac{1}{\rho}$, then Ω satisfies the ρ-ball condition.

Let F be as in Theorem 4.1. In the sequel, as soon as we say that F or $\widehat{F}$ satisfies the ball condition, we shall always mean the $\frac{1}{j_0}$-ball condition.

Remark 4.5. If C_i is convex for any $i \in \{0, \dots m\}$, $\sup_{\partial C_0} \kappa < j_0$ and

$$\operatorname{dist}(\partial C_i, \partial C_j) > \frac{2}{j_0} \qquad \forall (i,j) \in \{0, \dots k\},\ i \neq j,$$

then F satisfies the ball condition.

The following result, proved in [6], provides a necessary condition and a sufficient condition for the existence of a vector field $z : F \to \mathbb{R}^2$ satisfying (14).

Theorem 4.6. *Let F be as above and denote by κ the curvature of ∂F. Assume that there exists a vector field $z : F \to \mathbb{R}^2$ satisfying* (14)*. Then*

$$\sup_{\partial C_i} \kappa \le j_0 \le \inf_{\partial C_j}(-\kappa), \qquad i \in \{0, \dots, k\},\ j \in \{k+1, \dots, m\}. \tag{18}$$

Conversely, assume that $\widehat{F}$ satisfies the ball condition, that the inequalities at the right hand side of (18) *hold, and*

$$\operatorname{dist}(\partial C_i, \partial C_j) > \frac{2}{j_0} \qquad \forall (i,j) \in \{0, \dots, k\}^2 \cup \{k+1, \dots, m\}^2, i \neq j. \tag{19}$$

Then there exists a vector field $z : F \to \mathbb{R}^2$ satisfying (14).

4.2. Unbounded domains

Let us now consider the case $C_0 = \mathbb{R}^2$ and $C_1, \dots, C_m$ as above, which means the F is unbounded but $\mathbb{R}^2 \setminus F$ is bounded. In this case, problem (13) reduces to the existence of a vector field $z \in L^\infty(F, \mathbb{R}^2)$ such that

$$\begin{cases} -\operatorname{div} z = 0 & \text{on } F, \\ \|z\|_\infty \le 1, & \\ [z, \nu^F] = -1 & \mathcal{H}^1\text{-a.e. on } \partial C_i, i \in \{1, \dots k\}, \\ [z, \nu^F] = 1 & \mathcal{H}^1\text{-a.e. on } \partial C_j, j \in \{k+1, \dots m\}. \end{cases} \tag{20}$$

There holds a result analogous to Theorem 4.1.

Theorem 4.7. *The following conditions are equivalent*

(i) *Problem* (20) *has a solution.*

(ii) *There holds*

$$0 \le \int_F |Dw| + \sum_{i=1}^{k} \int_{\partial C_i} w - \sum_{j=k+1}^{m} \int_{\partial C_j} w \quad \forall w \in BV(F). \tag{21}$$

(iii) *For any $X \subseteq F$ of finite perimeter, there holds*

$$P(X, F) \ge \left| \sum_{j=k+1}^{m} \mathcal{H}^1(\partial^* X \cap C_j) - \sum_{i=1}^{k} \mathcal{H}^1(\partial^* X \cap C_i) \right|. \tag{22}$$

(iv) *let E_1 be a solution of the variational problem*

$$\min \left\{ P(E) : \bigcup_{j=k+1}^{m} C_j \subseteq E \subseteq \mathbb{R}^2 \setminus \bigcup_{i=1}^{k} C_i \right\}, \tag{23}$$

then there holds

$$P(E_1) = \sum_{j=k+1}^{m} P(C_j); \tag{24}$$

let E_2 be a solution of the variational problem

$$\min \left\{ P(E) : \bigcup_{i=1}^{k} C_i \subseteq E \subseteq \mathbb{R}^2 \setminus \bigcup_{j=k+1}^{m} C_j \right\}, \tag{25}$$

then there holds

$$P(E_2) = \sum_{i=1}^{k} P(C_i). \tag{26}$$

Notice that *(iv)* implies that each C_i is a convex set. Moreover, since any minimizer of problems (23) and (25) has a boundary (lying inside F) made of a finite number of segments which intersect tangentially ∂F (and there are only a finite number of such segments), the number of such minimizers is finite. Finally, conditions (24) and (26) are essentially distance conditions between sets C_i of

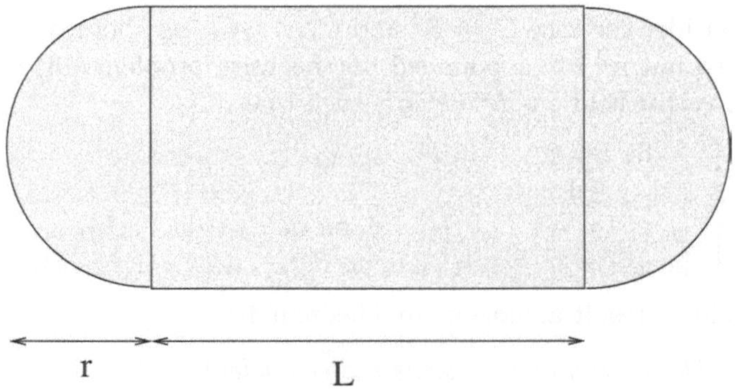

FIGURE 1. A bean-shaped admissible set

the same type, for example they are satisfied if $\mathrm{dist}(\partial C_i, \partial C_j) > P(C_l)$, for any $(i,j,l) \in \{1,\ldots,k\}^3 \cup \{k+1,\ldots,m\}^3$, $i \neq j$.

5. Some examples

In order to clarify the conditions given in Section 4, we shall discuss some explicit examples.

Example 1. Let $u(x,y) = \frac{2x}{x^2+y^2}$; then u is a solution of (11) with no jumps, satisfying condition (vi) of Proposition (3.4). Notice that u is not continuous at $(0,0)$.

Example 2. Let $F \subset \mathbb{R}^2$ be the set in Figure 1. It is easy to check that F satisfies the assumptions of Theorem 4.6, since it is a convex set with $C^{1,1}$ boundary and there holds

$$\sup_{\partial F} \kappa = \frac{1}{r} < \frac{2\pi r + 2L}{\pi r^2 + 2rL} = \frac{P(F)}{|F|}. \tag{27}$$

Moreover, since the inequality in (27) is strict, the solution of (1) starting from $\chi_{F'}$ remains a characteristic function for any convex set F' of class $C^{1,1}$ close enough to F in the $C^{1,1}$-norm.

Example 3. Let $F \subset \mathbb{R}^2$ be the union of two disjoint balls of radius r whose centers are at distance L. In this case, conditions (24), (26) of Theorem 4.7 become

$$L \geq \pi r.$$

It follows that, under this condition, the function $u = \frac{2}{r}\chi_F$ is a solution of (11).

Example 4. We shall now describe a class of radially symmetric tower solutions. Let $0 = R_0 < R_1 < \cdots < R_p < R_{p+1} = +\infty$, and let $\Omega_i := B_{R_i} \setminus \overline{B}_{R_{i-1}}$, $i = 1,\ldots,p+1$ (where we set $B_0 = \emptyset$, $B_{+\infty} = \mathbb{R}^2$). Let also $a_1,\ldots,a_{p+1}$ be real numbers such that $a_{p+1} = 0$ and $a_i \neq a_{i+1}$, for $i = 1,\ldots,p$. Then, there exist real numbers

$b_1, \ldots, b_{p+1}$, with $b_{p+1} = 0$ and $b_i \neq b_{i+1}$, such that $\mathrm{sign}(b_{i+1} - b_i) = \mathrm{sign}(a_{i+1} - a_i)$ for any $i \in \{1, \ldots, p\}$ (i.e. the numbers b_i have the same "qualitative ordering" of a_i) and the function $u = \sum_{i=1}^{p} b_i \chi_{\Omega_i}$ is a solution of (11). Indeed, applying Theorem 4.6 to the sets Ω_i, one can check that the function u is a solution of (11) when

$$b_i = -\frac{2}{\mathrm{sign}(a_i - a_{i-1}) R_{i-1} + \mathrm{sign}(a_{i+1} - a_i) R_i}.$$

Notice that, since $R_i > R_{i-1}$, it follows $\mathrm{sign}(b_i) = -\mathrm{sign}(a_{i+1} - a_i)$.

This example has already been discussed in [14].

6. Solutions of the denoising problem

In this last section, we will show how to obtain explicit solutions of the denoising problem (4), starting from solutions of (11), see [5, Proposition 7, Proposition 8].

Proposition 6.1. *Let $\lambda > 0$, $b \in \mathbb{R}$ and $u \in BV(\mathbb{R}^2)$ a solution of* (11). *Then the function $v = au$ is the solution of the variational problem* (4) *with $a := \mathrm{sign}(b)[(|b| - \lambda) \vee 0]$ and $f := bu$. Conversely, if $v = au \in BV(\mathbb{R}^2)$ is the solution of* (4) *with $f = bu$, for some $a, b \in \mathbb{R}$ with $b - a = \pm\lambda$, then the function u is a solution of* (11).

The following proposition, which extends the previous result, clarifies how to construct solutions of (4) from tower solutions of (11).

Proposition 6.2. *Let $C_0, C_1, \ldots, C_m \subset \mathbb{R}^2$ be open sets with boundary of class $C^{1,1}$, such that $C_i \cap C_j = \emptyset$ for $i \neq j$, $\bigcup_{i=0}^{m} \overline{C_i} = \mathbb{R}^2$ and C_i is bounded for any $i \in \{1, \ldots, m\}$. Assume that the function $u = \sum_{i=1}^{m} c_i \chi_{C_i}$ is a solution of* (11), *for some $c_i \in \mathbb{R}$. Let now $b_i \in \mathbb{R}$, $i \in \{1, \ldots, m\}$, $\lambda > 0$ and $a_i = \mathrm{sign}(b_i)[(|b_i| - \lambda) \vee 0]$. Then the function $u = \sum_{i=1}^{m} a_i c_i \chi_{C_i}$ is the solution of the variational problem* (4) *with $f = \sum_{i=1}^{m} b_i c_i \chi_{C_i}$.*

References

[1] L. Ambrosio, N. Fusco, and D. Pallara. *Functions of Bounded Variation and Free Discontinuity Problems.* Oxford Mathematical Monographs, 2000.

[2] F. Andreu, J.M. Mazón, S. Segura and J. Toledo. *Existence and uniqueness for a degenerate parabolic equation with L^1-data.* Trans. Amer. Math. Soc., 315:285–306, 1999.

[3] F. Andreu, V. Caselles, J.I. Diaz and J.M. Mazón. *Qualitative properties of the total variation flow.* Jour. Funct. Anal., 188:516–547, 2002.

[4] F. Andreu, V. Caselles and J.M. Mazón. *Existence and uniqueness of solutions for a parabolic quasilinear problem for linear growth functionals with L^1 data.* Math. Ann., 322:139–206, 2002.

[5] G. Bellettini, V. Caselles and M. Novaga. *The Total Variation Flow in $\mathbb{R}^N$.* Journal of Diff. Eqs., 184:475–525, 2002.

[6] G. Bellettini, V. Caselles and M. Novaga. *Self-similar solutions of the total variation flow.* In preparation.

[7] H. Brezis. *Operateurs Maximaux Monotones.* North-Holland, Amsterdam, 1973.

[8] J.T. Chen. *On the existence of capillary free surfaces in the absence of gravity.* Pacific J. Math., 88:323–361, 1980.

[9] A. Chambolle and P.-L. Lions. *Image recovery via total variation minimization and related problems.* Numer. Math., 76:167–188, 1997.

[10] S. Durand, F. Malgouyres, and B. Rougé. *Image de-blurring, spectrum interpolation and application to satellite imaging.* Control, Optimisation and Calculus of Variation, 5, 2000.

[11] E. Giusti. *On the equation of surfaces of prescribed mean curvature. Existence and uniqueness without boundary conditions.* Invent. Math., 46:111–137, 1978.

[12] S.N. Kruzhkov. *First order quasilinear equations in several independent variables.* Math. USSR-Sb., 10:217–243, 1970.

[13] L. Rudin, S. Osher, and E. Fatemi. *Nonlinear total variation based noise removal algorithms.* Physica D, 60:259–268, 1992.

[14] D.M. Strong and T. F. Chan. *Relation of regularization parameter and scale in total variation based image denoising.* CAM Report, 1996.

[15] G.H. Golub T. F. Chan and P. Mulet. *A nonlinear primal-dual method for total variation based image restoration.* SIAM J. Sci. Computing, 20:1964–1977, 1999.

[16] C. R. Vogel and M. E. Oman. *Fast total variation based image reconstruction.* Proceedings of the 1995 ASME Design Engineering Conferences, Vol. 3, pages 1009–1015, 1995.

Dipartimento di Matematica,
Università di Pisa,
via Buonarroti 2
56127 Pisa, Italy
E-mail address: `novaga@dm.unipi.it`

International Series of Numerical Mathematics, Vol. 147, 237–254

A Mathematical Model for Diffusion-induced Grain Boundary Motion

Oliver Penrose and J.W. Cahn

Abstract. For a film of metallic alloy immersed in a suitable vapour, we use a system of four coupled nonlinear differential equations to model the steady diffusion-induced motion of: a grain boundary, the surfaces of the two grains, and the triple junction where they all meet. One of the equations models diffusion along the moving grain boundary; another models the force balance which determines its speed. The remaining two equations model diffusion in the surfaces of the two grains. The equations are linked by boundary conditions at the triple junction. The resulting system of differential equations and boundary conditions is solved here for the case of 'trailing' grain boundaries (ones where the growing crystal grain develops as a layer beneath the surface of the specimen rather than filling up the entire space between the two surfaces) in a limit where the elastic driving force is very small. The main result is that for small values of Δc, defined as the (experimentally controllable) jump in mole fraction of solute at the triple junction, the growth velocity of the trailing grain is approximately proportional to $(\Delta c)^4$, but for large positive Δc the velocity is approximately proportional to $(\Delta c)^5$. The thickness of the trailing grain is approximately proportional to $(\Delta c)^{-2}$ for small Δc and to $(\Delta c)^{-8/3}$ for large. There is a negative value of Δc beyond which the model predicts that the velocity and thickness are independent of Δc, but this result should be treated with caution because the solution may be unstable.

1. Introduction

Diffusion-induced grain boundary motion (DIGM) was first observed by Hillert and Purdy in 1978 [9]. A thin foil of iron was placed in a vapour of zinc atoms. The zinc diffused along the boundaries between the different crystal grains in the foil, but not into the grains themselves, where the diffusivity is much lower. Parts of these grain boundaries were then observed to move, the grain on one side growing while the grain on the other side shrank. The zinc atoms which had diffused in along the grain boundary were left behind in the newly grown part of the crystal, thereby alloying it. Hillert and Purdy also demonstrated that DIGM occurred for de-alloying by placing a foil of an iron-zinc alloy in a vacuum. DIGM has been observed in many other alloy systems and with many geometries; for a review see ref. [7].

There has been controversy in the metallurgical community about the force that drives DIGM. With a view to helping resolve the controversy, Cahn et al [2] put forward a phase-field model[1] in which the consequences of different assumptions about the driving force could be worked out and compared with experiment. It was concluded that the most likely candidate was a mechanism suggested (in a slightly different context) by Sulonen [14] and first studied quantitatively by Hillert[8]; this mechanism is based on the energy of elastic deformations caused by mismatch between the sizes of the two different types of atom[2]. It remains the only mechanism consistent with all experiments and with some insightful *Gedanken* experiments.

The analysis given in ref. [2] is based on the assumption that the grain boundary is flat, but real grain boundaries are not flat: they can curve in the middle while still connecting the two sides of the foil, or they can curve so much that they separate into two parts, one close to each face of the foil. Both types of grain boundary were observed in the original experiments [9]. We shall call them connecting and trailing grain boundaries; they are illustrated in Fig. 1.

To model these curved grain boundaries, Fife et al.[5] developed the mathematics of the phase-field model further, deriving a sharp-interface approximation which led to a system of differential equations for the shape of the grain boundary (represented in this approximation as a geometrical surface). Some numerical studies of both phase-field and sharp-interface models were described at this conference by Vanessa Styles (see also ref. [6]).

2. The equations for the grain boundary

The sharp-interface equations obtained by Fife et al. [5] can also be derived by physical arguments which do not rely on the phase field model [1, 3]. There are two differential equations. One of them, describing diffusion within the grain boundary, can be written (in a specially chosen system of units, *v.i.*)

$$\frac{d^2 c_{gb}}{ds^2} = (c_{gb}(s) - c_a(s))v\cos\theta \qquad (-\tfrac{1}{2}\pi < \theta < \tfrac{1}{2}\pi) \tag{1}$$

Here c_{gb} denotes the local mole fraction of solute within the advancing grain boundary, c_a denotes the local mole fraction just in front of the advancing grain boundary, s denotes arc length measured along the grain boundary, v is the (positive) velocity of the grain boundary in the positive y direction (the y axis being taken parallel to the faces of the foil, as shown in Fig. 1) and θ is the angle between the grain boundary and the positive x-axis, which is taken to be perpendicular to the face of the foil. The local mole fraction in the grain left behind by the advancing grain boundary is equal to c_{gb}.

[1] For some existence and uniqueness results about the equations of this model, see ref. [4].

[2] For a detailed analysis of this interaction, arising out of discussions at the conference between J. Ockendon and one of the authors, see ref. [13].

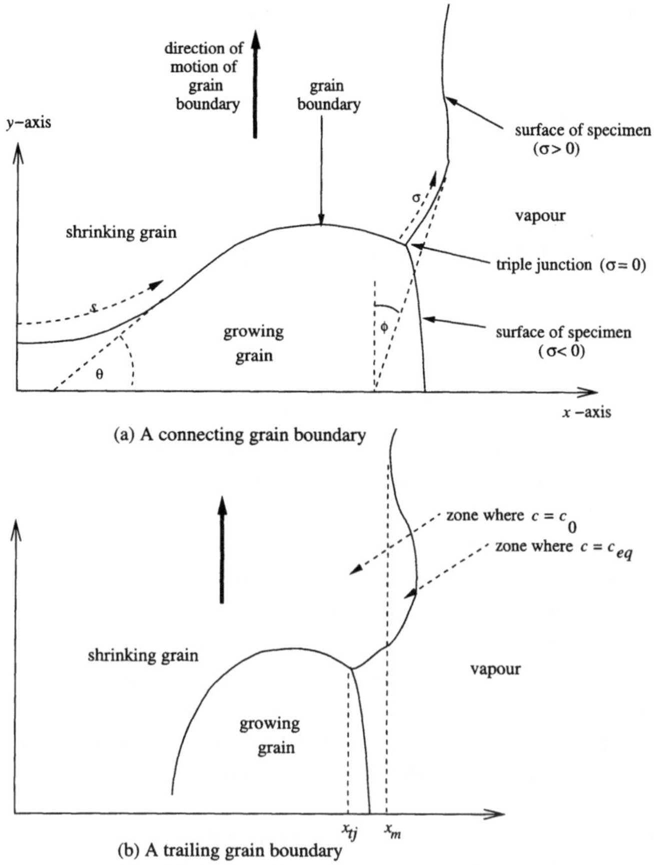

FIGURE 1. The two types of steadily moving grain boundary and the notation used in this paper. In this paper we solve the trailing case. Arc length in the grain boundary is denoted by s; in the solid-vapour surface, by σ (with $\sigma > 0$ in the shrinking grain, $\sigma < 0$ in the growing grain). The angle between the grain boundary and the x-axis is denoted by θ; the angle between the solid-vapour interface and the y-axis, by ϕ.

Usually c_a is equal to c_0, the concentration in the shrinking grain well ahead of the moving grain boundary, but near the surface there can be a thin zone in which the metal in front of the grain boundary arrived there as a result of deposition from the vapour; in this zone, c_a is approximately equal to c_{eq}, the concentration of solute at a flat interface in equilibrium with the vapour. Let x_m be the smallest local minimum value of x reached by the surface of the shrinking grain at points other than the triple junction, and let x_{tj} be the x-coordinate of the triple junction. If $x_m > x_{tj}$ (as is the case in Fig.1(b)) then the zone, if it exists, does not meet the grain boundary, but if $x_m \le x_{tj}$ then its inner boundary meets the grain boundary at a point whose x-coordinate is x_m. The rule determining c_a can be written

$$c_a = \begin{cases} c_0 & \text{if} \quad 0 \le x < \min(x_m, x_{tj}), \\ c_{eq} & \text{if} \quad x_m < x < x_{tj}. \end{cases} \tag{2}$$

The second differential equation, expressing the local normal velocity $v \cos\theta$ of the grain boundary in terms of the elastic driving force and the surface tension force due to the curvature of the grain boundary, is

$$v \cos\theta = d\theta/ds + \epsilon(c_{gb} - c_a)^2 \qquad \text{if } -\tfrac{1}{2}\pi < \theta < \tfrac{1}{2}\pi, \tag{3}$$

where

$$\epsilon := Y\eta^2 D\lambda / M\gamma^2 \tag{4}$$

is a dimensionless constant measuring the strength of the elastic interaction driving the DIGM. In this definition, Y is a certain elastic modulus, η is the rate of change of stress-free lattice spacing with respect to changes in mole fraction, D is the diffusivity in the grain boundary, λ is its thickness, M is its mobility (velocity per unit driving pressure) and γ is its energy per unit area.

The units of measurement have been chosen so that $D\lambda = 1$ and also $M\gamma = 1$. In a general system of units, s would be replaced by $(D\lambda/M\gamma)s$ and v by $[(M\gamma)^2/D\lambda]v$, but the formula (4) would not change. Although the values of both D and M are not known with any accuracy for real metals and are expected to vary greatly with composition and with temperature (as an Arrhenius exponential of the reciprocal of temperature) only their ratio, which is less sensitive to temperature, appears in (4); hence the order of magnitude of ϵ can be estimated from theory or experiment. Nevertheless, since η can vary widely (from near zero in silver-gold to 0.15 in nickel-gold) we expect ϵ also to vary widely. In the present paper we investigate the behaviour of this mathematical model in the limit of very small ϵ.

Eqns (1)–(3) are given here only in the form they take when $-\frac{1}{2}\pi < \theta < \frac{1}{2}\pi$ on the whole of the grain boundary. For values of θ outside this range the equations are different, reflecting the possibility of the grain boundary's passing a point in the metal more than once; this generalization is considered in section 8.

The solutions of eqns (1) and (3) satisfy boundary conditions at each end of the curve. One of these ends is the triple junction shown in Fig. 1, at which the grain boundary meets the external surfaces of the two crystal grains. For connecting grain boundaries, the other end is a similar triple junction at the corresponding point on the other side of the mid-plane of the foil. Assuming that the grain boundary is symmetrical about this plane, the concentration will be an even function of displacement from this plane and the angle will be an odd function, so that we need consider one half of the grain boundary only, with suitable conditions at the mid-point, which we also take as the origin of the arc length coordinate s:

$$\left.\begin{aligned} \theta &= 0 \\ dc_{gb}/ds &= 0 \end{aligned}\right\} \tag{5}$$

For trailing grain boundaries, there is no symmetry, but instead there are conditions in the far distance behind the triple junction, namely

$$\left.\begin{array}{rcl}\theta & \rightarrow & \frac{1}{2}\pi \\ c_{gb}-c_a & \rightarrow & 0\end{array}\right\} \tag{6}$$

where the location of the point at which $s = 0$ is now arbitrary.

In addition to the condition (5) or (6) there are two conditions at the triple junction itself. These conditions are the same for both cases. One of them (arising from the requirement that the chemical potentials must be continuous at the triple junction) can be written

$$c_{tj} = c_{eq} \tag{7}$$

where c_{tj} denotes the value of c_{gb} at the triple junction and c_{eq} is defined above. By varying the partial pressure of solute atoms in the vapour, the experimenter can control the value of c_{eq} and hence that of c_{tj}. The second boundary condition concerns the direction of the tangent to the grain boundary at the triple junction, *i.e.*, the value of $\theta_{tj} := \theta(s_{tj})$ where s_{tj} denotes the value of s at the triple junction. In refs [2, 5] the simplest possible approximation was used, namely that the grain boundary is perpendicular to the surface (i.e. $\theta_{tj} = 0$), but this approximation can obscure some interesting effects. To complete the specification of the boundary conditions, we need a condition from which θ_{tj} can be determined; this condition is considered in the following section.

3. The Mullins theory for surface motion by surface diffusion

In the presence of a grain boundary the surface of the specimen is not flat because there is a groove at the triple junction, where the angles are determined by the surface tensions (free energies per unit area) of the three interfaces meeting there. The resulting curvature affects the chemical potential at the surface, and gradients in this chemical potential cause matter to diffuse along the surface so that it is deposited in some places and removed from others.

A theory of this surface motion by diffusion in the case of a pure substance was developed by Mullins [10]. According to his ideas, the chemical potential varies linearly with the curvature, the material flux along the surface is proportional to the gradient of the chemical potential, and the deposition rate on the surface is minus the surface divergence of the flux; the resulting local normal velocity of the surface is therefore proportional to minus the surface Laplacian of surface curvature, in our case $d^3\phi/d\sigma^3$ where σ is arc length along the surface and ϕ is the angle between the surface and the y axis. If the surface profile of the metal moves like a wave travelling in the y direction with velocity v, this local deposition rate is $-v\sin\phi$, and so ϕ satisfies the differential equation

$$b\, d^3\phi/d\sigma^3 = v \sin\phi \tag{8}$$

where b is a dimensionless material constant. In a general system of units, b would be replaced by the material constant B used by Mullins, which has dimensions (length)^4 (time)$^{-1}$ and is related to b by the formula $B = b(D\lambda)^2/(M\gamma)$. Eq (8), which was given by Kanel, Novick–Cohen and Vilenkin [12], is the nonlinear generalization of an equation formulated by Mullins [11], which in our notation would read $b\, d^4x/dy^4 = v\, dx/dy$.

In ref. [11] Mullins found that the end of the moving grain boundary would make a fixed angle with the x axis, independent of the speed v (though it would change sign if the sign of v were reversed). His result can be written

$$\theta_{tj} = -\tfrac{1}{6}\delta \tag{9}$$

where δ is the discontinuity in ϕ, which is determined by the surface tension equilibrium at the triple junction. One of the results of the present paper is a generalization of (9), eqn (35) below. The extra term on the left side of (35) arises from the nonlinearity of eqn (8). The right side of (35) would be replaced by zero for the problems considered by Mullins and by Kanel et al.; this term arises from a further effect which we now consider.

Eqn (8) was formulated for a pure substance in equilibrium with its vapour. Here, however, we are concerned with an alloy whose concentration may vary along the surface. Furthermore the formulation of the boundary condition at the triple junction has to be amended to take account of flow into and out of the grain boundary. The relevant generalization of Mullins' theory has been considered by Brener and Temkin [1], who (using the same linearization approximation as Mullins [11]) obtain a pair of equations for the fluxes of both types of atom along the surface. The present paper applies to an alloy (*e.g.* iron-zinc) where the solute (zinc) is volatile while the solvent (iron) can be assumed non-volatile, in which case only one equation is necessary; for a travelling wave the equation is [3]

$$b(1 - c_{eq})\, d^3\phi/d\sigma^3 = (1 - c_a)v \sin\phi \tag{10}$$

where c_a means essentially the same thing as in eqn (1), i.e. $c_a = c_0$ at points which the surface is passing for the first time, and $c_a = c_{eq}$ if the surface has been there before. In the rest of this paper we confine ourselves to the case where c_{eq} and c_0 are close enough together to justify treating the factors $1 - c_{eq}$ and $1 - c_{de}$ as equal, so that (8) can be used in place of (10).

The boundary conditions on (8) at infinity are, naturally,

$$\phi(\sigma) \to 0 \text{ as } \sigma \to \pm\infty \tag{11}$$

At the triple junction (where we take $\sigma = 0$) the function $\phi(\sigma)$ is discontinuous. The angles there are related to each other and to the angle at the end of the grain boundary by

$$\begin{aligned} \phi(+0) - \phi(-0) &= \delta \\ \phi(+0) + \phi(-0) &= -2\theta_{tj} \end{aligned} \tag{12}$$

where in the second line we have assumed that the surface tensions of the two external surfaces are equal, so that at the triple junction the tangent to the grain boundary bisects the angle between the tangents to the two external surfaces. A second condition comes from requiring the chemical potential of the solute to be continuous at the triple junction; since this chemical potential depends on the curvature, the condition is

$$\phi'(+0) = \phi'(-0) \tag{13}$$

where the prime attached to ϕ denotes a derivative.

Finally, there is a condition arising from the fact that matter is conserved at the triple junction, so that the algebraic sum of the flows of solvent in the three surfaces meeting there is zero. Since the flows in the outer surfaces are proportional to the gradient of the curvature ϕ' while that in the grain boundary is proportional to the gradient of the concentration there, the conservation condition can be shown [3] to take the form

$$\phi''(+0) - \phi''(-0) = b^{-1}c'_{tj}. \tag{14}$$

where c'_{tj} denotes the value of dc_{gb}/ds at the outer end of the grain boundary, i.e. at the triple junction. In a general system of units the coefficient b^{-1} would be replaced by $M\gamma/D\lambda b = D\lambda/B$ where B is the material constant mentioned earlier, which is defined in Mullins' papers [10, 11].

For the problem considered in refs [11, 12], there is no flow of matter along the grain boundary, and so the right side of (14) is replaced there by 0.

4. Analysis of the nonlinear surface diffusion equation

In the phase space of the differential equation (8) there is a fixed point at $(\phi, \phi', \phi'') = (0, 0, 0)$. Near this fixed point the equation can be approximated as the linear system $(d/d\sigma)[\phi, \phi', \phi'']^T = A[\phi, \phi', \phi'']^T$, where T denotes a transpose and A is the matrix

$$A := \begin{bmatrix} 0 & 1 & 0 \\ 0 & 0 & 1 \\ v/b & 0 & 0 \end{bmatrix} \tag{15}$$

The expanding subspace of this fixed point is one-dimensional and is spanned by the vector $(1, \alpha, \alpha^2)^T$ where $\alpha := (v/b)^{1/3}$, so that on it we have $\phi' = \alpha\phi, \phi'' = \alpha^2\phi$. The corresponding expanding manifold can be (locally) parametrized by the variable ϕ : in some neighbourhood of the fixed point the other two coordinates can be expressed as power series in ϕ whose linear terms correspond to the linearized solution, namely

$$\phi' = \alpha\phi[1 - \tfrac{1}{78}\phi^2 + \tfrac{79}{628680}\phi^4 - \tfrac{1021}{6521926320}\phi^6 + \ldots] \tag{16}$$

$$\phi'' = \alpha^2\phi[1 - \tfrac{2}{39}\phi^2 + \tfrac{98}{78585}\phi^4 - \tfrac{5764}{407620395}\phi^6 + \ldots] \tag{17}$$

These series were computed by requiring them to satisfy the equations $\phi' d\phi'/d\phi = \phi'', \phi' d\phi''/d\phi = \alpha^3 \sin\phi$

The contracting subspace of the fixed point is two-dimensional, being spanned by the real and imaginary parts of the vector $(1, \alpha\omega, \alpha^2\omega^2)$ where ω is either of the complex cube roots of 1. Since $1 + \omega + \omega^2 = 0$, any vector (ϕ, ϕ', ϕ'') in the subspace satisfies $\alpha^2\phi + \alpha\phi' + \phi'' = 0$. The corresponding contracting manifold can be locally parametrized by the pair ϕ, ϕ'; in some neighbourhood of the fixed point the remaining coordinate ϕ'' can be expressed as a power series in these parameters whose linear term corresponds to the linearized solution:

$$\phi'' = -p - q + \tfrac{23}{273}p^3 + \tfrac{15}{182}p^2q + \tfrac{4}{91}pq^2 + \tfrac{1}{91}q^3 - \tfrac{125663}{73162635}p^5 \qquad (18)$$

$$+\tfrac{12793}{39020072}p^4q + \tfrac{36256}{14632527}p^3q^2 + \tfrac{3495}{1393574}p^2q^3 + \tfrac{5961}{4877509}pq^4 + \tfrac{184}{696787}q^5 + \dots$$

where $p := \alpha^2\phi, q := \alpha\phi'$. This series was computed by requiring it to satisfy $\phi'\partial\phi''/\partial\phi + \phi''\partial\phi''/\partial\phi' = \alpha^3 \sin\phi$.

5. The matching condition at the triple junction

For the shrinking grain, σ is positive and the function $\phi(\sigma)$ satisfies the boundary condition (11) for $\sigma \to +\infty$; it therefore belongs to the contracting manifold and satisfies (18). For the growing grain, σ is negative and the function $\phi(\sigma)$ satisfies the boundary condition (11) for $\sigma \to -\infty$; it therefore belongs to the expanding manifold and satisfies (16), (17). The two second derivatives in (14) can therefore be written

$$\begin{array}{llll} \text{shrinking grain}: & \phi''(+0) & = -\alpha^2\phi(+0) - \alpha\phi'(+0) + \dots & \text{by (18)} \\ & & = -\alpha^2\phi(+0) - \alpha\phi'(-0) + \dots & \text{by (13)} \\ & & = -\alpha^2\phi(+0) - \alpha^2\phi(-0) + \dots & \text{by (16)} \\ \text{growing grain}: & \phi''(-0) & = \alpha^2\phi(-0) + \dots & \text{by (17)} \end{array} \qquad (19)$$

so that (14) itself becomes

$$\begin{aligned} b^{-1}c'_{tj} &= \phi''(+0) - \phi''(-0) \\ &= -\alpha^2\phi(+0) - 2\alpha^2\phi'(-0) + \dots \end{aligned} \qquad (20)$$

Using (12), which implies $\phi(\pm 0) = -\theta_{tj} \pm \frac{1}{2}\delta$, we can write (20) in the form

$$b^{-1}\alpha^{-2}c'_{tj} = 3\theta_{tj} + \tfrac{1}{2}\delta + H(\theta_{tj}, \delta) \qquad (21)$$

where the higher terms in the series are incorporated in the function

$$H(\theta, \delta) := -\tfrac{2}{7}\theta^3 + \tfrac{3}{91}\theta^2\delta - \tfrac{8}{91}\theta\delta^2 - \tfrac{1}{273}\delta^3 + O(|\theta|^5 + |\delta|^5) \qquad (22)$$

6. Solving the equations

Only trailing grain boundaries will be studied in this paper. The connecting grain boundaries will be considered in a forthcoming paper [3]. To simplify the notation we write $c(s)$ or c in place of $c_{gb}(s)$ from here on.

The problem we consider is to find the speed v of the trailing grain boundary for a given value of the experimentally controllable parameter $\Delta c := c_{eq} - c_0$, which lies in the interval $[-1, +1]$. Our method is to find a family of solutions for eqns (1) and (3) satisfying the boundary condition (6), treating v as a parameter. Then we pick out the solution and the value of v so that the boundary conditions (7) and (21) are satisfied.

We shall make the approximation of assuming that $x_{tj} - x_m$, if positive, is negligibly small, so that (2) simplifies to $c_a = c_0$; this assumption will be justified *a posteriori* — see the discussion of eqn (40). It is convenient also to introduce the new variable $\psi := \frac{1}{2}\pi - \theta$, so that eqns (1), (3) become

$$\begin{aligned} d^2c/ds^2 &= (c-c_0)v\sin\psi && (23)\\ d\psi/ds &= \epsilon(c-c_0)^2 - v\sin\psi \qquad \text{if } 0<\psi<\pi) && (24)\end{aligned}$$

with the boundary conditions $c \to c_0, dc/ds \to 0, \psi \to 0$ as $s \to -\infty$. In the range where $0 < \psi < \pi$, the order of the system can be reduced to two by going over to ψ as independent variable:

$$(\epsilon(c-c_0)^2 - v\sin\psi)\frac{\partial}{\partial\psi}\left[(\epsilon(c-c_0)^2 - v\sin\psi)\frac{\partial c}{\partial\psi}\right] = (c-c_0)v\sin\psi \qquad (25)$$

with the boundary conditions $c \to c_0, [\epsilon(c-c_0)^2 - v\sin\psi]dc/d\psi \to 0$ as $\psi \to 0$.

The solutions of eqn (25) depend on the material parameter ϵ as well as on the experimentally controllable parameter Δc. We shall look for a linkage between these two parameters which leads to a family of solutions which are asymptotically self-similar in the limit $\epsilon \to 0$. To do this, we seek solutions whose asymptotic behaviour in this limit is

$$\left.\begin{aligned} c(\psi) - c_0 &\sim \epsilon^m\hat{c}(\psi)\\ v &\sim \epsilon^n\hat{v}\end{aligned}\right\} \quad \text{as} \quad \epsilon \to 0 \qquad (26)$$

where m and n are constants, to be chosen later so that a useful self-similar family is obtained. To begin the process of choosing m and n, let us require

$$2m + 1 < n; \qquad (27)$$

then it follows that $v\sin\psi$ becomes negligible in comparison with ϵc^2 as $\epsilon \to 0$, so that (25) simplifies to

$$\epsilon^{2+4m}\hat{c}\frac{d}{d\psi}\left(\hat{c}^2\frac{d\hat{c}}{d\psi}\right)(1+o(1)) = \epsilon^n\hat{v}\sin\psi \qquad (28)$$

with the boundary conditions $\hat{c} \to 0, \hat{c}^2 d\hat{c}/d\psi \to 0$ as $\psi \to 0$. To obtain a useful result, we require

$$2 + 4m = n \qquad (29)$$

so that eqn (28) simplifies further to

$$\hat{c}\frac{d}{d\psi}\left(\hat{c}^2\frac{d\hat{c}}{d\psi}\right) = \hat{v}\sin\psi \qquad (30)$$

The solution of this equation satisfying the boundary conditions at $\psi = 0$ can be computed as a series or by numerical integration. The series solution is

$$\hat{c} = \pm\left(\frac{16\hat{v}}{15}\right)^{1/4}\psi^{3/4}[1 - \tfrac{5}{472}\psi^2 + \tfrac{511}{6683520}\psi^4 + \tfrac{66797}{357734071296}\psi^6 + \ldots] \qquad (31)$$

To apply the concentration condition at the triple junction, eqn (7), we first evaluate (31) at the triple junction, obtaining

$$\hat{c}_{tj} = \pm \left(\frac{16\hat{v}}{15}\right)^{1/4} \psi_{tj}^{3/4}[1 - \tfrac{5}{472}\psi_{tj}^2 + \ldots] \tag{32}$$

where $\psi_{tj} := \frac{1}{2}\pi - \theta_{tj}$ is the value of ψ at the triple junction and $\hat{c}_{tj}$ is defined, in analogy with (26), by $c_{tj} - c_0 = \epsilon^m \hat{c}_{tj}$. Then (7) becomes $c_{eq} - c_0 = \epsilon^m \hat{c}_{tj}$, i.e.

$$\Delta c = \epsilon^m \hat{c}_{tj}. \tag{33}$$

To apply the triple junction angle condition (21), we need a formula for dc/ds; eqns (24), (26) and (31) give the asymptotic result

$$\frac{dc}{ds} = [\epsilon(c-c_0)^2 - v\sin\psi]\frac{dc}{d\psi} \sim \epsilon^{3m+1}\hat{c}^2\frac{d\hat{c}}{d\psi} = \pm\frac{3\epsilon^{3m+1}}{4}\left(\frac{16\hat{v}}{15}\right)^{3/4}\psi^{5/4}\times$$

$$\times[1 - \tfrac{85}{1416}\psi^2 + \tfrac{6305}{4010112}\psi^4 - \tfrac{7198697}{357734071296}\psi^6 + \ldots]. \tag{34}$$

The sign in (34) is the same as that of $c - c_0$. When (34) is substituted into the triple junction angle condition (21), using also (26), the result is

$$3\theta_{tj} + \tfrac{1}{2}\delta + H(\theta_{tj}, \delta) = b^{-1}(v/b)^{-2/3}c'_{tj}$$

$$= \pm b^{-1/3}(\epsilon^n\hat{v})^{-2/3}\frac{3\epsilon^{3m+1}}{4}\left(\frac{16\hat{v}}{15}\right)^{3/4}\psi_{tj}^{5/4}[1 - \tfrac{85}{1416}\psi_{tj}^2 + \ldots](1 + o(1)). \tag{35}$$

To make eqn (35) approach a useful limit as $\epsilon \to 0$, we require the right side to be formally independent of ϵ, *i.e.*,

$$-2n/3 + 3m + 1 = 0 \tag{36}$$

Then (35) simplifies (after dividing by 3 and taking the limit $\epsilon \to 0$) to

$$\theta_{tj} + \tfrac{1}{6}\delta + \tfrac{1}{3}H(\theta_{tj}, \delta)$$

$$= \pm\frac{b^{-1/3}}{4}\left(\frac{16}{15}\right)^{3/4}\hat{v}^{1/12}\psi_{tj}^{5/4}[1 - \tfrac{85}{1416}\psi_{tj}^2 + \ldots] \tag{37}$$

where the sign is the same as that of $\hat{c}_{tj}$. The solution of eqns (29), (36) is $m = 1, n = 6$, and these numbers also satisfy (27). Thus we can now be more specific about the ansatz (26) and the formula (33), replacing them by

$$\left.\begin{array}{rcl} c(\psi) - c_0 & \sim & \epsilon\hat{c}(\psi) \\ v & \sim & \epsilon^6\hat{v} \end{array}\right\} \quad \text{as} \quad \epsilon \to 0$$

$$\Delta c = \epsilon\hat{c}_{tj} \tag{38}$$

From these formulas we can obtain information about the sizes of the structures under study. The growing crystal grain forms a thin layer under the surface of the specimen. Its thickness is the sum of two parts; one part is the depth of its inner surface below the triple junction, given by

$$\begin{aligned} h_1 &:= \int_{-\infty}^{s_{tj}} \cos\theta \, ds = \int_0^{\psi_{tj}} \frac{\sin\psi \, d\psi}{\epsilon(c-c_0)^2 - v\sin\psi} && \text{by (24)} \\ &\sim \frac{1}{\epsilon^3}\int_0^{\psi_{tj}} \frac{\sin\psi \, d\psi}{\hat{c}^2} && \text{by (38)} \\ &= \frac{1}{\epsilon^3}\left(\frac{15}{16\hat{v}}\right)^{1/2} \int_0^{\psi_{tj}} \frac{\sin\psi \, d\psi}{\psi^{3/2}[1-\frac{5}{472}\psi^2+\ldots]^2}, \end{aligned} \tag{39}$$

by (31). The other contribution is the depth of the triple junction below the surface of the shrinking grain, given by

$$\begin{aligned} h_2 &:= -\textstyle\int_{-\infty}^{0} \sin\phi(\sigma)\, d\sigma \\ &= -(b/v)\textstyle\int_{-\infty}^{0} d^3\phi/d\sigma^3 \, d\sigma && \text{by (8)} \\ &= -(b/v)\phi''(-0) \\ &= -(b/v)(\alpha^2\phi(-0)+\ldots) && \text{by (19)} \\ &= (b/v)^{1/3}(\theta_{tj}+\tfrac{1}{2}\delta+\ldots) && \text{by (12)} \\ &\sim \epsilon^{-2}(b/\hat{v})^{1/3}(\theta_{tj}+\tfrac{1}{2}\delta+\ldots) && \text{by (38)} \end{aligned} \tag{40}$$

Since ϵ is small, the contribution h_1 dominates, and the thickness of the growing layer is of order ϵ^{-3}. Obviously this must be less than the overall thickness of the specimen, so it is a necessary condition for the applicability of these results that the thickness of the specimen should be of order ϵ^{-3} or greater. If the specimen is thinner than this, then DIGM is possible, if at all, only for connecting grain boundaries, which are outside the scope of this paper.

Formula (40) shows that the depth of the triple junction below the surface of the growing crystal is of order ϵ^{-2}. The sizes of other features of the outer boundary such as the depth of the triple junction below the surface of the growing crystal, the difference in thickness of the two crystals, and the thickness of the zone comprising the parts of the shrinking crystal that are deposited from the vapour as the grain boundary advances, are likewise of order ϵ^{-2}. From this it follows that $x_{tj} - x_m = O(\epsilon^{-2}) \ll x_{tj} = O(\epsilon^{-3})$, justifying our neglect of $x_{tj} - x_m$ in setting up eqn (23).

7. The speed of DIGM

To complete the calculation we want to find $\hat{v}$, which determines the speed of the grain boundary and enters into the formula (31) for $\hat{c}(\psi)$ from which the shape and size of the grain boundary can be calculated. We treat $\hat{c}_{tj}$ as known (from (4) and (38)), so it is a question of solving the pair of equations (32) and (37) for the two unknowns $\hat{v}$ and ψ_{tj}. A reasonably accurate method of solution is first to take the cube root of (32) and eliminate $v^{1/12}$ between the resulting equation and (37) to obtain

$$(K\hat{c}_{tj})^{1/3} = \frac{F(\psi_{tj},\delta)}{\psi_{tj}} \tag{41}$$

where

$$K := \frac{4}{225b} \tag{42}$$

$$F(\psi_{tj},\delta) := [\tfrac{1}{2}\pi - \psi_{tj} + \tfrac{1}{6}\delta + \tfrac{1}{3}H(\tfrac{1}{2}\pi - \psi_{tj},\delta)]\frac{[1-\frac{5}{472}\psi_{tj}^2+\ldots]^{1/3}}{[1-\frac{85}{1416}\psi_{tj}^2+\ldots]} \tag{43}$$

As ψ_{tj} increases over its allowed range $[0,\pi]$ at fixed δ the right side of (41) decreases monotonically from $+\infty$ to a negative value $F(\pi,\delta)/\pi$. Eqn (41) therefore has a unique solution for ψ_{tj} provided that $(K\hat{c}_{tj}) \geq -|F(\pi,\delta)/\pi|^3$. The value of $|F(\pi,\delta)/\pi|^3$ varies from $(0.671\ldots)^3 = 0.302\ldots$ at $\delta = 0$ to $(0.181\ldots)^3 = 0.0059\ldots$ at $\delta = \pi$.

An estimate of the value of $\hat{v}$ for a given $\hat{c}_{tj}$ can be derived from the following (unproved) bounds, which were obtained from graphs of $F(\psi_{tj},\delta) - \frac{1}{6}\delta$ against ψ_{tj} for various δ:

$$0.35 - 0.8\psi_{tj} < F(\psi_{tj},\delta) - \tfrac{1}{6}\delta < 1.5 - 0.8\psi_{tj} \tag{44}$$

Using these in (41) and solving for ψ_{tj} we find

$$\psi_{tj} = \frac{N_1 + \frac{1}{6}\delta}{0.8 + (K\hat{c}_{tj})^{1/3}} \tag{45}$$

where $0.35 < N_1 < 1.5$.

Solving (32) for $\hat{v}$ and then using (45) together with the fact that $0.903 < [1-\frac{5}{472}\psi_{tj}^2+\ldots] \leq 1$, we obtain

$$\begin{aligned}\hat{v} &= \frac{15\hat{c}_{tj}^4}{16\psi_{tj}^3}[1-\tfrac{5}{472}\psi_{tj}^2+\ldots]^{-4}\\ &= \frac{15\hat{c}_{tj}^4}{16}\left(\frac{0.8+(K\hat{c}_{tj})^{1/3}}{N_1+\frac{1}{6}\delta}\right)^3 N_2\\ \text{i.e. } v &= \frac{15N_2\epsilon\Delta c^4}{16}\left(\frac{0.8\epsilon^{1/3}+(K\Delta c)^{1/3}}{N_1+\frac{1}{6}\delta}\right)^3(1+o(1))\end{aligned} \tag{46}$$

where $1 \leq N_2 \leq (0.903)^{-4} = 1.504\ldots$. Thus, provided always that $K\hat{c} \geq -|F(\pi,\delta)/\pi|^3$, the speed of trailing DIGM is approximately quadratic in ϵ and proportional to the fourth power of the applied concentration difference Δc when Δc is small (compared with $(0.8)^3\epsilon/K$), but linear in ϵ and proportional to the fifth power of Δc, when Δc is large and positive.

Besides the speed, we can also calculate the thickness of the trailing layer. To leading order, the thickness is given by (39) as $\hat{v}^{-1/2}$ times an increasing function of ψ_{tj} which is approximately proportional to $\psi_{tj}^{1/2}$ at small ψ_{tj} and approaches a constant as $\psi_{tj} \to \pi$. Consequently, by (45) and (46), the thickness is approximately proportional to $\epsilon^{-3}\hat{c}_{tj}^{-2} = \epsilon^{-1}(\Delta c)^{-2}$ at small Δc, and to $\epsilon^{-3}\hat{c}_{tj}^{-5/2}\hat{c}_{tj}^{-1/6} = \epsilon^{-3}\hat{c}_{tj}^{-8/3} = \epsilon^{-1/3}(\Delta c)^{-8/3}$ at large Δc.

8. Generalizing the equations of the model

To treat the case $K\hat{c} < -|F(\pi,\delta)/\pi|^3$ we need to generalize the fundamental equations of our model to values of θ outside the range $-\frac{1}{2}\pi < \theta < \frac{1}{2}\pi$ used in (1) and (3). The new possibility is that $\cos\theta$ may be negative, describing a section of grain boundary along which s increases to the left instead of the right, so that the moving grain boundary may be traversing material points that it has already passed before. The formulation of the equations for such points does not involve any new (local) physics, just rewriting the old equations in a more general notation.

For points on the grain boundary where $\cos\theta \neq 0$, let us define a local unit vector $\mathbf{n}$ which is normal to the grain boundary and makes an acute angle with the positive y-axis. Because we are assuming a travelling wave, the local direction of motion of the grain boundary is in the direction of $\mathbf{n}$, and its normal velocity in this direction is the positive quantity $v|\cos\theta|$. The extension of (1) to general values of θ is therefore

$$\frac{d^2 c_{gb}}{ds^2} = (c_{gb}(s) - c_a(s))v|\cos\theta|, \tag{47}$$

where c_a, as before, represents the mole fraction c just in front of the moving grain boundary. If the relevant material point has not been passed before by the grain boundary, c_a is given by eqn (2); otherwise it is equal to the value of c_{gb} at the same value of x for the most recent previous passing, that is $c_a(x,y) = c_{gb}(x,y_1)$ for the smallest $y_1 > y$ at which (x,y_1) is on a grain boundary. Eqn (47) can also be used at the places where $\cos\theta = 0$ since the normal velocity there is zero regardless of the direction chosen for the unit normal vector.

The generalized force balance equation says that the normal velocity $v|\cos\theta|$ equals the sum of the curvature force and the elastic driving force, both reckoned as positive if they are in the direction of the above unit normal vector. The curvature force is $d\theta/ds$ in the direction of the more usual normal vector, which is on the left of a person walking along the curve in the direction of increasing s. This more usual normal vector is equal to $\mathbf{n}S(\theta)$, where

$$S(\theta) := \begin{cases} +1 & \text{if} \quad \cos\theta > 0 \\ -1 & \text{if} \quad \cos\theta < 0. \end{cases} \tag{48}$$

Therefore the curvature force in the direction of $\mathbf{n}$ is $S(\theta)d\theta/ds$. The elastic driving force always acts in the direction of motion, and is therefore positive under our sign convention and is given by the same expression as in (3). Combining the three terms, we obtain the following generalization of the force balance equation (3)

$$v|\cos\theta| = S(\theta)d\theta/ds + \epsilon(c_{gb}(s) - c_a(s))^2 \text{ if } \cos\theta \neq 0. \tag{49}$$

Here c_a, as in (47), denotes the concentration just in front of the advancing grain boundary. An equivalent form for (49), which we shall refer to later, is

$$d\theta/ds = v\cos\theta - \epsilon(c_{gb}(s) - c_a(s))^2 S(\theta) \text{ if } \cos\theta \neq 0. \tag{50}$$

Unlike (47), eqn (49) cannot be used at the places where $\cos\theta = 0$; at such places the normal velocity of the grain boundary is zero and so the elastic force on the grain boundary is not given by the formula $\epsilon(c_{gb}(s) - c_a(s))^2$, which applies only when the normal velocity is non-zero. Nevertheless, if we include the natural stipulation that (so as to exclude kinks) θ must vary continuously with s, eqn (49) does make a clear prediction about what happens at places where $\cos\theta = 0$, provided also that $\sin\theta < 0$ there. To obtain this prediction, note that eqn (49) implies, whenever $\cos\theta \neq 0$,

$$\begin{aligned} \frac{d}{ds}|\cos\theta| &= S(\theta)\frac{d}{ds}\cos\theta \\ &= -\sin\theta\{v|\cos\theta| - \epsilon(c_{gb} - c_a)^2\} \text{ by (49)} \\ &\leq v|\cos\theta| + \epsilon\sin\theta(c_{gb} - c_a)^2 \end{aligned} \tag{51}$$

If we define $f(s) := e^{-vs}|\cos\theta(s)|$, it follows that $df/ds \leq 0$ whenever $\cos\theta \neq 0$ and $\sin\theta < 0$. Now let s_0 be some value of s at which $\theta = -\frac{1}{2}\pi$. At this value we have $f(s) = 0$ and $\sin\theta(s) < 0$, and it follows[3] from the non-positivity of df/ds that $f(s) = 0$ for all $s > s_0$, and hence that θ remains at the value $-\frac{1}{2}\pi$ for all $s > s_0$.

This is a puzzling result. The model predicts that the grain boundary can have a straight part, as illustrated in Fig.2, on which there is no normal velocity and no curvature; yet the mole fractions at nearby points on the two sides of the straight part are different, which leads one to expect an elastic force on it, in violation of the force balance principle used to derive (49). Mathematically, the difficulty is related to the singular character of the differential equation (50), whose right side is not a continuous function of s as $\cos\theta$ passes through the value $-\frac{1}{2}\pi$. The way round the paradox may be connected with stability; for although the result shows that the straight section of the grain boundary is stable against small perturbations within the class of travelling wave solutions, it does not prove stability against perturbations which are outside this class. The theory of these more general perturbations, in which the shape of the curve in Fig. 2 would depend on time, is outside the scope of this paper; all we do here is to work out the travelling wave solution from which such a stability analysis would start.

[3] If there were an $s_2 > s_0$ with $f(s_2) > 0$, then by the continuity of $\theta(s)$ there would be an s_1 satisfying $s_0 < s_1 < s_2$ with $0 = f(s_0) < f(s_1) < f(s_2)$, but this last inequality is incompatible with the non-positivity of df/ds when $\cos\theta \neq 0$.

9. Predictions of the model for the case $K\hat{c} < -|F(\pi,\delta)/\pi|^3$

The result just derived implies that the grain boundary consists of two distinct parts, as illustrated in Fig.2 : a curved inner part on which $\theta > -\frac{1}{2}\pi$, smoothly joined to a straight outer part, parallel to the faces of the crystal, on which $\theta = -\frac{1}{2}\pi$. The outer straight part would be a solution to (50) if the last term were set to 0 by assumption when $\cos\theta = 0$.

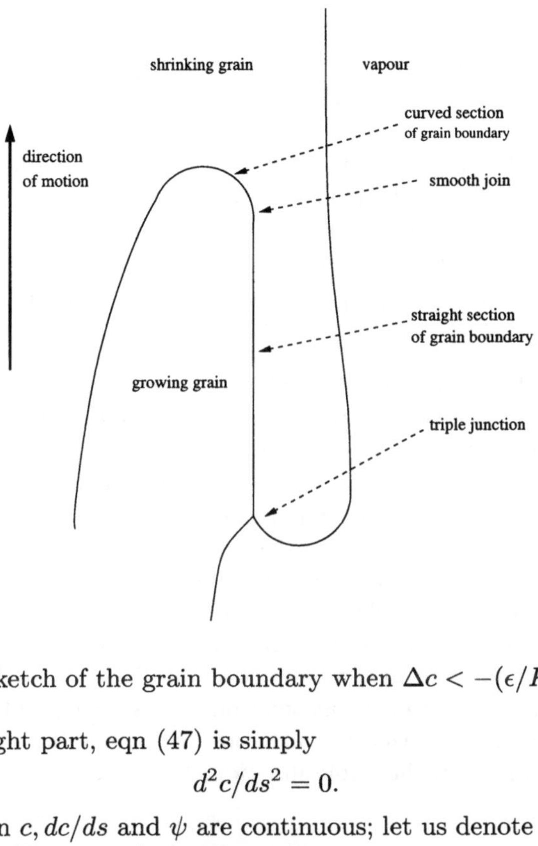

FIGURE 2. Sketch of the grain boundary when $\Delta c < -(\epsilon/K|F(\pi,\delta)/\pi|^3$

On the straight part, eqn (47) is simply

$$d^2c/ds^2 = 0. \tag{52}$$

At the smooth join $c, dc/ds$ and ψ are continuous; let us denote their values there by c_{sj}, c'_{sj} and ψ_{sj}. Then eqn (52) tells us that

$$\begin{aligned} c(s) &= c_{sj} + (s - s_{sj})c'_{sj} \\ c_{tj} &= c_{sj} + (s_{tj} - s_{sj})c'_{sj} \\ c'_{tj} &= c'_{sj} \neq 0 \end{aligned} \tag{53}$$

For the curved part of the grain boundary the method of solving the equations is nearly the same as in section 6; the only difference is that the places of c_{tj}, c'_{tj} and ψ_{tj} are now taken by c_{sj}, c'_{sj} and $\psi_{sj} = \pi$. The analogue of (41) can be written

$$\hat{c}_{sj} = \frac{1}{K}\left(\frac{F(\pi,\delta)}{\pi}\right)^3 = -\frac{1}{K}\left|\frac{F(\pi,\delta)}{\pi}\right|^3 \tag{54}$$

Putting this into the analogue of the first line of (46), we obtain

$$\begin{aligned}
\hat{v} &= \frac{15\hat{c}_{sj}^4}{16\pi^3[1-\frac{5}{472}\pi^2+\ldots]^4} = 0.0455\hat{c}_{sj}^4 \\
&= \frac{0.0455}{K^4}\left(\frac{F(\pi,\delta)}{\pi}\right)^{12} \text{ by (54)} \\
\text{i.e. } v &= \frac{0.0455\epsilon^6}{K^4}\left(\frac{F(\pi,\delta)}{\pi}\right)^{12}(1+o(1)) \qquad (55)
\end{aligned}$$

Provided Δc is less than the critical value $-(\epsilon/K)|F(\pi,\delta)/\pi|^3$, the velocity is independent of Δc : no matter how large and negative Δc is made, the velocity cannot be increased beyond the value given in (55). In a similar way it can be shown, using (39) and (40), that the thickness of the grain boundary is independent of Δc under these conditions. Increases of $|\Delta c|$ beyond $(\epsilon/K)|F(\pi,\delta)/\pi|^3$ serve only to increase the length of the straight part of the grain boundary, which according to (53) and (54) is the positive quantity

$$s_{tj} - s_{sj} = \frac{(c_0 - c_{tj}) - (c_0 - c_{sj})}{-c'_{sj}} = \frac{|\Delta c| - (\epsilon/K)|F(\pi,\delta)/\pi|^3}{|c'_{sj}|} \qquad (56)$$

Apart from the number 0.0455, which comes from the solution of the grain boundary equations, the scaled speed $\hat{v}$ is determined entirely by the non-linear Mullins theory, depending only on the angle δ at the triple junction and on the value of $K = 4/225b$. However, although $\hat{v}$ is independent of $\hat{c}$, it is extremely sensitive to the other parameters, particularly δ : for example, when $\delta = 0$, the coefficient of K^{-4} in equation (55) is 4×10^{-4}, but when $\delta = \pi$ this coefficient is 7×10^{-11} (the smallness of these numbers is deceptive, since K^{-4} could be large). Moreover, as we have mentioned earlier, the travelling wave solution underlying the derivation of the results in this section may well be unstable against perturbations which, if they move forward in the y direction at all, do so with speed less than v. The stability against such perturbations would presumably get worse as $|\Delta c|$ was made larger, since any localized perturbation would (by (56)) have more time to grow before colliding with the triple junction.

Acknowledgements

We thank Paul Fife and Charles Elliott for help in the formulation of the sharp-interface limit of the phase-field equations, and for information about the properties of these equations and about the connection between this paper and ref. [5]. We also thank Carol Handwerker for helpful information about DIGM experiments, and both her and Geoffrey McFadden for critical reviews of the manuscript. J.C. thanks Heriot-Watt University for hosting a workshop during which this work was initiated. O.P. thanks the Open University for a research grant, the Institut des Hautes Etudes Scientifiques, Bures-sur-Yvette, France and NIST for their hospitality during part of the gestation of the paper, DIMACS at Rutgers University for financial support during a visit to the USA, and the Royal Society of London for a grant towards travel expenses to Trento. DIMACS is an NSF Science and Technology Center, funded under contract STC-91-1999; it also receives support from the New Jersey Commission on Science and Technology.

References

[1] Brener, E. A. and Temkin, D. E., *Theory of diffusion induced grain boundary migration: is mass transport along free surfaces important?*, Acta Mater. **50**, 1707–1716 (2002).

[2] Cahn, J. W., Fife, P. C., and Penrose, O., *A phase-field model for diffusion-induced grain-boundary motion*, Acta Mater. **45**, 4397–4413 (1997).

[3] Cahn, J. W. and Penrose, O., in preparation.

[4] Deckelnick, K. and Elliott, C. M., *An existence and uniqueness result for a phase-field model of diffusion-induced grain-boundary motion*, Proc. Roy. Soc. Edin. **A131**, 1323–1344 (2001).

[5] Fife, P., Cahn, J. W., and Elliott, C. M., *A Free Boundary Model for DIGM*, Interfaces Free Bound. **3**, 291–336 (2001).

[6] Deckelnick, K., Elliott, C. M., and Styles, V., *Numerical diffusion-induced grain boundary motion*, Interfaces Free Bound. **3**, 393–414 (2001).

[7] Handwerker, C., *Diffusion-induced grain boundary migration in thin films*, in *Diffusion Phenomena in Thin Films and Microelectronic Materials*, ed. D. Gupta and P. S. Ho, Noyes Publications, Park Ridge, NJ, (1985) 245–322.

[8] Hillert, M., *On the driving force for diffusion induced grain boundary migration*, Scripta Metall. **17**, 237–240 (1983).

[9] Hillert, M. and Purdy, G. R., *Chemically induced grain boundary motion*, Acta Metall. **26**, 333–340 (1977).

[10] Mullins, W. W., *Theory of thermal grooving, J. Appl. Phys.* **28**, 333–339 (1957).

[11] Mullins, W. W., *The effect of thermal grooving on grain boundary motion*, Acta Metall. **6**, 414–427 (1958).

[12] Kanel, J., Novick-Cohen, A. and Vilenkin, A., *Coupled surface and grain boundary motion : nonclassical traveling wave solutions*, Preprint, 2002.

[13] Penrose, O., *On the elastic driving force in DIGM*, in preparation.

[14] Sulonen, M.S., *On the driving force of discontinuous precipitation and dissolution*, Acta Metall. **12**, 749–753 (1964).

(O. Penrose) Department of Mathematics,
Heriot-Watt University,
Riccarton, Edinburgh EH14 4AS,
Scotland, UK
E-mail address: O.Penrose@ma.hw.ac.uk

(J.W. Cahn) Materials Science and Engineering Laboratory,
NIST,
Gaithersburg, MD 20899-8555,
USA
E-mail address: john.cahn@nist.gov

International Series of Numerical Mathematics, Vol. 147, 255–264

Continuation of the Solution to the Chemotaxis Problem Beyond its Blow-up

Mario Primicerio and Boris Zaltzman

Abstract. In this paper we deal with classical solutions to the chemotaxis problem and characterize cases of blow-up. Next, we define continuation of the classical solutions beyond blow-up time T_0. This definition is in fact non-trivial because, although allowing finite mass to concentrate in the origin after T_0 (without prescribing its amount as a function of time), it identifies the solution uniquely. Moreover, two ways of regularizing the problem are provided: they look quite natural and they are shown to converge to the solution.

1. Introduction

Blow-up of solutions to the system of partial differential equations modelling chemotaxis has been investigated recently by several authors (see e.g. [1], [2]). But also in the case of the basic Keller–Segel model ([3]) some problems have received only partial answer. Referring to that model and confining to a two-dimensional radial-symmetric problem in an infinite domain, we will try to contribute to a better understanding of this mathematical phenomenon.

In all radial-symmetric problems, the Keller–Segel system can be reduced to a single equation for a suitable mass function; this equation shows clearly that the phenomenon is the combination of diffusion and nonlinear convection. When the latter prevails, blow-up occurs, whereas when diffusion provides sufficiently strong dissipation, no singularity appears for the concentration.

This balance is dependent on the number of dimensions, so that in one space dimension there is no blow-up and eventually the equation is a Burger's equation. In three dimensions blow-up occurs for any size of initial datum, while in the two-dimensional case we will see that either the diffusion term or the nonlinear convection can be dominant, depending on the size of initial datum. In case of blow-up (which will occur in the center for symmetry reasons) some additional questions arise. Will all the mass initially present concentrate instantaneously in a "delta function" in the origin? And if this is not the case, is there any way of continuing the solution after blow-up? Moreover, can a regularization be performed so that this continuation can be actually computed?

In Section 2 of this paper we deal with classical solutions and characterize cases of blow-up by a comparison technique using as a barrier the only nontrivial asymptotic solution. In Section 3 we define continuation of the classical solutions beyond blow-up time T_0. This definition, although analogous to the already used definition of weak solution, is in fact non-trivial because, although allowing finite mass to concentrate in the origin after T_0 (without prescribing its amount as a function of time), it identifies the solution uniquely. Moreover, two ways of regularizing the problem are provided: they look quite natural and they are shown to converge to the solution.

2. Classical solution

The two-dimensional radial-symmetric problem we will consider is the following,

$$ru_t = (\nu r u_r - r u p_r)_r\,,\ r > 0,\ t > 0, \tag{1}$$

$$(rp_r)_r = -ru,\ r > 0,\ t > 0, \tag{2}$$

with the initial condition

$$u(r,0) = u_0(r),\ r > 0, \tag{3}$$

where $u_0(r)$ will be assumed such that $\lim_{r\to\infty} u_0(r)r^{2+\alpha} = 0$, for some $\alpha > 0$, so that the total initial mass $\widehat{M} = 2\pi\int_0^\infty r u_0(r)dr$ is finite.

We define the mass function $\mathfrak{M}(r,t)$ as

$$\mathfrak{M}(r,t) \stackrel{def}{=} 2\pi \int_0^r \rho u(\rho,t)d\rho. \tag{4}$$

Then (1), (2) reduces to the single equation

$$\mathfrak{M}_t = \nu r \left[\frac{\mathfrak{M}_r}{r}\right]_r + \frac{\mathfrak{M}\mathfrak{M}_r}{2\pi r},\ r > 0,\ t > 0. \tag{5}$$

Next, define

$$s \equiv \frac{r^2}{2},\ M(s,t) = \mathfrak{M}(\sqrt{2s},t) = 2\pi\int_0^{\sqrt{2s}} \rho u(\rho,t)d\rho. \tag{6}$$

Then, (5) becomes

$$M_t = 2\nu s M_{ss} + \frac{1}{2\pi} M M_s,\ s > 0,\ t > 0, \tag{7}$$

and (3) gives

$$M(s,0) = M_0(s) \equiv 2\pi \int_0^{\sqrt{2s}} \rho u_0(\rho)d\rho, \tag{8}$$

so that $\lim_{s\to\infty} M_0(s) = \widehat{M}$. Our first aim is to discuss the solvability of the following

Problem 2.1. *Given a bounded nondecreasing function* $M_0(s) \in C^1([0,\infty))$, $M_0(0) = 0$, *find* $T > 0$ *and a bounded function* $M(s,t) \in C([0,\infty) \times [0,T)) \cap C^{2,1}((0,\infty) \times (0,T))$, *satisfying*

$$\begin{cases} M_t = 2\nu s M_{ss} + \frac{1}{2\pi} M M_s, & s > 0,\ t \in (0,T), \\ M(s,0) = M_0(s), & s > 0, \\ M(0,t) = 0, & t \in (0,T). \end{cases} \tag{9}$$

$M(s,t)$ *will be called the classical solution to the problem and the interval* $(0,T_0)$, $T_0 = \sup\{T\}$, *finite or infinite, will be called the maximal interval of existence.*

First we state the following

Proposition 2.2. *Any solution of Problem* 2.1 *is such that*

$$0 \le M \le \widehat{M},\ M_s(s,t) \ge 0,\ s > 0,\ t \in (0,T_0), \tag{10}$$

Proof. A straightforward application of the maximum principle yields (10). □

Next, we prove the following result on monotone dependence upon the data.

Lemma 2.3. *Let* M_1, M_2 *be classical solutions of* (9) *corresponding to the data* $M_{10}(s)$ *and* $M_{20}(s)$ *respectively, let* T_{01} *and* T_{02} *be their respective maximal intervals of existence, and let* $\widetilde{T}_0 = \min(T_{01}, T_{02})$. *Then, if*

$$M_{10}(s) \le M_{20}(s),\ s > 0, \tag{11}$$

then

$$M_1(s,t) \le M_2(s,t),\ s > 0,\ t \in (0,\widetilde{T}_0). \tag{12}$$

Proof. Consider

$$V(s,t) = M_2(s,t) - M_1(s,t),\ s > 0,\ t \in (0,\widetilde{T}_0),$$

and note that

$$V_t = 2\nu s V_{ss} + \frac{M_1 + M_2}{4\pi} V_s + \frac{V}{s}(M_1 + M_2)_s\,,\ s > 0,\ t \in (0,\widetilde{T}_0). \tag{13}$$

Note that V is nonnegative for $t = 0$, $s > 0$ because of (11) and vanishes on $s = 0$. Hence (12) follows by maximum principle (consider e.g. an approximation V^ε whose data are $V^\varepsilon(s,0) = V(s,0) + \varepsilon$, $V^\varepsilon(0,t) = \varepsilon$ and prove that V^ε can never reach the minimum value $\varepsilon/2$). □

Essentially by the same arguments, we can prove

Lemma 2.4. *Let* M *solve Problem* 2.1. *Then the following implications hold:*

$$M_0''(s) \le 0 \ \text{ in } R^+ \Longrightarrow M_{ss}(s,t) \le 0 \text{ in } R^+ \times (0,T), \tag{14}$$

$$4\pi\nu s M_0''(s) + M_0(s) M_0'(s) \ge 0 \text{ in } R^+ \Longrightarrow M_t(s,t) \ge 0 \text{ in } R^+ \times (0,T), \tag{15}$$

Investigating the stationary solutions to (7), we obtain the following

Proposition 2.5. *Equation* (7) *admits nonconstant smooth stationary solutions* $\overline{M}(s)$ *of finite mass* $\widehat{M}$ *only if* $\widehat{M} = 8\pi\nu$. $\overline{M}(s)$ *has the form*

$$\overline{M}(s) = 8\pi\nu \frac{s}{s+\beta}, \tag{16}$$

for arbitrary $\beta > 0$.

We also have

Proposition 2.6. *Let*

$$M_\infty(t) = 2\pi \int_0^\infty r u(r,t) dr = \lim_{s\to\infty} M(s,t). \tag{17}$$

Then it is

$$M_\infty(t) = M_\infty(0) = \widehat{M}, \quad \forall t \in (0, T_0). \tag{18}$$

Now, we state and prove the main result of this section.

Theorem 2.7. *If*

$$\widehat{M} \le 8\pi\nu, \tag{19}$$

then Problem 2.1 *is uniquely solvable and* $T_0 = \infty$.
If, on the contrary,

$$\widehat{M} > 8\pi\nu, \tag{20}$$

then Problem 2.1 *is uniquely solvable,* T_0 *is finite and*

$$\lim_{t\to T_0-} M(s,t) \ge 8\pi\nu, \ \forall s \in R^+. \tag{21}$$

Proof. Local existence is rather standard. Uniqueness for a given T is a corollary of Lemma 2.3. We also note that as long as $\lim_{\substack{s\to 0+,\\ t\to T-}} M(s,t) = 0$ (i.e. as long as $u(0,t)$ is finite) the solution can be continued beyond T. First, we prove Theorem when $\widehat{M} < 8\pi\nu$. Indeed, the stationary solution (16) is uniformly and monotonically convergent to $8\pi\nu$ in any closed subset of R^+ when $\beta \to 0$. Hence, for sufficiently small β, $M_0(s) \le \overline{M}(s)$. Therefore, Lemma 2.3 enables us to claim that

$$M(s,t) \le \overline{M}(s) \text{ in } R^+ \times (0,T) \tag{22}$$

and thus the solution can be continued beyond any $T > 0$. Next, we prove that the same is true also when $\widehat{M} = 8\pi\nu$. Let

$$\widetilde{u} = \max(8\pi\nu, \max_{[0,1]} M_0'(s)), \tag{23}$$

and define

$$v_0(s) = \begin{cases} \widetilde{u}s, & s \in [0, 8\pi\nu/\widetilde{u}], \\ 8\pi\nu, & s \in (8\pi\nu, +\infty]. \end{cases} \tag{24}$$

If $\widetilde{T}$, $v(s,t)$ is the solution of Problem 2.1 with initial datum $v_0(s)$, Lemma 2.3 ensures that $M(s,t) \le v(s,t)$ in their common interval of existence and that $T_0 \ge \widetilde{T}$. So, to complete the proof of the first statement of the theorem we only need to prove that $\widetilde{T} = +\infty$. To this end, let $\varepsilon > 0$ and let P_ε be the solution of the following problem

$$P_{\varepsilon t} = 2\nu\varepsilon P_{\varepsilon ss} + \frac{1}{2\pi} P_\varepsilon P_{\varepsilon s}, \ s > 0, \ t \in (\widetilde{T}/2, \widetilde{T}),$$
$$P_\varepsilon(s, \widetilde{T}/2) = v(s, \widetilde{T}/2), \quad s > 0, \ P_\varepsilon(0,t) = 0, \ t \in (\widetilde{T}/2, \widetilde{T}).$$

Since $v_{ss} \leq 0$ in $R^+ \times (0, \widetilde{T})$ (recall (14) after proper smoothing of $v_0(s)$), we have

$$v(s,t) \leq P_\varepsilon(s,t) \text{ in } R^+ \times (\widetilde{T}/2, \widetilde{T}). \tag{25}$$

Consider the nonlinear wave propagation equation

$$P_t = \frac{1}{2\pi} P P_s, \text{ in } R^+ \times (\widetilde{T}/2, \widetilde{T}),$$
$$P(s, \widetilde{T}/2) = v(s, \widetilde{T}/2), \quad s > 0, \ P_\varepsilon(0,t) = 0, \ t \in (\widetilde{T}/2, \widetilde{T}).$$

The problem is solvable, since $v'(s, \widetilde{T}/2) > 0$, and

$$P(s,t) < 8\pi\nu, \text{ in } R^+ \times (\widetilde{T}/2, \widetilde{T}). \tag{26}$$

Using (25) and its limit for $\varepsilon \to 0$ and (26) we have that, for any $\bar{t} \in (\widetilde{T}/2, \widetilde{T})$, it is $v(s, \bar{t}) < 8\pi\nu$, $s > 0$, and taking $\bar{t}$ as a new initial time we are back to the case studied above. To conclude the proof of the theorem we have to consider the case in which (20) holds. Define

$$G(t) = \int_0^\infty r^3 u(r,t) dr, \ t \in (0,T). \tag{27}$$

Multiply equations (1) and (2) by r^3 and integrate over R^+. After integration by parts we find

$$G_t = \frac{\widehat{M}}{2\pi}(8\pi\nu - \widehat{M}), \ t \in (0, T_0), \tag{28}$$

where (18) has been taken into account. Since $G(t) \geq 0$ in $(0, T_0)$, it is clear that

$$T_0 \leq \frac{2\pi G(0)}{\widehat{M}(8\pi\nu - \widehat{M})}, \tag{29}$$

thus excluding global existence of classical solutions when (20) holds. To prove (21), assume that for a given $s_0 > 0$, there exists $\delta > 0$ such that

$$M(s_0, T_0) = \lim_{t \to T_0 -} M(s_0, t) = 8\pi\nu - 2\delta. \tag{30}$$

Therefore, for some $\sigma > 0$, $M(s_0, t) < 8\pi\nu - \delta$, $t \in (T_0 - \sigma, T_0)$, and, hence, (recall $M_s \geq 0$)

$$M(s, T_0 - \sigma) \leq 8\pi\nu - \delta, \ s \in [0, s_0]. \tag{31}$$

But this would imply the possibility of defining $\overline{M}(s)$ as in (16) with sufficiently small $\beta > 0$ in such a way that

$$M(s, T_0 - \sigma) \leq \overline{M}(s), \ s \in R^+. \tag{32}$$

By Lemma 2.3

$$M(s,T_0) \le \overline{M}(s) \tag{33}$$

and thus we find that contradicting (21) would imply that $(0,T_0)$ is not the maximal interval of existence of classical solution. □

3. Global in time solutions

We state

Problem 3.1. *Given a bounded nondecreasing function $M_0(s) \in C^1([0,\infty))$ with $M_0(0) = 0$, such that $\widehat{M} > 8\pi\nu$, let T_0 be the maximal interval of existence of the classical solution. We look for a bounded nondecreasing function M defined on $R^+ \times (0,\infty)$ and such that for all $T > 0$*

$$\begin{cases} M_t = 2\nu s M_{ss} + \frac{1}{2\pi} M M_s, & s>0,\ t\in(0,T), \\ M(s,0) = M_0(s), & s>0, \\ M(0,t) = \begin{cases} 0, & t<T_0, \\ \ge 0, & t>T_0, \end{cases} & t \in (0,T). \end{cases} \tag{34}$$

$M(s,t)$ will be called global-in-time solution *to our problem.*

Remark 3.2. The global solution is a "gluing" of a smooth classical solution and a singular (at $r=0$) "weak" solution at the blow-up instant $t = T_0$.

As a preliminary, we note that classical results on parabolic equations yield

Proposition 3.3. *For any global-in-time solution we have $M \in C^\infty(R^+ \times (0,T))$.*

Let us prove that blow-up singularities never "melt", in the sense that

Lemma 3.4. *For any weak solution we have*

$$M(s,t) \ge 8\pi\nu \text{ for all } s>0,\ t\ge T_0. \tag{35}$$

Proof. Using Proposition 3.3 we find

$$M_t = 2\nu s M_{ss} + \frac{1}{2\pi} M M_s,\ s>0,\ t>T_0, \tag{36}$$

$$M(s,T_0) \ge 8\pi\nu,\ s>0, \tag{37}$$

$$M(0,t) \ge 0,\ t \ge T_0. \tag{38}$$

Applying Lemma 2.3, we obtain that

$$M(s,t) \ge \overline{M}(s) \text{ in } R^+ \times (T_0,\infty) \tag{39}$$

where $\overline{M}(s)$ is any stationary solution defined by (16). Since the stationary solutions are uniformly and monotonically convergent to $8\pi\nu$ in any closed subset of R^+ when $\beta \to 0$, (39) yields the estimate (35). □

Next, we prove uniqueness of the global-in-time solutions.

Theorem 3.5. *Let M_1 and M_2 be continuations of the same classical solution beyond T_0. Then $M_1 = M_2$ in $R^+ \times (T_0, T)$.*

Proof. Integrating the equation

$$(sM_s)_s = \frac{1}{2\nu}\left(M_t - \frac{1}{4\pi}\left[M^2 - 8\pi\nu M\right]_s\right)$$

in s and t, we find for any global-in-time solution

$$sM_s \in L^\infty(0,\infty; L^1(0,T)) \text{ for all } T > 0. \tag{40}$$

Define $N(s,t) = M_1(s,t) - M_2(s,t)$. Because of Proposition 3.3,

$$N_t = 2\nu(sN_s)_s + \frac{1}{4\pi}\left(N(M_1 + M_2 - 8\pi\nu)\right)_s \tag{41}$$

is satisfied in $R^+ \times (T_0, T)$. Moreover, $N(s, T_0) = 0$ in R^+. Multiply (41) by $f_\gamma(N(s,t))$, where the functions $f_\gamma(N)$ are smooth approximations of $sgn(N)$:

$$f_\gamma(N) = \frac{N}{\sqrt{N^2+\gamma}}, \quad \gamma > 0. \tag{42}$$

Then, for any $\delta > 0$, integrate on $(\delta, +\infty)$. After integration by parts the following inequality is obtained

$$\frac{d}{dt}\int_\delta^\infty \frac{sN^2}{\sqrt{N^2+\gamma}}ds \le 2\nu\left[-s^2 f_\gamma N_s\right]|_{s=\delta} + 2\nu\int_\delta^\infty f_\gamma N ds- \tag{43}$$

$$-\frac{1}{4\pi}\left[Nf_\gamma\left(M_1 + M_2 - 8\pi\nu\right)\right]|_{s=\delta} - \frac{1}{4\pi}\int_0^\infty N\left(M_1 + M_2 - 8\pi\nu\right)\frac{\partial f_\gamma}{\partial s}sds-$$

$$-\frac{1}{4\pi}\int_0^\infty N\left(M_1 + M_2 - 8\pi\nu\right) f_\gamma ds.$$

Now, integrate in t, take the limit $\gamma \to 0$, let δ tend to zero and apply (40) to obtain

$$\int_0^\infty s|N(s,t)|ds \le -\frac{1}{4\pi}\int_0^\infty |N|\left(M_1 + M_2 - 16\pi\nu\right) ds, \ t \in (T_0, T), \tag{44}$$

and making use of Lemma 3.3 completes the proof. □

To prove the existence, we will use a monotonicity argument based on two different types of regularization.

(i) First, we regularize the Keller–Segel model as we did in [4], [5], i.e. by assuming that the coefficient of chemotactic response vanishes when the concentration of the cells exceeds a threshold value. The regularized problem takes the form

$$ru_t^\varepsilon = \left(\nu r u_r^\varepsilon - H\left(\frac{1}{\varepsilon} - u^\varepsilon\right) r u^\varepsilon p_r^\varepsilon\right)_r, \ r > 0, \ t > 0, \tag{45}$$

$$\left(rp_r^\varepsilon\right)_r = -ru^\varepsilon, \ r > 0, \ t > 0, \tag{46}$$

$$u^\varepsilon(r, 0) = u_0(r), \ r > 0. \tag{47}$$

In (45), H is the Heaviside graph.

(ii) Alternatively, regularization will be based on the assumption that the coefficient of chemotactic response vanishes for $r < \varepsilon$:

$$ru_t^\varepsilon = (\nu r u_r^\varepsilon - H(r-\varepsilon) r u^\varepsilon p_r^\varepsilon)_r, \; r > 0, \; t > 0, \tag{48}$$

$$(rp_r^\varepsilon)_r = -ru^\varepsilon, \; r > 0, \; t > 0, \tag{49}$$

$$u^\varepsilon(r,0) = u_0(r), \; r > 0. \tag{50}$$

In both cases, by sending ε to zero we will obtain the global-in-time solution to the original problem.

We start with approach (i). Through the same steps used to obtain (7), we get for $\varepsilon < \varepsilon_0 = 2\pi/\sup M_0'$

$$M_t^\varepsilon = 2\nu s M_{ss}^\varepsilon + \frac{1}{2\pi} H\left(\frac{2\pi}{\varepsilon} - M_s^\varepsilon\right) M^\varepsilon M_s^\varepsilon, \; s > 0, \; t > 0, \tag{51}$$

$$M^\varepsilon(s,0) = M_0(s), \; s > 0. \tag{52}$$

The cutting effect on the nonlinear term provides the existence of a unique global solution (48), (49) such that $M^\varepsilon(0,t) = 0$, $t > 0$.

Next, we prove

Lemma 3.6. *Assume*

$$M_0'' \le 0, \; s > 0. \tag{53}$$

Then

$$M_{ss}^\varepsilon \le 0, \; a.e. \; s > 0, \; t > 0. \tag{54}$$

Moreover, for two different initial data M_{10}, M_{20} satisfying (53) *it is*

$$M_{10}(s) \ge M_{20}(s) \Longrightarrow M_1^\varepsilon(s,t) \ge M_2^\varepsilon(s,t). \tag{55}$$

Proof. We apply the argument of Lemma 2.4 in the domain where $M_s^\varepsilon < 2\pi/\varepsilon$. To prove (55), note that when (54) holds then

$$H\left(\frac{2\pi}{\varepsilon} - M_s^\varepsilon\right) = H\left(\frac{2\pi s}{\varepsilon} - M^\varepsilon\right). \tag{56}$$

Then , we approximate H in L^2 by smooth monotonic functions H_n and we are reduced to prove (55) for the corresponding smooth solutions $\widetilde{M_1^\varepsilon}$, $\widetilde{M_2^\varepsilon}$. But this follows from the maximum principle applied to the difference $\widetilde{N} = \widetilde{M_1^\varepsilon} - \widetilde{M_2^\varepsilon}$, $\widetilde{N}(s,0) \ge 0$ that satisfies

$$\widetilde{N}_t = 2\nu\nu\widetilde{N}_{ss} + \frac{1}{2\pi} H_n^{(1)} M_1^\varepsilon \widetilde{N}_s + \frac{1}{2\pi} H_n^{(1)} M_{2s}^\varepsilon \widetilde{N} + \frac{1}{2\pi} H_n(v) \widetilde{N} M_2^\varepsilon M_{2s}^\varepsilon, \; s > 0, \; t > 0, \tag{57}$$

where

$$H_n^{(1)} = H_n\left(\frac{2\pi s}{\varepsilon} - M_1^\varepsilon\right) \text{ and } v \in \left(M_1^\varepsilon - \frac{2\pi s}{\varepsilon}, M_2^\varepsilon - \frac{2\pi s}{\varepsilon}\right). \qquad \square$$

Now we consider the following nonlinear wave propagation problem

$$P_t = PP_s, \text{ in } R^+ \times R^+, \tag{58}$$

$$P(s,0) = M_0(s), \ s > 0. \tag{59}$$

Note that $P(s,t)$ is the limit of the solutions to the following family of problems for Burgers equations

$$P_{nt} = \frac{1}{n^2} P_{nss} + P_n P_{ns} \text{ in } R^+ \times R^+, \tag{60}$$

$$P_n(s,0) = M_0(s), \ s > 0, \ P_n(0,t) = 0, \ t > 0. \tag{61}$$

We can prove the following comparison lemma

Lemma 3.7. *If* (53) *holds, then*

$$M^\varepsilon(s,t) \le P(s,t), \ s > 0, \ t > 0. \tag{62}$$

We also have

$$(sM_s^\varepsilon)_s = M_s^\varepsilon + \frac{1}{2\nu}\left(M_t^\varepsilon - \frac{1}{2\pi} HM^\varepsilon M_s^\varepsilon\right),$$

and hence, for any $T > 0$

$$\int_0^T sM_s^\varepsilon dt \le \widehat{M}\left(1 + \frac{1}{4\pi\nu}\right)T + \frac{s\widehat{M}}{2\nu} \tag{63}$$

and condition (i) in Problem 3.1 is fulfilled uniformly with respect to $\varepsilon > 0$.

Now, we are able to prove

Theorem 3.8. *Assume* (53) *holds. Then the limit of the solutions of* (51)–(52) *as* ε *tends to zero coincides with the classical solution in* $(0, T_0)$ *and provides its unique continuation beyond* T_0.

Proof. Because of (56), H is nonincreasing both with respect to ε and to M, so that

$$\varepsilon_1 < \varepsilon_2 \Longrightarrow M^{\varepsilon_1} \le M^{\varepsilon_2} \text{ in } R^+ \times R^+. \tag{64}$$

From (64) we have that there exists a function $\mu(x,t)$ such that

$$\mu(s,t) = \lim_{\varepsilon \to 0} M^\varepsilon(s,t), \ s > 0, \ t > 0. \tag{65}$$

It is easy to check that μ fulfills the conditions in the statement of Problem 3.1 and that it is smooth in $R^+ \times R^+$. Next, compare M^ε with the (classical) solution $M(s,t)$ of Problem 2.1 in its maximal interval of existence. Noting that $H \le 1$, we have $M^\varepsilon(s,t) \le M(s,t)$, $s > 0$, $t \in (0, T_0)$ and passing to the limit as $\varepsilon \downarrow 0$

$$\mu(s,t) \le M(s,t), \ s > 0, \ t \in (0, T_0). \tag{66}$$

But this means that $\mu(0,t) = 0$ in $(0, T_0)$ and hence (T_0, μ) solve Problem 2.1 and, since the classical solution is unique,

$$\mu(s,t) = M(s,t), \ s > 0, \ t \in (0, T_0). \tag{67}$$

The use of the uniqueness beyond the blow-up (Theorem 3.5) completes the proof. □

Now we consider the regularization (ii) with the aim of avoiding assumption (52). We denote by M^ε the solution of

$$M_t^\varepsilon = 2\nu s M_{ss}^\varepsilon + \frac{1}{2\pi} H\left(s - \frac{\varepsilon^2}{2}\right) M^\varepsilon M_s^\varepsilon, \ s > 0, \ t > 0, \tag{68}$$

$$M^\varepsilon(s,0) = M_0(s), \ s > 0, \text{ and } M^\varepsilon(0,t) = 0, \ t > 0. \tag{69}$$

We proceed as in proving Lemmas 3.5 and 3.7 and obtain the uniform estimates

Lemma 3.9. *The approximating solutions are bounded from below by the diffusion-dominated problem and from above by the nonlinear wave propagation problem:*

$$Q(s,t) \leq M^\varepsilon(s,t) \leq P(s,t), \ s > 0, \ t > 0. \tag{70}$$

Moreover M^ε depends monotonically on the initial datum.

Proof. No major changes are needed in the proofs of Lemmas 3.5 and 3.7 to obtain the above statement. Let us remark that in the present case we do need assumption (53), since the "cutting" in (68) does not depend on M^ε. □

Theorem 3.10. *The limit of the solutions to the approximating problems* (68)–(69) *is the global-in-time solution to Problem* 3.1.

Proof. The proof is a slight modification of the argument of Theorem 3.8, and it is based on the use of the monotonic dependence of the solutions on $\varepsilon > 0$. □

References

[1] M.A. Herrero and J.J.L. Velazquez, *Chemotactic collapse for the Keller–Segel model.* J. Math. Biol. **35** (1996), 177–196.

[2] W. Jager and S. Luckhaus, *On explosions of solutions to a system of partial differential equations modelling chemotaxis.* Trans. Amer. Math. Soc. **329**, 2 (1992), 819–824.

[3] E.F. Keller and L.A. Segel, *Initiation of slime mold aggregation viewed as an instability.* J. Theor. Biol. **26** (1970), 399–415.

[4] M. Primicerio and B. Zaltzman, *Free boundary problems in radial symmetric chemotaxis,* in Proceedings of WASCOM 2001, World Scientific, 2001.

[5] M. Primicerio and B. Zaltzman, *A free boundary problem arising in chemotaxis.* Adv. Math. Sci. Appl. **12** (2002), 685–708.

(M. Primicerio) Dipartimento di Matematica, "U.Dini", Università di Firenze, Viale Morgagni 67/a, 50134, Firenze, Italy

(B. Zaltzman) DSEEP, Blaustein Institute for Desert Research, Ben-Gurion University of the Negev, Sede-Boqer Campus, 84990, Israel.

International Series of Numerical Mathematics, Vol. 147, 265–280

Shape Deformations and Analytic Continuation in Free Boundary Problems

Fernando Reitich

Abstract. In this paper we present a general approach to the analysis of stability and bifurcation of steady states and travelling waves for a class of free-boundary problems. The approach is based on the explicit consideration of series solutions in a parameter measuring the variations of the domains (e.g. an initial configuration or a specific steady state) from separable geometries (e.g. planes, spheres, etc). The method relies on the derivation of formulae for the recursive evaluation of (Taylor) approximations of arbitrary order in the variation parameters, and on the iterative estimation of the growth, in appropriately defined spaces, of the resulting functional coefficients. We describe a variety of instances where this procedure leads to general results on analyticity of solutions, on existence of complex steady-state and travelling-wave solutions and on nonlinear stability of equilibria. We further show that the relevance of these results goes beyond the theoretical. Actually, it also leads to an understanding of the mechanisms behind the observed performance of a class of numerical algorithms, based on shape-perturbation theory, which have been extensively used in simulations of free-boundary (and boundary value) models. Indeed, our research shows that the standard implementations of these schemes can suffer from severe ill-conditioning as a result of pronounced cancellations in the underlying recursions. Moreover, our work further suggests alternative implementations with greatly improved properties of numerical stability and convergence, enhanced by methods of analytic continuation. As we demonstrate, these modifications can have a dramatic effect on the accuracy and applicability of perturbative numerical approaches.

1. Introduction

The issues of stability and bifurcation of steady states or other distinguished solutions (e.g. travelling waves) of free-boundary models are of central importance in applications, both in process design (e.g. solidification) as well as in their control (e.g. tumor growth). A classical approach to these issues is based on the analysis of *linearized equations*, leading to "linear stability analysis" or to "bifurcation equations". Of course, these analyses are only *suggestive* of the behavior of solutions, as any definitive assertion would entail the study of the full non-linear system. These latter studies, on the other hand, are significantly more complex than their linearized counterparts, and they typically involve a completely different approach. In particular, nonlinear analyses seldom shed direct light on the *order* and nature of possible instabilities and, in addition, their highly abstract nature may render them inadequate for numerical simulations. In contrast, linearized approaches can clearly identify first order instabilities, and they also deliver linear systems whose simulation can be readily implemented. A most natural question then relates to the possibility of upgrading classical linear analyses to full nonlinear studies, by iteratively computing and analyzing higher-order corrections to the linearized model. In this paper we tackle this very question with a view to providing a general, direct setting for the theoretical and numerical analysis (stability, bifurcation) of steady states and travelling waves of free-boundary models.

A most interesting aspect of our program is our demonstration that a direct approach to analyzing higher-order corrections to the linearized models will typically fail. Indeed, the higher order systems obviously correspond to inhomogeneous instances of the linearized equations, where the source terms depend on solutions to lower-order systems. Unfortunately, as we show in the next section, these source terms arise as sums that entail significant *cancellations of singularities*, as the most singular part of each term in the sum exactly cancels out to yield an appropriately regular right-hand-side. This simple fact has rather dramatic consequences, both in theory and simulation. On the theoretical side, these cancellations preclude a straightforward iterative estimation of the higher-order corrections, as any use of the triangle inequality destroys these, rendering a useless bound (see §2). With regards to numerical simulations, on the other hand, the cancellations can be identified with the source of numerical instabilities that have been observed in standard algorithms based of shape deformations [3, 4, 5, 11, 12, 16, 17]; see §4.

As we have found, however, these theoretical and numerical impediments can be overcome. Indeed, as we show in §3, a simple change of independent variables that collapses the perturbed domains onto the unperturbed geometry has the beneficial effect of implicitly accounting for cancellations. This observation then can be used to advantage in both theory and practice. In the theory, the lack of significant cancellations in the new variables allows for a direct estimation of the higher-order terms in the perturbation expansion; and, from a practical standpoint, this realization suggests a number of alternative implementations of perturbative schemes that exhibit enhanced stability and convergence properties. The rest of the paper is devoted to substantiating these assertions. First, in §2, we provide a description of the origins and nature of the cancellations that arise as we consider higher-order terms in the perturbation series of a field quantity with respect to variations of its domain of definition. There, we further show how these cancellations conspire against a direct estimation of the terms in the series. Then, in §3, we briefly explain how cancellations are mollified by a change of variables, thereby allowing for iterative estimates. We close, finally, with a description, in §4, of some algorithmic improvements on numerical methods based on shape perturbations that are motivated by the theory.

2. Cancellations in higher-order corrections

As will be clear from the presentation, our observations, results and numerical algorithms will be relevant to rather general free-boundary problems, just as first-order linearized stability analyses are. Specific examples, in d spatial dimensions ($d = 2, 3$), include the classical (e.g. quasi-steady, one-phase) Stefan problem [8]

$$\begin{aligned} &\Delta C = 0 && \text{in} \quad \Omega(t) \subset \mathbb{R}^d,\ t > 0 \\ &\frac{\partial C}{\partial n} = -V_n && \text{on} \quad \partial\Omega(t),\ t > 0 \\ &C = \gamma\kappa && \text{on} \quad \partial\Omega(t),\ t > 0 \end{aligned} \tag{1}$$

for the evolution of a particle growing by diffusion or heat flow; a model for a capillary drop [9, 10]

$$\begin{aligned} &\Delta\vec{u} - \nabla p = 0 && \text{in} \quad \Omega(t),\ t > 0 \\ &\operatorname{div}(\vec{u}) = 0 && \text{in} \quad \Omega(t),\ t > 0 \\ &\left(\nabla\vec{u} + (\nabla\vec{u})^T - pI\right)\vec{n} = -\gamma\kappa\vec{n} && \text{on} \quad \partial\Omega(t),\ t > 0 \\ &V_n = \vec{u}\cdot\vec{n} && \text{on} \quad \partial\Omega(t),\ t > 0; \end{aligned} \tag{2}$$

models of tumor growth [6, 7] of the form

$$\begin{aligned} &c\frac{\partial\sigma}{\partial t} = \Delta\sigma - \lambda\sigma && \text{in} \quad \Omega(t),\ t > 0 \\ &\Delta p = -\mu\left(\sigma - \tilde{\sigma}\right) && \text{in} \quad \Omega(t),\ t > 0 \\ &\sigma = \bar{\sigma} && \text{on} \quad \partial\Omega(t),\ t > 0 \\ &\frac{\partial p}{\partial n} = -V_n && \text{on} \quad \partial\Omega(t),\ t > 0 \\ &p = \gamma\kappa && \text{on} \quad \partial\Omega(t),\ t > 0; \end{aligned} \tag{3}$$

and the classical model of gravity-capillary water waves [13]

$$\begin{aligned} &\Delta\varphi(x,y,t) = 0 && \text{for } x \in \mathbb{R}^{d-1},\ -h < y < \eta(x,t) \\ &\partial_y\varphi(x,-h,t) = 0 && \\ &\partial_t\varphi + \frac{1}{2}\left|\nabla\varphi\right|^2 + g\eta = \gamma\kappa && \text{on } y = \eta(x,t) \\ &-\partial_t\eta - \nabla_x\eta\cdot\nabla_x\varphi + \partial_y\varphi = 0 && \text{on } y = \eta(x,t) \end{aligned} \tag{4}$$

to describe but a few. In all cases, linearized stability analyses are rather straightforward. For instance, the version of (1) linearized about the spherical steady state

$$\Omega(t) = S^{d-1} \equiv \left\{\omega \in \mathbb{R}^d \ :\ |\omega| = 1\right\}, \quad C(x,t) \equiv \gamma \tag{5}$$

reads

$$
\begin{aligned}
&\Delta C_1 = 0 && \text{in} \quad |x| < 1 \\
&\frac{\partial C_1}{\partial r} + \frac{\partial \Lambda_1}{\partial t} = 0 && \text{on} \quad S^{d-1} \\
&C_1 + \gamma\left(\Lambda_1 + \frac{1}{d-1}\Delta_\omega \Lambda_1\right) = 0 && \text{on} \quad S^{d-1} \\
&\Lambda_1(\omega, 0) = \lambda(\omega) && \text{on} \quad S^{d-1}
\end{aligned}
\tag{6}
$$

where (C_1, Λ_1) denote the first order corrections to the steady state (5). More precisely,

$$
C = \gamma + \varepsilon C_1(x,t) \quad \text{and} \quad \Omega(t) = \left\{ x \in \mathbb{R}^d \ : \ |x| = 1 + \varepsilon \Lambda_1(\omega, t), \ \omega = \frac{x}{|x|} \right\} \tag{7}
$$

solve (1) with the initial conditions

$$
\Omega(0) = \left\{ x \in \mathbb{R}^d \ : \ |x| = 1 + \varepsilon \lambda(\omega), \ \omega = \frac{x}{|x|} \right\} \tag{8}
$$

to *first order* in ε. From (6), a simple calculation reveals the (linear) stability of all perturbations except for those with the least oscillations (Fourier/Spherical harmonics of order zero and one).

As we mentioned in the Introduction, we wish to capitalize on the simplicity of the first-order, linear stability analysis through a simple extension to higher orders in ε, in the hope of attaining a direct approach to *non-linear* stability. More precisely, in keeping with example (1), we seek a solution to (1), (8) in the form

$$
C = C(x,t;\varepsilon) = \gamma + \textstyle\sum_{n \geq 1} C_n(x,t)\varepsilon^n \quad \text{and} \tag{9}
$$

$$
\Omega(t) = \left\{ x \in \mathbb{R}^d \ : \ |x| = 1 + \Lambda(\omega, t; \varepsilon) \equiv 1 + \textstyle\sum_{n \geq 1} \Lambda_n(\omega, t)\varepsilon^n, \ \omega = \frac{x}{|x|} \right\}.
$$

As easily verified, the n-th order correction (C_n, Λ_n) for $n \geq 2$ solves a problem of the form

$$
\begin{aligned}
&\Delta C_n = 0 && \text{in} \quad |x| < 1 \\
&\frac{\partial C_n}{\partial r} + \frac{\partial \Lambda_n}{\partial t} = G_n^1 && \text{on} \quad S^{d-1} \\
&C_n + \gamma\left(\Lambda_n + \frac{1}{d-1}\Delta_\omega \Lambda_n\right) = G_n^2 && \text{on} \quad S^{d-1} \\
&\Lambda_n(\omega, 0) = 0 && \text{on} \quad S^{d-1}
\end{aligned}
\tag{10}
$$

where the functions G_n^1 and G_n^2 are given as *sums* of terms that depend upon $\left(\{C_k\}_{k=1}^{n-1}, \{\Lambda_k\}_{k=1}^{n-1}\right)$. Symbolically, equations (10) can be written as

$$\begin{aligned} \mathbf{L}(C_n, \Lambda_n) &= \Phi_n\left(\{C_k\}_{k=1}^{n-1}, \{\Lambda_k\}_{k=1}^{n-1}\right) \\ &= \sum_{n=1}^{K_n} \Psi_n^m\left(\{C_k\}_{k=1}^{n-1}, \{\Lambda_k\}_{k=1}^{n-1}\right) \end{aligned} \tag{11}$$

where each Ψ_n^m is simply a monomial in the functions $\left(\{C_k\}_{k=1}^{n-1}, \{\Lambda_k\}_{k=1}^{n-1}\right)$. At this stage, a natural approach suggests itself whereby the functions (C_n, Λ_n) are estimated inductively using (11). The complexity of the right-hand-sides in this equation (whose number of terms K_n increases with n), however, conspires against the use of any bound other than the triangle inequality

$$\| (C_n, \Lambda_n) \|_* \leq \sum_{m=1}^{K_n} \|\Psi_m\left(\{C_k\}_{k=1}^{n-1}, \{\lambda_k\}_{k=1}^{n-1}\right)\|_{**} \tag{12}$$

where the norms are appropriately defined. As we mentioned, the sum in (11) entails significant *cancellations* among its constitutive terms, which are destroyed upon use of (12). As a result, the bound obtained in this manner *does not* guarantee the convergence of the series in (9).

In fact, this observation is not related to the free-boundary character of the problems: an entirely analogous situation arises on an attempt at solving a boundary value problem via a boundary perturbation approach. To see this, consider the simple problem

$$\begin{aligned} &\Delta C = 0 && \text{in} \quad \Omega_\varepsilon = \left\{x \in \mathbb{R}^2 \,:\, |x| = r < 1 + \varepsilon\lambda(\theta)\right\} \\ &C = C(r,\theta)\big|_{r=1+\varepsilon\lambda(\theta)} = f(\theta) && \text{for } \theta \in [0, 2\pi] \end{aligned} \tag{13}$$

and propose a solution $C = C(r, \theta; \varepsilon)$ as in (9)

$$C = C_0(r,\theta) + \sum_{n\geq 1} C_n(r,\theta)\varepsilon^n \tag{14}$$

where now $C_0(r, \theta)$ is the solution of the unperturbed problem (13) with $\varepsilon = 0$. More precisely, if

$$f(\theta) = \sum_{p=-M}^{M} f_p e^{ip\theta}, \tag{15}$$

we have

$$C_0(r,\theta) = \sum_{p=-M}^{M} f_p r^{|p|} e^{ip\theta}. \tag{16}$$

The analog of (10) is then

$$\Delta C_n = 0 \qquad \text{in} \quad \Omega_0 = \{x \in \mathbb{R}^2 \,:\, |x| < 1\}$$

$$C_n(1,\theta) = \delta_{n,0} f(\theta) - \sum_{l=0}^{n-1} \frac{\lambda(\theta)^{n-l}}{(n-l)!} \partial_r^{n-l} C_l(1,\theta) \quad \text{for } \theta \in [0, 2\pi]. \tag{17}$$

The *cancellations* in formula (17) are now rather apparent: for instance, the n-th order formula entails increasingly high-order derivatives of the corrections C_l and, therefore, high-order derivatives of the perturbation $\lambda(\theta)$ (upon iteration). On the other hand, classical results on analyticity of potentials, such as those that can be derived from the work of Calderón [1] and Coifman and Meyer [2] on the theory of singular integrals, guarantee that the series (14) converges, for $r < 1$ and ε sufficiently small, for perturbations $\lambda(\theta)$ that are only Lipschitz continuous. This apparent contradiction between the high-order of differentiations in (17) and the low degree of regularity necessary for the convergence of the series can only be reconciled if these high-order derivatives actually cancel out in (17), yielding an appropriately regular right-hand side [13, 14].

A particularly clear verification of this assertion, which also shows how the bound (12) fails to accurately estimate the functions C_n, can be derived if we consider equation (17) in Fourier space. Indeed, letting

$$C_n(r,\theta) = \sum_{p=-(n+1)M}^{(n+1)M} d_{n,p} r^{|p|} e^{ip\theta}$$

and

$$\frac{\lambda(\theta)^k}{k!} = \sum_{p=-kM}^{kM} \lambda_{k,p} e^{ip\theta} \tag{18}$$

the recurrence (17) can be written as

$$d_{n,p} = \delta_{n,0} f_p - \sum_{l=0}^{n-1} \sum_q \lambda_{n-l,p-q} \frac{|q|!}{(|q|-(n-l))!} d_{l,q}.$$

In particular, for $p = (n+1)M$ we have

$$D_n = \delta_{n,0} f_M - \sum_{l=0}^{n-1} \lambda_{n-l,(n-l)M} \frac{((l+1)M)!}{((l+1)M-(n-l))!} D_l \tag{19}$$

where we have set

$$D_k = d_{k,(k+1)M}.$$

In Figure 1, we show some numerical results obtained from implementations of (19) in the most favorable case of an *analytic* boundary

$$\lambda(\theta) = 2\cos(\theta) = e^{i\theta} + e^{-i\theta} \qquad (M = 1).$$

Figure 1(a) shows the effect of cancellations on the accuracy that can be attained: at sufficiently high-orders, round-off errors completely overcome the values of interest. These errors, of course, arise from computing smaller values (i.e. the values of D_n) as *differences* of larger numbers (i.e. the individual terms in the right-hand side of (19)). The fact that these large numbers do indeed significantly cancel out is further demonstrated in Figure 1(b), which also exemplifies the effect of estimates of the form (12). For this, we introduce the *majorizing* sequence Δ_n, defined by the recurrence

$$\Delta_n = \delta_{n,0}|f_M| + \sum_{l=0}^{n-1} |\lambda_{n-l,(n-l)M}| \frac{((l+1)M)!}{((l+1)M-(n-l))!} \Delta_l. \tag{20}$$

Clearly,

$$|D_n| \leq \Delta_n$$

as (20) can be obtained from (19) upon use of the triangle inequality. As we show in Figure 1(b), this straightforward estimate, which destroys the cancellations in (19), leads to a series with a vanishing radius of convergence.

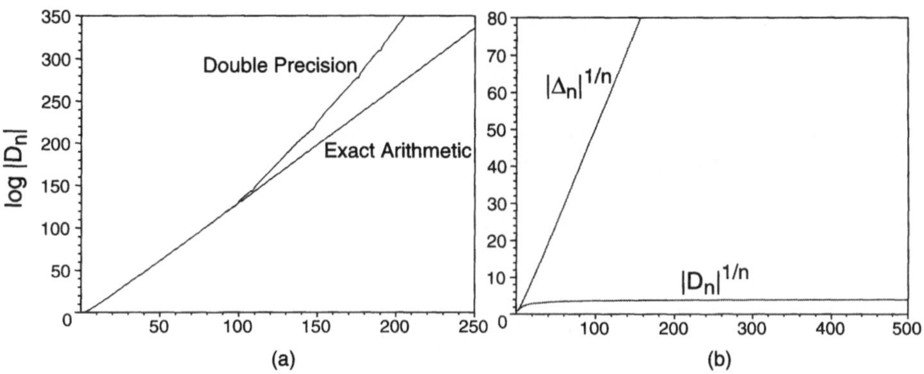

FIGURE 1. (a) Results of the implementation of the recursion (19) in exact and finite precision arithmetic. (b) Root test for the sequences D_n and Δ_n in (19), (20).

3. Recursive estimation of higher-order corrections

As we stated in the Introduction, these cancellations can be avoided with a change of independent variables that maps the perturbed domains onto the unperturbed geometry. For instance, for the geometry of (4) a natural transformation is

$$x' = x, \quad y' = \frac{y-\eta}{h+\eta}; \tag{21}$$

similar transformations can be derived for other arrangements, such as those arising in connection with problems (1)–(3). As it can be easily verified, the recurrences that arise when proposing solutions in the form of perturbation series for these problems in the new variables preserve the structure of that obtained in the original variables. However, substantial differences appear in the right-hand sides, most significantly the absence of high-order derivatives of lower-order terms (which, in turn, would entail high-order derivatives of the data and the boundary itself).

For example, let us consider the travelling wave solutions

$$\varphi(x+\mathbf{c}t, y), \quad \eta(x+\mathbf{c}t)$$

of (4): we are thus led to the system

$$\begin{aligned}
&\Delta\varphi(x,y) = 0 && \text{for } x \in \mathbb{R}^{d-1},\ -h < y < \eta(x) \\
&\partial_y \varphi(x,-h) = 0 && \\
&[\mathbf{c}\cdot\nabla_x]\,\varphi + \frac{1}{2}\left|\nabla\varphi\right|^2 + g\eta = \gamma\kappa && \text{on } y = \eta(x) \\
&-[\mathbf{c}\cdot\nabla_x]\,\eta - \nabla_x\eta\cdot\nabla_x\varphi + \partial_y\varphi = 0 && \text{on } y = \eta(x)
\end{aligned} \tag{22}$$

which obviously supports the trivial solutions

$$\varphi \equiv 0, \quad \eta \equiv 0, \quad \mathbf{c} = \mathbf{c}_0.$$

A natural approach to the derivation of nontrivial solutions to (22) is then to seek these in the form of perturbation series

$$\varphi(x,y;\varepsilon) = \sum_{n\geq 1} \varphi_n(x,y)\varepsilon^n, \quad \eta(x;\varepsilon) = \sum_{n\geq 1} \eta_n(x)\varepsilon^n \quad \text{and} \tag{23}$$

$$\mathbf{c}(\varepsilon) = \mathbf{c}_0 + \sum_{n\geq 1} \mathbf{c}_n\varepsilon^n \tag{24}$$

that *bifurcate* from the trivial solution. In these original variables, the equations for the n-th order correction $(\varphi_n, \eta_n, \mathbf{c}_{n-1})$ take on the form

$$\begin{aligned}
&\Delta\varphi_n(x,y) = 0 \qquad\qquad \text{for } x \in \mathbb{R}^{d-1},\ -h < y < 0 \\
&\partial_y\varphi_n(x,-1) = 0 \\
&[\mathbf{c}_0\cdot\nabla_x]\,\varphi_n(x,0) + [g-\gamma\Delta_x]\,\eta_n(x) + [\mathbf{c}_{n-1}\cdot\nabla_x]\,\varphi_1(x,0)(1-\delta_{n,1}) = P_n(x) \\
&-[\mathbf{c}_0\cdot\nabla_x]\,\eta_n(x) + \partial_y\varphi_n(x,0) - [\mathbf{c}_{n-1}\cdot\nabla_x]\,\eta_1(x)(1-\delta_{n,1}) = Q_n(x)
\end{aligned} \tag{25}$$

where, as in (10) and (17), the inhomogeneous terms involve derivatives of lower order corrections of increasingly high order. In contrast, if we seek a solution in the form

$$\varphi(x',y';\varepsilon) = \sum_{n\geq 1} \varphi_n(x',y')\varepsilon^n, \quad \eta(x';\varepsilon) = \sum_{n\geq 1} \eta_n(x')\varepsilon^n, \tag{26}$$

the analog of (25) reads

$$\begin{aligned}
&\Delta\varphi_n(x',y') = F_n(x',y') = \nabla_{x'}\cdot F_n^1(x',y') + \partial_{y'}F_n^2(x',y') + F_n^3(x',y') \\
&\partial_{y'}\varphi_n(x',-1) = 0 \qquad (27)\\
&[\mathbf{c}_0\cdot\nabla_{x'}]\,\varphi_n(x',0) + [g-\gamma\Delta_{x'}]\,\eta_n(x') + [\mathbf{c}_{n-1}\cdot\nabla_{x'}]\,\varphi_1(x',0)(1-\delta_{n,1}) = R_n(x')\\
&-[\mathbf{c}_0\cdot\nabla_{x'}]\,\eta_n(x') + \partial_{y'}\varphi_n(x',0) - [\mathbf{c}_{n-1}\cdot\nabla_{x'}]\,\eta_1(x')(1-\delta_{n,1}) = S_n(x')
\end{aligned}$$

where now the right-hand sides *do not* contain high-order derivatives of $\{\varphi_k\}_{k=1}^{n-1}$. For instance,

$$F_n^1(x',y') = \sum_{m=0}^{n-3}\sum_{l=0}^{m}\eta_{l+1}\eta_{m+1-l}\nabla_{x'}\varphi_{n-2-m} - 2\sum_{l=0}^{n-2}\eta_{l+1}\nabla_{x'}\varphi_{n-1-l} + (1+y')$$
$$\times\sum_{m=0}^{n-3}\sum_{l=0}^{m}\eta_{l+1}\partial_{y'}\varphi_{n-2-m}\nabla_{x'}\eta_{m+1-l} + (1+y')\sum_{l=0}^{n-2}\partial_{y'}\varphi_{n-1-l}\partial_{y'}\varphi_{n-1-l}\nabla_{x'}\eta_{l+1}$$

with similar expressions for F_n^2, F_n^3, R_n and S_n. At this stage, it can be shown that the absence of these high-order derivatives allows for the inductive estimation of the terms in the recurrence (27) in appropriate functional spaces. The underlying reason is now perhaps rather clear: the main difference between the approaches that deliver (25) and (27) resides in the boundary differentiations (with respect to ε) that must be performed to derive them. Indeed, while for (25) these differentiations must be performed *across* the boundary of the domains of definition of the field quantities in order to justify the use of the "chain rule", this justification is provided by the change of variables in the case of (27). In fact, the change of variables guarantees that differentiations are always performed in the *interior* of the domain of definition of the fields. In this manner, the iterative use of recurrences such as that in (27) can lead to general results on non-linear stability of steady states for systems like (1) or (2), and to interesting new insights on bifurcation of solutions (steady states, travelling waves) for (3) and (4); for details, we refer to [6, 7, 8, 9, 10].

4. Numerical methods based on boundary variations: consequences and improved algorithms

In this last section, we discuss the consequences of the cancellations described in §2 and the estimates discussed in §3 on the stability and convergence properties of numerical methods based on boundary perturbations. For the sake of brevity, we shall concentrate on the popular "Operator Expansion" method [3, 11, 12, 16, 17], although entirely analogous analyses hold for most of the other classical implementations of these methods (see e.g. [13, 14, 15]).

The Operator Expansion (OE) method is an approach which has been shown to be well-suited to the theoretical and numerical analysis of rather general boundary-value and free-boundary problems, and specifically of those where the governing equations in the "bulk" are relatively simple (e.g. systems of linear equations with constants coefficients). The basic strategy consists in posing the problem entirely in terms of unknowns defined *on* the (fixed or free) boundary of the domain of definition of the field quantities. In general, of course, these quantities will include derivatives of the bulk fields at the boundary which, in turn, depend on their behavior *inside* their domain of definition. On the other hand, if the field equations are relatively simple (as is the case for a wide variety of boundary-value and free-boundary problems from mathematical physics), these derivatives can be obtained from suitably defined (non-local) *boundary* operators, thus reducing the overall approximation problem to a problem of approximation of these maps. To exemplify the details behind the procedure, let us consider a classical example of application, namely to the equations (4) modeling the evolution of gravity water waves ($\gamma = 0$). Clearly, letting

$$\xi(x,t) \equiv \varphi(x,\eta(x,t),t) \quad \text{and}$$
$$\nu(x,t) \equiv \frac{\partial \varphi}{\partial n} \equiv -\nabla_x \eta(x,t) \cdot \nabla_x \varphi(x,\eta(x,t),t) + \partial_y \varphi(x,\eta(x,t),t),$$

equations (4) can be entirely rewritten in terms of the quantities $\eta(x,t)$, $\xi(x,t)$ and $\nu(x,t)$. In fact, introducing the Dirichlet-to-Neumann Operator (DNO) $G(\eta)$,

$$G(\eta)[\xi] = \nu, \tag{28}$$

equations (4) are equivalent to the (Hamiltonian) system [18]

$$\partial_t \eta = \delta_\xi H$$
$$\partial_t \xi = -\delta_\eta H,$$

where the Hamiltonian H is given by

$$H(\eta,\xi) = \frac{1}{2}\int \xi \; G(\eta)[\xi] + g \; \eta^2 \, dx.$$

Thus, from a numerical standpoint, the full problem is essentially reduced to finding an efficient and accurate strategy for the approximation of the DNO in (28). In this connection, note that the operator can be *explicitly* found for a flat surface $\eta = 0$. Indeed, for instance, if periodic two-dimensional solutions to (4) are to be found

$$\eta(x+2\pi,t) = \eta(x,t), \quad \xi(x+2\pi,t) = \xi(x,t)$$

on an tank of infinite depth ($h = \infty$), the operator on a flat surface is simply given by

$$G(0)[\sum_p \hat{\xi}_p e^{ipx}] = \sum_p |p| \hat{\xi}_p e^{ipx}. \tag{29}$$

In turn, this fact suggests a perturbation method to approximate the DNO on general surfaces η, wherein the operator is sought in the form of a series

$$G(\eta)[\xi] = \sum_{n\geq 0} G_n(\eta)[\xi] \tag{30}$$

with $G_n(\eta) = O(\eta^n)$. Letting $D = -i\nabla_x$, the ansatz in (30) readily leads to the recursion [14]

$$G_n(\eta)[\xi] = |D|^{n-1} D \frac{\eta^n}{n!} D\xi - \sum_{l=0}^{n-1} |D|^{n-l} \frac{\eta^{n-l}}{(n-l)!} G_l(\eta)[\xi] \tag{31}$$

which can be easily implemented in the frequency domain (cf. (29)).

As in the case of (14), the convergence of the series (30) follows from the work in [1, 2] under the assumption that η is Lipschitz (and e.g. $\xi \in H^1$). On the other hand, as in (17), the right-hand sides of (31) involve high-order derivatives of the boundary η (and of the Dirichlet data ξ) which must, therefore, cancel out. As we explained in §2, this is a "recipe" for ill-conditioning which, in the present context, we further demonstrate in Figure 2.

These figures display relative mean-squared errors in the Neumann data $G(\eta)[\xi]$ calculated for some specific solutions in a tank of unit depth ($h = 1$) where a formula similar to (31) holds [14]. Indeed, if $h = 1$, the functions

$$\varphi_p(x, y) = \cosh(|p|(y + 1)) \cos(p \cdot x) \tag{32}$$

solve the first two equations in (4), and therefore

$$G(\eta)[\cosh(|p|(\eta + 1)) \cos(p \cdot x)] = (\partial_y - \nabla_x \eta \cdot \nabla_x) \varphi_p\big|_{y=\eta} \tag{33}$$

can be explicitly computed for any η and compared with the approximate numerical solutions. The errors in Figure 2(a)–(c) are displayed as functions of n, the number of terms retained in the Taylor series, and they correspond to three different profiles η with decreasing regularity: analytic (Figure 2(a)), four times continuously differentiable (Figure 2(b)) and Lipschitz continuous (Figure 2(c)). In addition to the OE results, we include the corresponding outcome for two additional perturbative approximations, which we refer to as "Field Expansions" (FE) and "Transformed Field Expansion" (TFE). In contrast with the OE formulation, in the FE approach the DNO is not approximated directly as an operator, but rather a posteriori of a perturbation expansion of the *fields* themselves, as in (23). The TFE scheme, on the other hand, is similar to the FE procedure, but now the fields are expanded in the transformed variables, as in (26).

The figure confirms the theoretical predictions of §§2 and 3. Indeed, the effects of the ill-conditioning of the OE and FE formulae, arising as a result of cancellations, is evidenced here in the form of an explosive divergence of the series beyond a few terms. As the theory would predict, the onset of this divergence is precipitated by a profile's roughness, as each term in the right-hand side of the recurrences becomes more singular with decreasing boundary regularity. Indeed, the figures show that as the profile is varied from smooth to rough to Lipschitz,

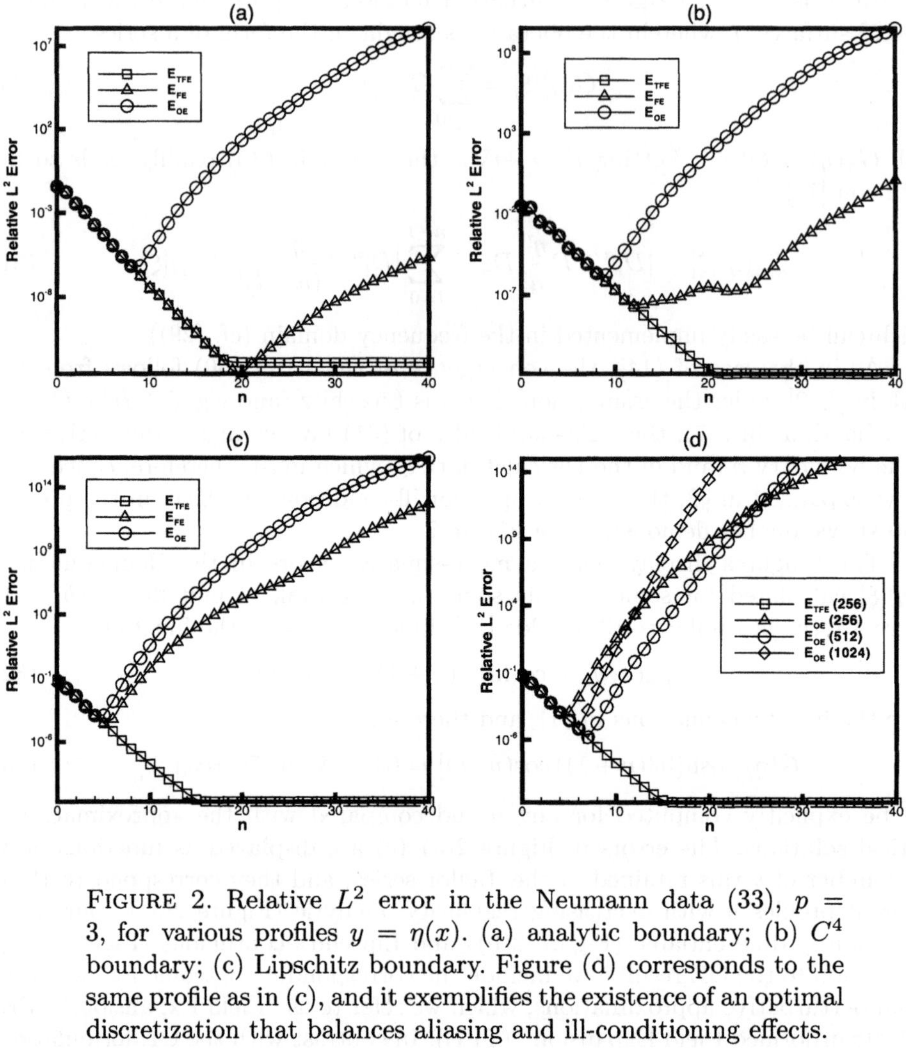

FIGURE 2. Relative L^2 error in the Neumann data (33), $p = 3$, for various profiles $y = \eta(x)$. (a) analytic boundary; (b) C^4 boundary; (c) Lipschitz boundary. Figure (d) corresponds to the same profile as in (c), and it exemplifies the existence of an optimal discretization that balances aliasing and ill-conditioning effects.

the onset of divergence for the OE and FE methods changes from $n = 9$ to 8 to 4 and from $n = 20$ to 13 to 5 respectively. In contrast, and again as expected, the results of TFE are consistently stable and only limited by the numerical resolution (N_y = number of discretization points in the y-direction = 64, in these experiments). In all cases, a number $N_x = 256$ of discretization points in the x-direction were used, which typically provided an optimal compromise between aliasing effects — for which N_x large is preferred — and ill-conditioning — which accentuates with increasing N_x. Indeed, a further refinement (to $N_x = 512$), results in a loss of accuracy in the OE and FE results for both the smooth and rough profiles, indicating that conditioning errors overcome those that may arise from aliasing. For the Lipschitz profile, on the other hand, a discretization with $N_x = 512$ actually produces slightly better results (for both OE and FE), but they again deteriorate at $N_x = 1024$; see Figure 2(d). For comparison, this latter figure also includes the results of TFE with $N_x = 256$ which demonstrate a further and unexpected advantage of TFE over the OE implementation, as the latter appears to contain more substantial information at high wavenumbers in each term of the perturbation series. Indeed, this figure shows that at a resolution of $N_x = 256$ the calculation of terms beyond $n = 5$ with OE is significantly aliased, while that of TFE is not (as demonstrated by the higher resolution results of OE).

The examples above clearly demonstrate the beneficial effect of the newly derived recursions underlying the TFE approach. However, these improvements come at a cost, namely increased computational effort. Indeed, the appearance of "volumetric sources" in the new recurrence (cf. the first equation in (27)) precludes the use of the exact solutions (32) as a basis for a spectral method, as it can be done within the OE and FE schemes. As it can be readily verified, the use of this basis translates into a reduction of dimensionality for the OE and FE procedures, as the dependence in the transverse direction y can be analytically derived; in contrast, this dependence must be numerically resolved within the TFE framework. As we have shown above (see also [14, 15]), however, this increased cost may be compensated by a substantial increase in accuracy that can, in fact, allow for computations beyond the reach of the OE and FE algorithms.

Still, if we insist on computational times that are exactly comparable to those associated with the OE or FE methods, the theory we have presented also suggests another possible improvement. Indeed, as we mentioned (see the last paragraph in §3), the instabilities in the OE recursion (31) can be associated with the *boundary* differentiations that must be performed to derive it; the change of variables (21), on the other hand, guarantees that differentiations are performed in the *interior* of the domains. In turn, this suggests an alternative implementation of the OE method, whereby the DNO $G(\eta)$ is computed by analytic continuation through a family of operators $\widetilde{G}(\varepsilon\eta)$ rather than by the more natural $G(\varepsilon\eta)$ as in (30). In accordance with the foregoing remarks, a good candidate for the family $\widetilde{G}(\varepsilon\eta)$ is one of "Dirichlet-to-Interior Neumann Operators" (DIDO) defined as

$$\widetilde{G}(\varepsilon\eta)[\xi](x) = (\partial_y - \varepsilon\nabla_x\eta\cdot\nabla_x)\,\varphi(x, b(1-\varepsilon) + \varepsilon\eta(x)))$$

where b is to be chosen, and $b > 0$ so as to ensure *interior* differentiations. We conclude with an example of application of such a scheme. In Figure 3 we present a plot of maximum errors in the Neumann data (33) ($p = 5$) for the case of a sinusoidal profile

$$\eta(x) = \frac{H}{2}\cos(x) \tag{34}$$

with $H = 0.1$ and for several values of the parameter b. The significant gains in stability of the DIDO approach over the standard OE method ($b = 0$) are evident in Figure 3(a), which again shows the error as a function of the number n of terms in the Taylor series approximation. Indeed, the figure shows that this enhanced stability allows for the accurate calculation of the Taylor coefficients over a significantly larger range for n. And this, in turn, translates into substantial improvements in the accuracy of the calculations, particularly when appropriate re-summation (analytic continuation) mechanisms are introduced as in Figure 3(b). This last figure displays the corresponding errors in the approximations of the Taylor sum via (diagonal) Padé functions, showing that DIDO can deliver results that are several orders of magnitude better than those of classical OE at a comparable computational cost.

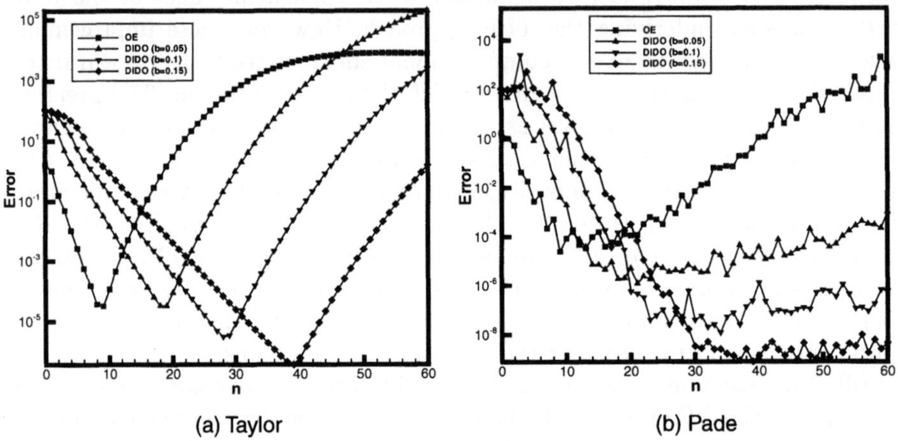

FIGURE 3. L^∞ error in the Neumann data (33), $p = 5$, for the sinusoidal profile (34). (a) Taylor series approximation. (b) Approximation via $[n/2, n/2]$ Padé approximants.

Acknowledgments. This work was supported in part by NSF through grant No. DMS-9971379, AFOSR through contract No. F49620-02-1-0052 and by the Army High Performance Computing Research Center (AHPCRC) under Army Research Laboratory cooperative agreement number DAAD19-01-2-0014.

References

[1] A. P. Calderón, *Cauchy integrals on Lipschitz curves and related operators*, Proc. Nat. Acad. Sci. USA **75** (1977), 1324–1327.

[2] R. Coifman and Y. Meyer, *Nonlinear harmonic analysis and analytic dependence*, in Pseudodifferential operators and applications, Amer. Math. Soc., 1985, 71–78.

[3] W. Craig and C. Sulem, *Numerical simulation of gravity waves*, J. Comput. Phys. **108** (1993), 73–83.

[4] D. G. Dommermuth and D. K. P. Yue, *A high-order spectral method for the study of nonlinear gravity waves*, J. Fluid Mech. **184** (1987), 267–288.

[5] T. Elfouhaily, D. R. Thompson, D. Vandemark and B. Chapron, *Truncated Hamiltonian versus surface perturbation in nonlinear wave theories*, Waves Random Media **10** (2000), 103–116.

[6] A. Friedman and F. Reitich, *On the existence of spatially patterned dormant malignancies in a model for the growth of non-necrotic vascular tumors*, Math. Models and Meth. Appl. Sci. **11** (2001), 601–625.

[7] A. Friedman and F. Reitich, *Symmetry-breaking bifurcation of analytic solutions to free-boundary problems: an application to a model of tumor growth*, Trans. Amer. Math. Soc. **353** (2001), 1587–1634.

[8] A. Friedman and F. Reitich, *Nonlinear stability of a quasi-static Stefan problem with surface tension: a continuation approach*, Ann. Scuola Norm. Sup. Pisa Cl. Sci. (4) **30** (2001), 341–403.

[9] A. Friedman and F. Reitich, *Quasi-static motion of a capillary drop, I: the two-dimensional case*, J. Diff. Equations **178** (2002), 212–263.

[10] A. Friedman and F. Reitich, *Quasi-static motion of a capillary drop, II: the three-dimensional case*, J. Diff. Eqs. **186** (2002), 509–557.

[11] H. Hasimoto and H. Ono, *Nonlinear modulation of gravity waves*, J. Phys. Soc. Japan **33** (1972), 805–811.

[12] D. M. Milder, *The effects of truncation on surface-wave Hamiltonians*, J. Fluid Mech. **216** (1990), 249–262.

[13] D. Nicholls and F. Reitich, *A new approach to analyticity of Dirichlet–Neumann operators*, Proc. R. Soc. Edinburgh A **131** (2001), 1411–1433.

[14] D. Nicholls and F. Reitich, *Stability of high-order perturbative methods for the computation of Dirichlet–Neumann operators*, J. Comput. Phys. **170** (2001), 276–298.

[15] D. Nicholls and F. Reitich, *Analytic continuation of Dirichlet–Neumann operators*, Numer. Math., to appear.

[16] R. A. Smith, *An operator expansion formalism for nonlinear surface waves over variable depth*, J. Fluid Mech. **363** (1998), 333–347.

[17] K. M. Watson and B. J. West, *A transport-equation description of nonlinear ocean surface wave interactions*, J. Fluid Mech. **70** (1975), 815–826.

[18] V. Zakharov, *Stability of periodic waves of finite amplitude on the surface of a deep fluid*, J. Appl. Mech. Tech. Phys. **9** (1968), 190–194.

School of Mathematics
University of Minnesota
206 Church St. S. E.
Minneapolis, MN 55455, USA
E-mail address: reitich@math.umn.edu

International Series of Numerical Mathematics, Vol. 147, 281–291

Error Estimates for Dissipative Evolution Problems

Giuseppe Savaré

Abstract. We present a quick overview of the general problem of finding optimal *a priori* and *a posteriori* error estimates for the approximation of dissipative evolution equations in Hilbert and metric spaces by means of a variational formulation of the implicit Euler scheme. We shall discuss what are the intrinsic metric arguments which are involved in the derivation of the estimates and we will present an elementary proof in a simplified finite dimensional case. An application to the porous medium equation in the new framework of the Wasserstein distance is briefly sketched.

1. Introduction

In this paper we are interested in optimal *a priori* and *a posteriori* error estimates for the approximation of the solution $u : [0,T] \to \mathscr{S}$ of the gradient flow equation

$$\begin{cases} u'(t) = -\nabla\phi(u(t)) & t \in [0,T], \\ u(0) = u^0, \end{cases} \tag{1}$$

by means of the implicit Euler scheme

$$\frac{U_\tau^n - U_\tau^{n-1}}{\tau_n} = -\nabla\phi(U_\tau^n), \quad n = 1, \cdots, N; \quad U_\tau^0 \approx u^0 \text{ given.} \tag{2}$$

Here $\phi : \mathscr{S} \to \mathbb{R}$ is a given functional defined in a suitable ambient space $\mathscr{S}$, $\tau := \{\tau_n\}_{n=1}^N$ denotes the variable step sizes induced by a partition $\mathcal{P}_\tau$ of the time interval $[0,T]$ into N subintervals $I_\tau^n := (t_\tau^{n-1}, t_\tau^n]$,

$$\mathcal{P}_\tau := \{0 = t_\tau^0 < t_\tau^1 < \ldots < t_\tau^{N-1} < t_\tau^N = T\}, \quad \tau_n = t_\tau^n - t_\tau^{n-1} = |I_\tau^n|, \tag{3}$$

and the equation (2) should be understood as a recursive algorithm which indicates how to find U_τ^n once U_τ^{n-1} is known.

The *discrete solution* $U_\tau(t_\tau^n) := U_\tau^n, n = 1, \cdots, N$, will thus provide an approximation of the exact solution u at the nodes t_τ^n of the partition $\mathcal{P}_\tau$. We want

This work was partially supported by grants of M.I.U.R. (cofin2002) and of the Institute of Applied Mathematics and Information Technology (I.M.A.T.I.) of the C.N.R., Pavia, Italy.

to find estimates of the *error* $E_{\boldsymbol{\tau}}$ between $U_{\boldsymbol{\tau}}$ and the exact solution u

$$E_{\boldsymbol{\tau}} := \max_{t\in\mathcal{P}_{\boldsymbol{\tau}}} d(U_{\boldsymbol{\tau}}(t), u(t)), \quad d \text{ being a suitable } \textit{distance} \text{ defined on } \mathscr{S},$$

addressing the following fundamental issues.

- The estimates are *a posteriori*, i.e. they are expressed in terms of computable quantities which depend solely on the time-steps, the discrete solution, and the data.
- The *a posteriori* bounds should converge to zero as the maximum of the time steps
$$|\boldsymbol{\tau}| := \max_{1\le n\le N} \tau_n \tag{4}$$
goes to 0 with an *optimal rate* w.r.t. the regularity of the solution u.
- The estimates should depend on the structural properties (*dissipativity*) of $\mathscr{S}$ and ϕ which ensure the well-posedness of (1): in particular, they should not depend on the space dimension of $\mathscr{S}$ and they are explicitly determined without need of solving any auxiliary problem. No extra regularity assumptions should be assumed.
- No *a priori* constraints should be imposed between consecutive time-steps, which could just be tailored to the *a posteriori* error estimators alone.

Four possible frameworks

We will focus on frameworks where (1) is, in some sense to be better specified, *dissipative* and its dependence with respect to (w.r.t.) perturbations of the initial datum can be explicitly controlled in terms of the distance d.

a) The simplest case: $\mathscr{S} := \mathbb{R}^m$, d is the usual Euclidean distance, ϕ is a (quadratic perturbation of a) C^1 convex function: (1) is a system of ODE's.

b) Infinite dimensional Hilbert spaces $\mathscr{S} := \mathscr{H}$*, the distance being induced by the scalar product* $\langle\cdot,\cdot\rangle$. The most interesting case in view of applications to nonlinear evolutionary PDE's [18] is provided by lower semicontinuous (l.s.c.) convex functionals ϕ (continuous quadratic perturbations are still allowed), which take their values in the *extended* real line $(-\infty, +\infty]$. A crucial feature of the convexity of ϕ is the variational formulation of both (1) and (2) by means of evolution inequalities. In fact, the solution of (1) can also be characterized by means of the system

$$\langle u'(t), u(t) - v\rangle + \phi(u(t)) \le \phi(v) \quad \forall\, v \in \mathscr{H}, \tag{5}$$

which, at the discrete level, corresponds to

$$\langle \tau_n^{-1}(U_{\boldsymbol{\tau}}^n - U_{\boldsymbol{\tau}}^{n-1}), U_{\boldsymbol{\tau}}^n - V\rangle + \phi(U_{\boldsymbol{\tau}}^n) \le \phi(V) \quad \forall\, V \in \mathscr{H}; \tag{6}$$

they both lead to the nonexpansivity property of the continuous (and discrete) trajectories starting from two initial values u^0, v^0 (resp. $U_{\boldsymbol{\tau}}^0, V_{\boldsymbol{\tau}}^0$)

$$d(u(t), v(t)) \le d(u^0, v^0), \quad d(U_{\boldsymbol{\tau}}^n, V_{\boldsymbol{\tau}}^n) \le d(U_{\boldsymbol{\tau}}^0, V_{\boldsymbol{\tau}}^0) \quad t \in [0,T],\ n = 1, \cdots, N. \tag{7}$$

In the framework of this well developed theory (see e.g. [4]), the same approximation results of case *a)* can be reproduced. We refer to [14] for a detailed account of the various contributions and for a presentation of the estimates and of the related applications; here we simply quote the optimal *a priori* estimates of [3], [15], and the linear theory of [8] (the last of an important series of papers on this subject).

c) $\mathscr{S}$ *is a Riemannian manifold, endowed with its Riemannian distance d.* In this case equation (1) should be imposed in the tangent space of $\mathscr{S}$ at the point $u(t)$ and we could interpret the notion of convexity as *convexity along geodesics.*

Conversely, (2) requires more attention, since now the difference $U^n_\tau - U^{n-1}_\tau$ is no more defined. One possible way to circumvent this difficulty is to rewrite (2) in a variational form, which is already suggested in the Euclidean case. We simply notice that among the solutions of (2) we can select the minima of the functionals

$$V \mapsto \Phi^n_\tau(V) := \frac{1}{2\tau_n} d^2(V, U^{n-1}_\tau) + \phi(V), \tag{8a}$$

i.e. U^n_τ, $n = 1, \cdots, N$, solve recursively the family of minimum problems

$$\Phi^n_\tau(U^n_\tau) = \min_{V \in \mathscr{S}} \Phi^n_\tau(V) \quad n = 1, \cdots, N. \tag{8b}$$

(8a,b) still make sense in a Riemannian manifold, and they can be used to approximate (1). Of course, we expect the error estimates to be affected by the curvature of $\mathscr{S}$; nevertheless, it is interesting to note that, at least in the case of a simply connected manifold with *nonpositive* sectional curvature, we can reproduce the same result of the Euclidean framework. It is one of the most interesting contribution of [11], [2] to show that, as in the Euclidean case, these estimates do not directly rely on the finite dimension or on the local differentiable structure of the manifold, but on simple geometric properties of its Riemannian distance.

d) $(\mathscr{S}, d)$ *is a metric space.* In this case is no more clear what is the meaning of (1), whereas the recursive variational scheme (8a,b) still applies and provides a family of (hopefully) approximate solutions to (1). This fact has led several authors to use this approximation procedure to define a solution of (1), when the differential problem does not exhibit a useful linear structure. In an abstract metric framework the contributions of E. De Giorgi and his collaborators [7], [6], [1], show how to give a meaning to (1) and provide general conditions for the convergence (up to the extraction of a subsequence) of the variational scheme with uniform step sizes.

Before discussing the delicate question of convergence estimates in this abstract setting, let us briefly recall a particular example, which was also one of the main motivation of the present investigation: it arises from the recent papers of F. Otto [9, 16] on a new variational interpretation of the Fokker–Planck and the porous medium equations as gradient flows w.r.t. suitable functionals in the space of probability measures metricized by the so called Wasserstein distance. For the sake of simplicity, here we will consider only measures which are absolutely continuous w.r.t. the Lebesgue measure, and we will identify them with their densities.

An example of the metric setting: the porous medium equation and the Wasserstein distance

Let us introduce the set $\mathscr{S} = \mathscr{P}^2_a(\mathbb{R}^k)$

$$\mathscr{P}^2_a(\mathbb{R}^k) := \Big\{ u \in L^1(\mathbb{R}^k) : u \geq 0, \int_{\mathbb{R}^k} u(x)\,dx = 1, \int_{\mathbb{R}^k} |x|^2 u(x)\,dx < +\infty \Big\}, \tag{9a}$$

endowed with the *Wasserstein distance* (we refer to the recent book [17] for a comprehensive and up-to-date overview of this and related topics)

$$\begin{aligned} W^2(v,w) := \min\Big\{ \int_{\mathbb{R}^k} |\boldsymbol{r}(x) - x|^2\, v(x)\,dx : \quad & \boldsymbol{r} : \mathbb{R}^k \to \mathbb{R}^k \text{ is a Borel map,} \\ \int_{\mathbb{R}^k} \zeta(y) w(y)\,dy = \int_{\mathbb{R}^k} \zeta(\boldsymbol{r}(x)) v(x)\,dx \quad & \text{for all functions } \zeta \in C^0_c(\mathbb{R}^k) \Big\}. \end{aligned} \tag{9b}$$

The map $\boldsymbol{r}$ attaining the minimum in (9b) is usually called *the optimal transportation* between v and w. For $m > 1$, we also introduce the functional on $\mathscr{P}^2_a(\mathbb{R}^k)$

$$\varphi(u) := \frac{1}{m-1} \int_{\mathbb{R}^k} u^m(x)\,dx. \tag{9c}$$

In [16] F. Otto showed that the nonlinear (*porous medium*) diffusion equation

$$\begin{cases} \partial_t u - \Delta(u^m) = 0 & \text{in } \mathbb{R}^k \times (0,T), \\ u(x,0) = u^0(x) & \text{in } \mathbb{R}^k, \quad u^0 \in \mathscr{P}^2_a(\mathbb{R}^k), \end{cases} \tag{9d}$$

can be considered as the *gradient flow* of the functional φ in $\mathscr{P}^2_a(\mathbb{R}^k)$ w.r.t. the distance W: here, the meaning of "*gradient flow*" has to be understood exactly as the limit of the discrete solutions $U_{\boldsymbol{\tau}}$ obtained by the recursive minimization algorithm (8a,b) with the choice $d := W$ and $\phi := \varphi$ corresponding to (9a,b,9c).

Even if $\mathscr{P}^2_a(\mathbb{R}^k)$ is a convex subset of $L^1(\mathbb{R}^k)$, segments do not provide the shortest path (w.r.t. W) connecting two elements, since they do not have even finite length. However, it is worth mentioning that the functional φ defined by (9c) is still convex along the (constant speed) geodesics of $\mathscr{P}^2_a(\mathbb{R}^k)$, as showed by [12].

Geodesic convexity is not enough for error estimates in metric spaces

Let us recall that the well-posedness (and therefore also the problem to find optimal error estimates) for gradient flows of convex functionals in general *Banach* spaces (even finite dimensional) is still completely open. This remark points out, at least heuristically, that severe restrictions should be imposed to the general metric setting, and that the geometric properties of the distance should play a crucial role.

A first step in this direction has been obtained by U. Mayer [11], who considered gradient flows of geodesically convex functionals on *nonpositively curved metric spaces*: these are length spaces where the distance from any given point $w \in \mathscr{S}$ satisfies the inequality

$$d^2(v_t, w) \leq (1-t)d^2(v_0, w) + t d^2(v_1, w) - t(1-t) d^2(v_0, v_1) \quad t \in [0,1], \tag{10}$$

along any constant speed geodesic curve $\{v_t\}_{t\in[0,1]}$ connecting two arbitrary points $v_0, v_1 \in \mathscr{S}$. This property, which was introduced by Aleksandrov on the basis of the analogous inequality satisfied in Euclidean spaces (in fact an equality) and in Riemannian manifolds of nonpositive sectional curvature [10, §2.3], allows to prove (7), and to obtain the convergence of the approximation scheme (and a suboptimal *a priori* error estimate) by following the same ideas of the celebrated Crandall–Liggett generation Theorem [5]. It is interesting to note that if d is induced by a norm in a vector space and it satisfies (10), then the norm obeys the parallelogram rule (simply choose $t = 1/2$ in (10)) and therefore it is induced by a scalar product.

These assumptions, however, do not cover the case of the Wasserstein distance (9a,b). In fact, it is possible to prove (cf. [2]) that the distance of this space satisfies the opposite inequality of (10), thus providing a *positively* curved space, as formally suggested also by [16].

A new geometric condition

The main idea to overcome this difficulty and to derive better error estimates is to allow more flexibility in the choice of the connecting curves, which do not need to be geodesics but could depend on the distance and on the functional. This approach relies on a careful analysis of the techniques of [14] and it is completely developed in an intrinsic metric framework in [2], where a particular attention to the case of the Wasserstein distance and its applications is devoted. It is surprising that all the estimates are based only on the following geometrical property:

for every triple $w, v_0, v_1 \in \mathscr{S}$ there exists a curve $\{v_t\}_{t\in[0,1]}$ connecting v_0 to v_1 which satisfies

$$\phi(v_t) \le (1-t)\phi(v_0) + t\phi(v_1) \quad t \in [0,1], \tag{11a}$$

$$d^2(w, v_t) \le (1-t)d^2(w, v_0) + td^2(w, v_1) - t(1-t)d^2(v_0, v_1). \tag{11b}$$

In the next section, following the ideas of [2], we will present a new elementary derivation of the estimates of [14] in the easiest finite dimensional framework (thus avoiding any technical tools related to the theory of maximal monotone operators), trying to use only the *metric properties* of the Euclidean spaces related to (11a,b). We shall obtain estimates independent of the dimension and of the vectorial structure, which therefore will be immediately applicable to general Hilbert and metric spaces. An application to the porous medium equation will be briefly sketched at the end of the paper.

2. Main result

In this section we shall deal with the simplest situtation, which is provided by a

convex C^1 functional ϕ defined in a finite dimensional space $\mathscr{S} := \mathbb{R}^m$,

endowed with the usual euclidean distance $d(u, v) := |u - v|$.

Since the elementary proof of this result will be based only on the fact that for each $w, v_0, v_1 \in \mathscr{S}$

the segment $v_t := (1-t)v_0 + tv_1$, $t \in [0,1]$, *satisfies (11a,b),*

we will keep the notation $d(u,v)$ instead of $|u-v|$ for the distance.

Theorem 2.1 (Optimal error estimate). *Let u be the solution of* (1), *let U_τ be the discrete solution obtained by solving* (2) *or, equivalently,* (8a,b). *Then*

$$E_\tau^2 = \max_{t\in\mathcal{P}_\tau} d^2(U_\tau(t), u(t)) \le d^2(u^0, U_\tau^0) + \sum_{n=1}^N \tau_n^2 \mathscr{E}_\tau^n, \tag{12}$$

where $\mathscr{E}_\tau^n$ is the a posteriori error estimator *defined by*

$$\mathscr{E}_\tau^n := \frac{\phi(U_\tau^{n-1}) - \phi(U_\tau^n)}{\tau_n} - \frac{d^2(U_\tau^{n-1}, U_\tau^n)}{\tau_n^2} \ge 0. \tag{13}$$

Moreover, if ϕ is bounded from below, then

$$\sum_{n=1}^N \tau_n^2 \mathscr{E}_\tau^n \le |\tau| \big(\phi(U_\tau^0) - \inf_{\mathscr{S}} \phi\big). \tag{14}$$

Finally, the following optimal a priori *bound in terms of the gradient of ϕ holds*

$$\sum_{n=1}^N \tau_n^2 \mathscr{E}_\tau^n \le \tfrac{1}{2}|\tau|^2 \,|\nabla\phi(U_\tau^0)|^2. \tag{15}$$

We will divide the proof in some steps.

A "metric variational inequality" for u. Our first remark is that (5) can also be written as

$$\tfrac{d}{dt}\tfrac{1}{2} d^2(u(t), v) + \phi(u(t)) \le \phi(v) \quad \forall v \in \mathscr{S},\ t \in [0,T]. \tag{16}$$

Observe that, being u regular, (16) contains all the information of (1); in particular it is easy to derive from (16) the contraction property for continuous trajectories (7). In fact, if w is another solution of (16), which we rewrite by introducing another variable s

$$\tfrac{d}{ds}\tfrac{1}{2} d^2(w(s), v) + \phi(w(s)) \le \phi(v) \quad \forall v \in \mathscr{S},\ t \in [0,T], \tag{17}$$

then, choosing $v := w(s)$ in (16), $v := u(t)$ in (17), and summing up the two inequalities, we find

$$\tfrac{\partial}{\partial t}\tfrac{1}{2} d^2(u(t), w(s)) + \tfrac{\partial}{\partial s}\tfrac{1}{2} d^2(u(t), w(s)) \le 0 \quad s,t \in [0,T]. \tag{18}$$

Evaluating this relation for $s = t$ we obtain

$$\tfrac{d}{dt}\tfrac{1}{2} d^2(u(t), w(t)) \le 0 \quad \text{i.e.} \quad d(u(t), w(t)) \le d(u^0, w^0) \quad t \in [0,T]. \tag{19}$$

A "metric variational inequality" for $U^n_{\boldsymbol\tau}$. The second step consists in writing an analogous variational inequality for the discrete solution: here, we will use (11a) and (11b) in a crucial way. In fact, choosing $w := U^{n-1}_{\boldsymbol\tau}$, it is easy to see that they yield the following strong convexity property for the functionals $\Phi^n_{\boldsymbol\tau}(\cdot)$:

$$\Phi^n_{\boldsymbol\tau}(v_t) \le (1-t)\Phi^n_{\boldsymbol\tau}(v_0) + t\Phi^n_{\boldsymbol\tau}(v_1) - \tfrac{1}{2\tau_n}t(1-t)d^2(v_0,v_1). \tag{20}$$

Starting from the minimum property (8b) and applying (20) with $v_0 := U^n_{\boldsymbol\tau}$, $v_1 := V$, we get

$$\Phi^n_{\boldsymbol\tau}(U^n_{\boldsymbol\tau}) \le \Phi^n_{\boldsymbol\tau}(v_t) \le (1-t)\Phi^n_{\boldsymbol\tau}(U^n_{\boldsymbol\tau}) + t\Phi^n_{\boldsymbol\tau}(V) - \tfrac{1}{2\tau_n}t(1-t)d^2(U^n_{\boldsymbol\tau},V) \quad \forall V \in \mathscr{S}.$$

The minimum condition says that the right derivative at $t = 0$ of the right hand side is nonnegative; thus, we find

$$\Phi^n_{\boldsymbol\tau}(V) - \Phi^n_{\boldsymbol\tau}(U^n_{\boldsymbol\tau}) - \tfrac{1}{2\tau_n}d^2(U^n_{\boldsymbol\tau},V) \ge 0 \quad \forall V \in \mathscr{S}, \tag{21}$$

which can also be written as

$$\tfrac{1}{\tau_n}\Big(\tfrac12 d^2(U^n_{\boldsymbol\tau},V) - \tfrac12 d^2(U^{n-1}_{\boldsymbol\tau},V)\Big) + \tfrac{1}{2\tau_n}d^2(U^n_{\boldsymbol\tau},U^{n-1}_{\boldsymbol\tau}) + \phi(U^n_{\boldsymbol\tau}) \le \phi(V). \tag{22}$$

A continuous formulation of (22). We want to write (22) as a true differential evolution inequality for the discrete solution $U_{\boldsymbol\tau}$, in order to compare it with the continuous one (16) and to try to reproduce the same comparison argument which we used for (18) and (19). A first idea, which was exploited in [14], is to use a linear interpolation between the consecutive values of the discrete solution. But an even simpler choice is possible (as in [13]); we interpolate the values of the functions of $U^n_{\boldsymbol\tau}$ instead of interpolating the arguments, since this does not even require an underlying linear structure. Therefore, we set

$$\phi_{\boldsymbol\tau}(s) := \text{“the linear interpolant of } \phi(U^{n-1}_{\boldsymbol\tau}) \text{ and } \phi(U^n_{\boldsymbol\tau})\text{” if } s \in (t^{n-1}_{\boldsymbol\tau}, t^n_{\boldsymbol\tau}], \tag{23}$$

i.e.

$$\phi_{\boldsymbol\tau}(s) := \frac{t^n_{\boldsymbol\tau} - s}{\tau_n}\phi(U^{n-1}_{\boldsymbol\tau}) + \frac{s - t^{n-1}_{\boldsymbol\tau}}{\tau_n}\phi(U^n_{\boldsymbol\tau}) \quad s \in (t^{n-1}_{\boldsymbol\tau}, t^n_{\boldsymbol\tau}]. \tag{24}$$

Analogously, for any $V \in \mathscr{S}$ we set

$$d^2_{\boldsymbol\tau}(s;V) := \frac{t^n_{\boldsymbol\tau} - s}{\tau_n}d^2(U^{n-1}_{\boldsymbol\tau},V) + \frac{s - t^{n-1}_{\boldsymbol\tau}}{\tau_n}d^2(U^n_{\boldsymbol\tau},V) \quad s \in (t^{n-1}_{\boldsymbol\tau}, t^n_{\boldsymbol\tau}]. \tag{25}$$

Since

$$\tfrac{d}{ds}d^2_{\boldsymbol\tau}(s;V) = \tfrac{1}{\tau_n}\Big(d^2(U^n_{\boldsymbol\tau},V) - d^2(U^{n-1}_{\boldsymbol\tau},V)\Big) \quad s \in (t^{n-1}_{\boldsymbol\tau}, t^n_{\boldsymbol\tau}],$$

(22) becomes

$$\tfrac{d}{ds}\tfrac12 d^2_{\boldsymbol\tau}(s;V) + \phi_{\boldsymbol\tau}(s) \le \phi(V) + \tfrac12\mathscr{R}_{\boldsymbol\tau}(s) \quad \forall s \in (0,T)\setminus \mathcal{P}_{\boldsymbol\tau}, \tag{26}$$

where we set, for $s \in (t_{\boldsymbol{\tau}}^{n-1}, t_{\boldsymbol{\tau}}^n]$, $t_{\boldsymbol{\tau}}^{n-1/2} := \frac{1}{2}(t_{\boldsymbol{\tau}}^n + t_{\boldsymbol{\tau}}^{n-1})$, and

$$\begin{aligned}\tfrac{1}{2}\mathscr{R}_{\boldsymbol{\tau}}(s) &:= \phi_{\boldsymbol{\tau}}(s) - \phi(U_{\boldsymbol{\tau}}^N) - \tfrac{1}{2\tau_n} d^2(U_{\boldsymbol{\tau}}^n, U_{\boldsymbol{\tau}}^{n-1}) \\ &= \frac{t_{\boldsymbol{\tau}}^n - s}{\tau_n}\Big(\phi(U_{\boldsymbol{\tau}}^{n-1}) - \phi(U_{\boldsymbol{\tau}}^n)\Big) - \frac{d^2(U_{\boldsymbol{\tau}}^n, U_{\boldsymbol{\tau}}^{n-1})}{2\tau_n} \\ &= (t_{\boldsymbol{\tau}}^n - s)\mathscr{E}_{\boldsymbol{\tau}}^n + (t_{\boldsymbol{\tau}}^{n-1/2} - s)\frac{d^2(U_{\boldsymbol{\tau}}^n, U_{\boldsymbol{\tau}}^{n-1})}{\tau_n^2}.\end{aligned} \tag{27}$$

The comparison argument. Taking a convex combination w.r.t. the variable $s \in I_{\boldsymbol{\tau}}^n$ of (16) written for $v := U_{\boldsymbol{\tau}}^{n-1}$ and $v := U_{\boldsymbol{\tau}}^n$, we easily get

$$\tfrac{\partial}{\partial t}\tfrac{1}{2} d_{\boldsymbol{\tau}}^2(s, u(t)) + \phi(u(t)) \le \phi_{\boldsymbol{\tau}}(s) \quad t, s \in [0, T], \tag{28}$$

whereas choosing $V := u(t)$ in (26) we find

$$\tfrac{\partial}{\partial s}\tfrac{1}{2} d_{\boldsymbol{\tau}}^2(s; u(t)) + \phi_{\boldsymbol{\tau}}(s) \le \phi(u(t)) + \tfrac{1}{2}\mathscr{R}_{\boldsymbol{\tau}}(s) \quad t \in [0, T],\ s \in [0, T] \setminus \mathcal{P}_{\boldsymbol{\tau}}. \tag{29}$$

Summing up (28) and (29), we end up with

$$\tfrac{\partial}{\partial t} d_{\boldsymbol{\tau}}^2(s, u(t)) + \tfrac{\partial}{\partial s} d_{\boldsymbol{\tau}}^2(s; u(t)) \le \mathscr{R}_{\boldsymbol{\tau}}(s) \quad t \in [0, T],\ s \in [0, T] \setminus \mathcal{P}_{\boldsymbol{\tau}}. \tag{30}$$

Choosing $s = t$, we eventually find

$$\tfrac{d}{dt} d_{\boldsymbol{\tau}}^2(t, u(t)) \le \mathscr{R}_{\boldsymbol{\tau}}(t) \quad t \in [0, T] \setminus \mathcal{P}_{\boldsymbol{\tau}}, \tag{31}$$

and therefore, being $t \mapsto d_{\boldsymbol{\tau}}(t, u(t))$ continuous,

$$d_{\boldsymbol{\tau}}^2(T, u(T)) \le d_{\boldsymbol{\tau}}^2(0, u^0) + \int_0^T \mathscr{R}_{\boldsymbol{\tau}}(t)\, dt. \tag{32}$$

Since $d_{\boldsymbol{\tau}}(t, V) = d(U_{\boldsymbol{\tau}}(t), V)$ if $t \in \mathcal{P}_{\boldsymbol{\tau}}$, a simple evaluation through (27) of the contribution of each subinterval $I_{\boldsymbol{\tau}}^n$ to the integral gives

$$d^2(U_{\boldsymbol{\tau}}(T), u(T)) \le d^2(U_{\boldsymbol{\tau}}^0, u^0) + \sum_{n=1}^N \tau_n^2 \mathscr{E}_{\boldsymbol{\tau}}^n. \tag{33}$$

Cancellation effects and rate of convergence. Choosing $V := U_{\boldsymbol{\tau}}^{n-1}$ in (22), we find

$$\tfrac{1}{\tau_n} d^2(U_{\boldsymbol{\tau}}^n, U_{\boldsymbol{\tau}}^{n-1}) + \phi(U_{\boldsymbol{\tau}}^n) \le \phi(U_{\boldsymbol{\tau}}^{n-1}); \tag{34}$$

this shows that $\mathscr{E}_{\boldsymbol{\tau}}^n \ge 0$ and that $n \mapsto \phi(U_{\boldsymbol{\tau}}^n)$ is decreasing; we get

$$\sum_{n=1}^N \tau_n^2 \mathscr{E}_{\boldsymbol{\tau}}^n \le \sum_{n=1}^N \tau_n\Big(\phi(U_{\boldsymbol{\tau}}^{n-1}) - \phi(U_{\boldsymbol{\tau}}^n)\Big) \le |\boldsymbol{\tau}|\Big(\phi(u_{\boldsymbol{\tau}}^0) - \phi(U_{\boldsymbol{\tau}}^N)\Big) \tag{35}$$

which proves (14). (15) requires a more refined argument. We start by writing (22) at the time step $n-1$ with the choice $V := U_{\boldsymbol{\tau}}^n$

$$\begin{aligned}&\phi(U_{\boldsymbol{\tau}}^{n-1}) - \phi(U_{\boldsymbol{\tau}}^n) \\ &\qquad \le \tfrac{1}{2\tau_{n-1}}\Big(d^2(U_{\boldsymbol{\tau}}^{n-2}, U_{\boldsymbol{\tau}}^n) - d^2(U_{\boldsymbol{\tau}}^{n-1}, U_{\boldsymbol{\tau}}^n) - d^2(U_{\boldsymbol{\tau}}^{n-1}, U_{\boldsymbol{\tau}}^{n-2})\Big).\end{aligned} \tag{36}$$

Now, we observe that the triangle and the Young inequalities yield

$$\begin{aligned} d^2(U_\tau^{n-2}, U_\tau^n) &\le \Big(d(U_\tau^{n-2}, U_\tau^{n-1}) + d(U_\tau^{n-1}, U_\tau^n)\Big)^2 \\ &\le \big(1 + \tfrac{\tau_n}{\tau_{n-1}}\big) d^2(U_\tau^{n-2}, U_\tau^{n-1}) + \big(1 + \tfrac{\tau_{n-1}}{\tau_n}\big) d^2(U_\tau^{n-1}, U_\tau^n). \end{aligned} \tag{37}$$

By substituting this inequality into (36), we obtain, after dividing by τ_n,

$$\frac{\phi(U_\tau^{n-1}) - \phi(U_\tau^n)}{\tau_n} \le \frac{d^2(U_\tau^{n-1}, U_\tau^n)}{2\tau_n^2} + \frac{d^2(U_\tau^{n-1}, U_\tau^{n-2})}{2\tau_{n-1}^2}. \tag{38}$$

If we insert (38) into the expression of $\mathscr{E}_\tau^n$ we get

$$\mathscr{E}_\tau^n \le \frac{d^2(U_\tau^{n-1}, U_\tau^{n-2})}{2\tau_{n-1}^2} - \frac{d^2(U_\tau^{n-1}, U_\tau^n)}{2\tau_n^2} \tag{39}$$

and therefore, by (13) and the positivity of $\mathscr{E}_\tau^n$, we eventually find

$$\sum_{n=1}^N \tau_n^2 \mathscr{E}_\tau^n \le |\tau|^2 \sum_{n=1}^N \mathscr{E}_\tau^n \le |\tau|^2 \Big(\frac{\phi(U_\tau^0) - \phi(U_\tau^1)}{\tau_1} - \frac{d^2(U_\tau^1, U_\tau^0)}{2\tau_1^2}\Big). \tag{40}$$

The modulus of the gradient of a convex function. For convex functions the modulus of the gradient can also be characterized as

$$|\nabla\phi(w)| = \sup\Big\{\frac{\phi(w) - \phi(v)}{d(w,v)} : v \in \mathscr{S}, v \ne w, \phi(v) \le \phi(w)\Big\}; \tag{41}$$

thus, since $\phi(U_\tau^1) \le \phi(U_\tau^0)$, we get

$$\tfrac{\phi(U_\tau^0) - \phi(U_\tau^1)}{\tau_1} - \tfrac{d^2(U_\tau^1, U_\tau^0)}{2\tau_1^2} \le |\nabla\phi(U_\tau^0)| \tfrac{d(U_\tau^1, U_\tau^0)}{\tau_1} - \tfrac{d^2(U_\tau^1, U_\tau^0)}{2\tau_1^2} \le \tfrac{1}{2}|\nabla\phi(U_\tau^0)|^2.$$

General metric space and the application to the porous medium equation

It should be clear from the above proof that

Corollary 2.2. *The statement of Theorem* 2.1 *holds even if* $(\mathscr{S}, d)$ *is a metric space,* ϕ *is a l.s.c. (extended) real functional defined on* $\mathscr{S}$ *whose gradient norm is defined by* (41), u *is a solution of the "metric evolution inequality"* (20), U_τ *solves the variational algorithm* (8a,b), *and* (11a,b) *hold.*

If we want to apply the previous result to the example of (9a,b,c,d), for a triple $w, v_0, v_1 \in \mathscr{P}_a^2(\mathbb{R}^k)$, we should exhibit a connecting curve v_t, $t \in [0,1]$, which satisfies (11a,b). Let us denote by $\boldsymbol{r}_0, \boldsymbol{r}_1$ the optimal transportation maps between w and v_0, v_1 respectively (cf. (9b)); we define the curve v_t by duality with continuous functions

$$\int_{\mathbb{R}^k} \zeta(y) v_t(y)\, dy := \int_{\mathbb{R}^k} \zeta((1-t)\boldsymbol{r}_0(x) + t\boldsymbol{r}_1(x)) w(x)\, dx \quad \forall \zeta \in C_c^0(\mathbb{R}^k), \tag{42}$$

and it is possible to prove [2] that (11a,b) are satisfied. Thus we obtain

Corollary 2.3. *Let u be the solution of the porous medium equation* (9d) *and let $U_{\boldsymbol{\tau}}$ be the discrete solution obtained by solving the recursive minimization scheme* (8a,b) *with the choices of d, ϕ corresponding to* (9b,c). *Then the* a posteriori *estimates of Theorem* 2.1 *hold; in particular, if $U^0_{\boldsymbol{\tau}} = u^0$*

$$\sup_{t\in\mathcal{P}_{\boldsymbol{\tau}}} W^2(u(t), U_{\boldsymbol{\tau}}(t)) \leq |\boldsymbol{\tau}| \frac{1}{m-1} \int_{\mathbb{R}^k} (u^0(x))^m \, dx, \tag{43}$$

and, if $u^{m-1/2} \in H^1(\mathbb{R}^k)$,

$$\sup_{t\in\mathcal{P}_{\boldsymbol{\tau}}} W^2(u(t), U_{\boldsymbol{\tau}}(t)) \leq \frac{|\boldsymbol{\tau}|^2}{2} \frac{m^2}{(m-1/2)^2} \int_{\mathbb{R}^k} \left|\nabla[(u^0(x))^{m-1/2}]\right|^2 dx. \tag{44}$$

References

[1] L. Ambrosio, *Minimizing movements*, Rend. Accad. Naz. Sci. XL Mem. Mat. Appl. (5) **19** (1995), 191–246.

[2] L. Ambrosio, N. Gigli, and G. Savaré, *Gradient flows of probability measures*, in preparation (2003).

[3] C. Baiocchi, *Discretization of evolution variational inequalities*, Partial differential equations and the calculus of variations, Vol. I (F. Colombini, A. Marino, L. Modica, and S. Spagnolo, eds.), Birkhäuser Boston, Boston, MA, 1989, pp. 59–92.

[4] H. Brézis, *Opérateurs maximaux monotones et semi-groupes de contractions dans les espaces de Hilbert*, North-Holland Publishing Co., Amsterdam, 1973, North-Holland Mathematics Studies, No. 5. Notas de Matemática (50).

[5] M. G. Crandall and T. M. Liggett, *Generation of semi-groups of nonlinear transformations on general Banach spaces*, Amer. J. Math. **93** (1971), 265–298.

[6] E. De Giorgi, *New problems on minimizing movements*, Boundary Value Problems for PDE and Applications (C. Baiocchi and J. L. Lions, eds.), Masson, 1993, pp. 81–98.

[7] E. De Giorgi, A. Marino, and M. Tosques, *Problems of evolution in metric spaces and maximal decreasing curve*, Atti Accad. Naz. Lincei Rend. Cl. Sci. Fis. Mat. Natur. (8) **68** (1980), no. 3, 180–187.

[8] K. Eriksson, C. Johnson, and S. Larsson, *Adaptive finite element methods for parabolic problems VI: Analytic semigroups*, SIAM J. Numer. Anal. **35** (1998), 1315–1325.

[9] R. Jordan, D. Kinderlehrer, and F. Otto, *The variational formulation of the Fokker–Planck equation*, SIAM J. Math. Anal. **29** (1998), no. 1, 1–17 (electronic).

[10] J. Jost, *Nonpositive curvature: geometric and analytic aspects*, Lectures in Mathematics ETH Zürich, Birkhäuser Verlag, Basel, 1997.

[11] U. F. Mayer, *Gradient flows on nonpositively curved metric spaces and harmonic maps*, Comm. Anal. Geom. **6** (1998), no. 2, 199–253.

[12] R. J. McCann, *A convexity principle for interacting gases*, Adv. Math. **128** (1997), no. 1, 153–179.

[13] R. H. Nochetto and G. Savaré, *Nonlinear evolution governed by accretive operators: error control and applications*, to appear (2003).

[14] R. H. Nochetto, G. Savaré, and C. Verdi, *A posteriori error estimates for variable time-step discretizations of nonlinear evolution equations*, Comm. Pure Appl. Math. **53** (2000), no. 5, 525–589.

[15] J. Rulla, *Error analysis for implicit approximations to solutions to Cauchy problems*, SIAM J. Numer. Anal. **33** (1996), 68–87.

[16] F. Otto, *The geometry of dissipative evolution equations: the porous medium equation*, Comm. Partial Differential Equations **26** (2001), no. 1-2, 101–174.

[17] C. Villani, *Topics in optimal mass transportation*, Graduate Studies in Mathematics, **58**, AMS (2003).

[18] A. Visintin, *Models of phase transitions*, Progress in Nonlinear Differential Equations and Their Applications, vol. 28, Birkhäuser, Boston, 1996.

Department of Mathematics,
Università di Pavia,
via Ferrata, 1
I-27100 Pavia, Italy
E-mail address: `savare@imati.cnr.it`
URL: `http://www.imati.cnr.it/~savare`

International Series of Numerical Mathematics, Vol. 147, 293–301

A Multi-mesh Finite Element Method for 3D Phase Field Simulations

Alfred Schmidt

Abstract. In phase field models, different components of the solution (temperature and phase variable) exhibit a strongly different local behaviour. Especially in 3D, the high resolution needed for the phase variable in the transition region calls for specially adapted methods for discretization in order to make computations with as good a resolution as possible. An optimal discretization should use a separate mesh for each component. We present a general framework for the adaptive solution of coupled systems and its application to phase field simulations, making 3D simulations possible even on desktop computers.

Free boundary problems give rise to coupled systems of partial differential equations, which are prominent examples of systems where a careful numerical discretization is needed in order to resolve the solution behaviour. Usually, a high resolution is essential near the free boundary, while coarser meshes are sufficient in the bulk. A reasonable numerical method should use locally adapted meshes to fulfil both the needs of accuracy and efficiency. Especially for 3D problems, accurate simulations may be impossible without appropriate discretization concepts. For coupled systems like the phase field system, where the components of the solution show strongly different local behaviour, an optimal discretization should use specially adapted meshes for each component.

1. Models for solidification

An undercooling or oversaturation of a liquid leads to a rapid solidification of the material. Models for this behaviour include energy and/or mass transport by diffusion and/or convection as well as energy/mass conservation conditions across the interface, like the Stefan condition. On a certain (meso-)scale, additional surface effects at the phase boundary play an important role, depending on the curvature C_Γ or the velocity V_Γ of the moving interface $\Gamma(t)$, usually described by a Gibbs–Thomson relation like

$$\varepsilon_c C_\Gamma + \varepsilon_v V_\Gamma + \theta = 0 \qquad \text{on } \Gamma(t).$$

Anisotropic surface effects, where the 'curvature' is the variation of an anisotropic surface energy which, as well as ε_c and ε_v, depends on the direction of the surface normal, lead to dendritic growth.

Mathematical models for solidification with surface effects differ mainly in the treatment of the free boundary, defining it as a sharp interface, a level set, or a diffuse interface (phase field). Numerical methods for solidification simulations are developed following the same lines.

1.1. Sharp interface models

In a sharp interface model, the phase boundary is a (smooth) hyper-surface of the underlying space (a surface in 3D or a curve in 2D). The advantage of the treatment of the interface as a smooth surface is its lower dimension and that curvature, which appears in the Gibbs–Thomson law, is well defined. Drawbacks are that the standard model allows no changes in interface topology, no nucleation. The notion of varifolds allows for some possibilities in this direction.

The free boundary problem leads to a degenerate parabolic equation for the interface motion, like an anisotropic mean curvature flow equation, coupled with heat or mass diffusion in solid and liquid phases. Additionally, convection can be added by coupling to Navier–Stokes' equations in the time dependent liquid phase.

NUMERICAL METHODS FOR SHARP INTERFACE SIMULATIONS. The sharp interface model leads to numerical methods with separate discretizations of bulk (3D) and interface (2D) (or 2D & 1D).

In the bulk, a fine mesh is essential near the interface to sufficiently resolve the behaviour of solutions (temperature, concentration). Adaptively refined meshes in finite element methods are crucial to get enough resolution, especially in 3D. A numerical sharp interface finite element method with adaptive meshing in 2D and 3D was presented in [7, 8], with additional convection in 2D in [2].

Especially in 3D, the discretization of the moving interface, which substantially changes shape during the growth process, is a delicate task, which makes repeated remeshing or at least mesh modifications necessary.

1.2. Diffuse interface models

Phase field models can be formulated either with a double obstacle potential or a smooth double well potential. We want to restrict ourself here to an obstacle formulation, as introduced by Blowey and Elliott [3].

The model describes the evolution of temperature θ and phase variable χ in a domain Ω and it is a system of two (degenerate) parabolic equations:

$$\partial_t(\theta + \lambda\chi) - \kappa\Delta\theta = f$$
$$\varepsilon\,\partial_t\chi - \varepsilon\,\mathrm{div}(a(\nabla\chi)) + \Lambda(\chi) - \tfrac{1}{\varepsilon}\beta\chi \ni \gamma\theta$$

in $\Omega \times (0,T)$ plus boundary and initial conditions. The function a may include anisotropic solidification parameters, and Λ is a (set valued) maximal monotone graph, the subdifferential of the double obstacle potential,

$$\Lambda(s) = \begin{cases} (-\infty, 0] & \text{if } s = -1 \\ 0 & \text{if } s \in (-1,1) \\ [0, +\infty) & \text{if } s = +1 \end{cases}$$

with the effect that values of χ are in the interval $[-1,+1]$. Further, f is a given heat source density, λ, κ, and β are non-negative, material-dependent coefficients, and $\varepsilon > 0$ a small parameter. The phase variable χ is equal to -1 (solid) or $+1$ (liquid) everywhere but in a narrow transition region of width $O(\varepsilon)$. For $\varepsilon \to 0$, the solution converges to a solution of the sharp interface model with a corresponding Gibbs–Thomson relation. Figure 1 shows the evolution of the phase boundary during the solidification of an undercooled liquid from a 2D simulation.

FIGURE 1. Solid-liquid interface at different times

At a first glance, the introduction of a phase variable makes mathematical analysis of long-time solutions and numerical methods much easier. The sharp moving interface is approximated by the level set of a smooth function, thus even allowing geometrical 'catastrophes' like topological changes of the interface, nucleation, etc., without big problems in the mathematical formulation (by introducing the notion of viscosity solutions, e.g.).

Both components of the solution, the temperature and the phase variable, show a strongly different local behaviour, see Figure 2.

- The phase variable is constant outside of a moving narrow strip of width $O(\varepsilon)$, where the phase transition occurs. Here, $|\nabla\chi| = O(\varepsilon^{-1})$.
- The temperature satisfies the heat equation outside this strip, thus it is smooth. Inside the moving strip, the gradient of temperature changes rapidly. In the limit $\varepsilon \to 0$, the Stefan condition holds at the sharp interface,

$$(\kappa\nabla\theta_{\text{liquid}} - \kappa\nabla\theta_{\text{solid}}) \cdot n = -\lambda V.$$

In order to resolve this behaviour with a finite element approximation, different requirements to the discretization hold for both components.

- The mesh for the discretization of χ must have a local mesh width such that $h_S < c\varepsilon$. Outside this strip, where χ is constant, the mesh might be arbitrarily coarse. In order to be able to track nucleations, the phase field

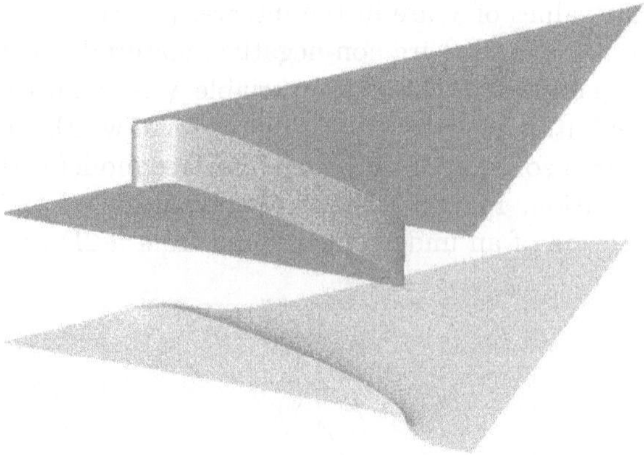

FIGURE 2. Graphs of phase variable and temperature on 1/8 domain

system should be solved in the whole domain, not only near the current interface.

- To resolve the temperature behaviour, a much coarser mesh is sufficient in the strip than the mesh needed for the phase variable. On the other hand, the mesh must have a sufficient fineness also in the remainder of the domain.

A separate discretization for both temperature and phase variable is needed in order to meet all these requirements in an efficient numerical method.

2. Adaptive finite element methods for coupled systems of PDE

Besides phase transitions, many physical problems lead to coupled systems of partial differential equations, too. Especially in case of nonlinear phenomena, the components u_i, $i = 1, \dots, n$ may show a strongly different behaviour (smoothness of solutions etc.) in the common underlying domain $\Omega \subset \mathbb{R}^d$.

The usual adaptive discretization for systems of PDEs uses the same mesh for all components, locally refined on the basis of a posteriori error indicators for the sum of error contributions on each mesh element.

An optimal adaptive discretization should use different locally refined meshes for the different components of the solution, controlled by separate error indicators for each component.

In [6] we present a general concept for adaptive finite element methods for stationary or time dependent coupled problems.

The solution components u_i are discretized in finite element spaces X_i, which are based on different locally refined simplicial grids $\mathcal{S}_i$, where all the meshes $\mathcal{S}_i$ are refinements of the same macro triangulation $\mathcal{S}_0$ of Ω, see Figure 3 for a simple example.

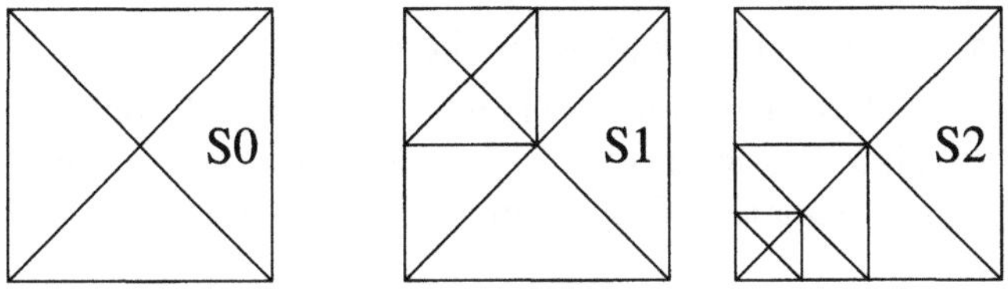

FIGURE 3. Two different refinements of a macro triangulation $\mathcal{S}_0$

The advantages of this approach are

- separately adapted mesh for each component of the solution;
- altogether fewer degrees of freedom, thus more efficient;
- by the common macro mesh, a direct local hierarchy of the triangulations $\mathcal{S}_i$ is given. This allows for an *exact* evaluation of each other component U_j on the elements of mesh $\mathcal{S}_i$.

A somewhat similar idea for phase field simulations was proposed by Elliott and Gardiner [5], who used two separate meshes for the temperature and the phase variable. A fine mesh with mesh size $h \ll \varepsilon$ for the phase variable is used only near the transition region (selected by a 'mask'), a coarse mesh with mesh size $H = 4h$ is used for the temperature. Use of the 'mask' reduces computations for the phase variable significantly, but it does not allow to track nucleations, e. g. Both meshes are aligned, but no error estimators or local mesh refinements are used.

2.1. Aspects of implementation

The adaptive multi-mesh method is implemented in the finite element toolbox ALBERT, a joint development with K. G. Siebert [9, 10]. The toolbox uses simplicial meshes in 2D (triangles) and 3D (tetrahedra) and local refinement by bisection of elements, which induces a hierarchical structure of the meshes and the finite element spaces.

The multi-mesh concept introduces the traversal of a common, *virtually refined* mesh $\mathcal{S}_M$, see Figure 4, for the calculation of coupling terms, which involve several components or functions from finite element spaces defined on different meshes, like $\int_\Omega U_i \Phi$ with $U_i \in X_i$, $\Phi \in X_j$.

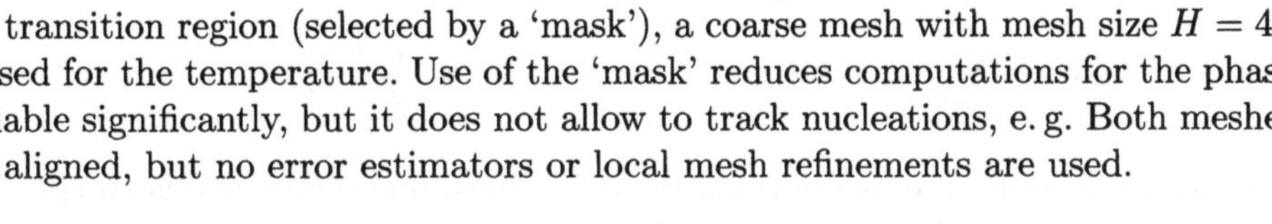

FIGURE 4. Virtual mesh $\mathcal{S}_M$ with locally maximal refinement from $\mathcal{S}_1, \mathcal{S}_2$

2.2. Numerical analysis and adaptive methods

We recall the standard a posteriori error estimates and adaptive finite element methods for a scalar elliptic problem. The error $\|u - U\|$ between the solution u and a discrete approximation U is estimated by a sum of local error indicators, e. g.

$$\|u - U\|^2 \leq \sum_{S\in\mathcal{S}} \eta_S^2,$$

where the indicators η_S must be computable from the discrete solution U and the given data of the problem (*a posteriori*).

On an quasi-optimal mesh for a given error tolerance tol, the local error indicators are equally distributed over all mesh elements,

$$\eta_S^2 \approx \frac{tol^2}{\#\mathcal{S}} \qquad \text{for all } S \in \mathcal{S},$$

compare [1], e. g. A similar approach is used in the case of coupled systems. The goal is an estimate of the errors $\|u_i - U_i\|$ by a sum of local error indicators on separate meshes, like

$$\sum_{i=1}^{n} \|u_i - U_i\|^2 \leq \sum_{S\in\mathcal{S}_1} \eta_{1,S}^2 + \cdots + \sum_{S\in\mathcal{S}_n} \eta_{n,S}^2.$$

In order to describe a quasi-optimal set of meshes, one possibility is to split the total tolerance tol to the separate meshes

$$tol_1^2 + \cdots + tol_n^2 = tol^2$$

and optimize every single mesh. This leads to the decoupled equidistribution conditions

$$\eta_{i,S}^2 \approx \frac{tol_i^2}{\#\mathcal{S}_i} \quad \text{for all } S \in \mathcal{S}_i, i = 1, \ldots, n.$$

3. Adaptive method for phase field models

In a joint paper with Z. Chen and R.H. Nochetto [4], we derive error estimates and adaptive methods for the double obstacle phase field system. Denoting by $u = \theta + \lambda\chi$ the energy density, and again the discrete (finite element) functions by uppercase letters, the estimate looks like

$$\begin{aligned} &\|u - U\|_{L^\infty(0,T;H^{-1}(\Omega))} + \sqrt{\varepsilon}\|\chi - X\|_{L^\infty(0,T;L^2(\Omega))} \\ &+ \left(\int_0^T \|\theta - \Theta\|_{L^2(\Omega)}^2 + \alpha\varepsilon \,\|\nabla\chi - \nabla X\|_{L^2(\Omega)}^2 \,dt\right)^{1/2} \\ \leq\ & \text{computable terms, localizable to mesh elements.} \end{aligned}$$

In [4], we use a common mesh for the discretization of the temperature and the phase variable. The local error indicators η_S contain contributions from both components,

$$\text{Estimated error } \leq \eta_0 + \max_m \sum_{S \in \mathcal{S}^m} \eta_S^2,$$

where m is the time step index. A quasi-optimal common mesh fulfils the equidistribution condition for the local indicators

$$\eta_S^2 \approx \frac{tol^2}{\#\mathcal{S}^m} \quad \text{for all } S \in \mathcal{S}^m.$$

Here, the total mesh element count $\#\mathcal{S}^m$ is very large because of the fine resolution in the strip, thus the local temperature error must be very small also in elements far from the interface.

Adaptive FE method with separate meshes

A separation of contributions from temperature and phase variable in the error indicator is possible:

$$\text{Estimated error } \leq \eta_0 + \max_m \Big(\sum_{S \in \mathcal{S}_\theta^m} \eta_{\theta,S}^2 + \sum_{S \in \mathcal{S}_\chi^m} \eta_{\chi,S}^2 \Big).$$

A split of the tolerance to both meshes, as described above, leads to a smaller element count $\#\mathcal{S}_\theta^m$, thus to larger local error tolerances for the temperature error. It follows that the temperature mesh contains considerably fewer elements than a common mesh. Also the mesh for the phase variable contains less elements, as a fine resolution is needed only in the transition region.

Finally, we show a few numerical results. Figure 6 presents pictures from a 3D simulation where the phase boundary is a growing sphere. Figure 5 depicts the corresponding numbers of mesh elements over time. As expected, the temperature mesh has only few elements whose number grows slowly, while the phase variable mesh is much finer and grows nearly linearly in time in order to resolve the growing free boundary.

With our method, we expect to be able to efficiently and accurately simulate three dimensional free boundary problems.

Acknowledgements.

This article reports partly joint work with Z. Chen (Beijing), R. H. Nochetto (College Park), and K. G. Siebert (Freiburg).

References

[1] I. Babuška and W. Rheinboldt, *Error estimates for adaptive finite element computations*, SIAM J. Numer. Anal. **15** (1978), 736–754.

[2] E. Bänsch and A. Schmidt, *Simulation of dendritic crystal growth with thermal convection*, Interfaces and Free Boundaries **2** (2000), 95–115.

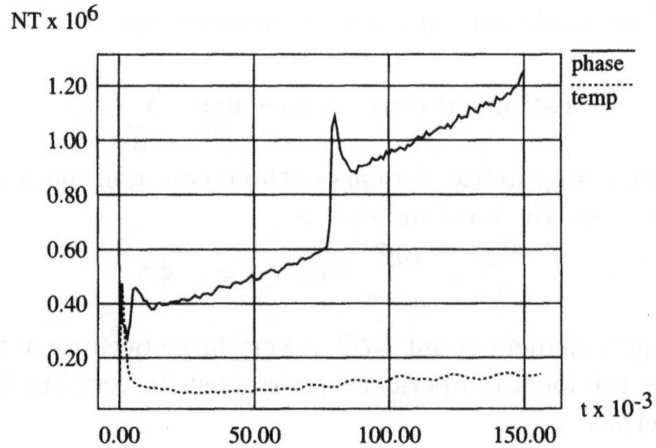

FIGURE 5. Numbers of 3D tetrahedral mesh elements for phase and temperature meshes over time.

[3] J. Blowey and C. Elliott, *Curvature dependent phase boundary motion and parabolic double obstacle problems.*, in Ni, Wei-Ming (ed.) et al., Degenerate diffusions. IMA Vol. Math. Appl. **47**, 19–60 , 1993.

[4] Z. Chen, R. H. Nochetto, and A. Schmidt, *Adaptive finite element methods for diffuse interface models.* In preparation.

[5] C. Elliott and A. Gardiner, *Double obstacle phase field computations of dendritic growth.* Report 96/19, University of Sussex.

[6] A. Schmidt, *Adaptive methods for coupled systems.* In preparation.

[7] ——, *Computation of three dimensional dendrites with finite elements*, J. Comput. Phys. **125** (1996), 293–312.

[8] ——, *Approximation of crystalline dendrite growth in two space dimensions*, Acta Math. Univ. Comenianae **67** (1998), 57–68.

[9] A. Schmidt and K. G. Siebert, *ALBERT: An adaptive hierarchical finite element toolbox.* Preprint 06/2000 Freiburg. Documentation.

[10] ——, *ALBERT - Software for scientific computations and applications*, Acta Math. Univ. Comenianae. **70** (2001), 105–122.

Zentrum für Technomathematik, Universität Bremen,
Bibliothekstrasse 1, 28359 Bremen, Germany
E-mail address: `schmidt@math.uni-bremen.de`

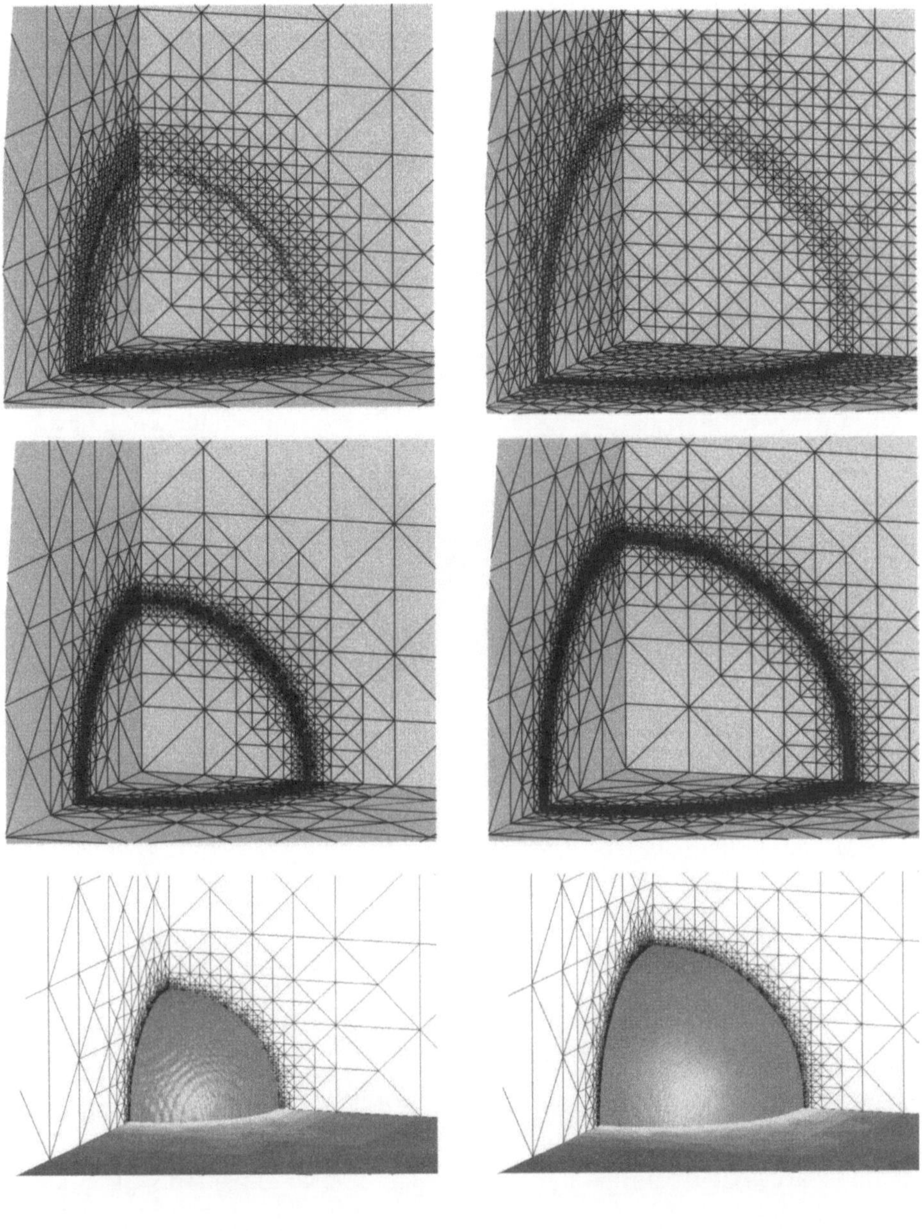

FIGURE 6. 3D simulation of growing sphere: Temperature mesh (top), phase variable mesh (middle) and phase variable zero level set and mesh (bottom) from different time steps.

International Series of Numerical Mathematics, Vol. 147, 303–312

Morse Description and Geometric Encoding of Digital Elevation Maps

Andrés Solé, Vicent Caselles, Guillermo Sapiro, and Francisco Arándiga

Abstract. Two complementary geometric structures for the topographic representation of an image are developed in this work. The first one computes a description of the Morse structure of the image, while the second one computes a simplified version of its drainage structure. The topographic significance of the Morse and drainage structures of Digital Elevation Maps (DEM) suggests that they can been used as the basis of an efficient encoding scheme. As an application we combine this geometric representation with an interpolation algorithm and lossless data compression schemes to develop a compression scheme for DEM. This algorithm permits to obtain compression results while controlling the maximum error in the decoded elevation map, a property that is necessary for the majority of applications dealing with DEM. We present the underlying theory and compression results for standard DEM data.

1. Introduction

The use of a topographic description of images, surfaces or $3D$ data has been introduced and motivated in different areas of research: image processing, computer graphics and geographic information systems (GIS) [BPS97, CK98, FS96, KK94, SKK91]. In this paper we propose a geometric approach to compress Digital Elevation Models (DEM). DEM data are a digital representation of a surface terrain where each cell corresponds to a point (x, y, z) in 3D space, (x, y) being the image coordinates and z its height. Obviously, this kind of data requires a large amount of bytes to be stored, thus some compression is needed. In the case of DEM data it seems reasonable to store only those geometric structures which are of special relevance (such as its Morse and drainage structures) and interpolate the rest from this non linear subsampling. The object of this paper is to compute these structures and to use them to efficiently compress DEM data, while controlling the L_∞ norm of the error.

Key words and phrases. Image compression, digital elevation maps, Morse structure, drainage structure, geometric encoding, geometric sampling.

This paper is structured as follows. Section 2 is devoted to the computation of the Morse structure of an image. Some basic notions, like the notion of *monotone section* of a topographic map, are introduced, and an algorithm to compute the Morse structure is proposed. We also briefly comment on the mathematical justification of the proposed algorithms. In Section 3 we describe an algorithm to compute a simplified drainage structure. Section 4 describes the coding and interpolation strategy and we summarize the compression algorithm used in our experiments. Section 5 shows comparative results between the proposed algorithm and the standards JPEG Lossless (JLS) method and JPEG-2000. Finally, in Section 6 we present the main conclusions of this work.

2. Morse theory and monotone sections

The aim of Morse theory is to describe the topological changes of the (iso)level sets of a function in terms of its critical points. We will consider here the case of 2D real valued data (gray level images, or height data) and we will describe a simple and efficient algorithm to compute its Morse structure from its upper and lower level sets.

To make the ideas precise, let us first recall some definitions. Let D be a subset of $\mathbb{R}^N$. Given a function $u : D \to \mathbb{R}$, we call upper (lower) level set of u any set of the form $[u \geq \lambda] := \{x \in D : u(x) \geq \lambda\}$ or $[u > \lambda] := \{x \in D : u(x) > \lambda\}$ ($[u \leq \lambda] := \{x \in D : u(x) \leq \lambda\}$ or $[u < \lambda] := \{x \in D : u(x) < \lambda\}$), where $\lambda \in \mathbb{R}$. The (upper) topographic map of a function u is the family of the connected components of the level sets of u, $[u \geq \lambda]$, $\lambda \in \mathbb{R}$, the connected components being understood in the relative topology of D. We shall assume that our image is a continuous function and each upper or lower level set has a finite number of connected components which is true for a discrete image or for a continuous image after some basic filtering. In [BCM01] the authors studied the Morse structure of the topographic map for continuous functions (a similar study can be done for bounded upper semicontinuous functions). They defined a notion of nonsingular region of the topographic map trying to express the fact that the level lines of the topographic map in a nonsingular region do not change topology, as it happens for smooth functions where singularities are understood in the usual way.

Let $u : D \to \mathbb{R}$ be a function. For each $\lambda, \mu \in \mathbb{R}$, $\lambda \leq \mu$ we define

$$U_{\lambda,\mu} = \{x \in D : \lambda \leq u(x) \leq \mu\}.$$

The connected components of a set $X \subseteq \mathbb{R}^N$ will be denoted by $CC(X)$. If $x \in X$, the connected component of X containing x will be denoted by $cc(X, x)$

Definition 2.1. Let $u : D \to \mathbb{R}$ be a continuous function. A *monotone section* of the topographic map of u is a set of the form $X_{\lambda,\mu} \in CC(U_{\lambda,\mu})$, for some $\lambda, \mu \in \mathbb{R}$ with $\lambda \leq \mu$, such that for any $\lambda', \mu' \in [\lambda, \mu]$, $\lambda' \leq \mu'$ the set $\{x \in X_{\lambda,\mu} : \lambda' \leq u(x) \leq \mu'\}$ is a connected component of $U_{\lambda',\mu'}$.

Due to small oscillations in the image, its Morse structure is too complex, i.e., there are many criticalities. To simplify the structure of the topographic map while preserving its main features we filter the image with the Vicent–Serra filters. If u has been filtered with the extrema filters with an area threshold ϵ and X is a connected component of $[u \geq \lambda]$ or $[u < \lambda]$, then $|X| > \epsilon$. The resulting image has a simplified topographic map structure. Indeed, for any $\lambda \leq \mu$, there are a finite number of connected components of $U_{\lambda,\mu}$. In particular, there are a finite number of connected components of $[u \geq \lambda]$ and $[u < \lambda]$. In the sequel, we shall assume that

(**H**) u has been filtered with the extrema filters.

Assuming property (**H**), if $X_{\lambda,\mu}$ is a monotone section, then the family of sets $X \cap [u = \alpha]$, $\alpha \in [\lambda, \mu]$, is a family of nested connected sets [CSS02]. Moreover, the union of monotone sections which intersect is a monotone section [CSS02]. This allows to define the notion of maximal monotone section containing a given point. Let $x \in D$ and $\lambda = u(x)$. For each $\eta \geq \lambda$, let $X_{\lambda,\eta} = cc(U_{\lambda,\eta}, x)$. We define

$$\eta_+(x,\lambda) = \sup\{\eta : \eta \geq \lambda \quad \text{s. t. } X_{\lambda,\eta} \text{ is a monotone section}\}.$$

Similarly, we define $\eta_-(x,\lambda) = \inf\{\eta : \eta \leq \lambda \quad \text{s. t. } X_{\eta,\lambda} \text{ is a monotone section}\}$. Note that both numbers are well defined since $X_{\lambda,\lambda}$ is always a monotone section. Note that, by definition, $\eta_-(x,\lambda) \leq \eta_+(x,\lambda)$. The (open, closed, half open, or half closed) interval $I(x,\lambda)$ containing λ whose end-points are $\eta_-(x,\lambda), \eta_+(x,\lambda)$ determines a monotone section containing x maximal with respect to inclusion, which we denote by $X_{I(x,\lambda)}$.

Maximal monotone sections represent the largest sections of the topographic map containing no topological changes and our purpose is to compute them. Intuitively monotone sections are topologically equivalent to truncated cones and the maximal monotone ones are the largest truncated cones contained in the graph of the image. The interpolation algorithms described in Section 4 are able to re-interpolate these truncated cones from the curves bounding them. To compute the maximal monotone sections we require some additional definitions.

Definition 2.2. Let $M \subseteq \overline{\Omega}$. We say that M is a *zonal maximum* (resp., *minimum*) of u at height λ if M is a connected component of $[u = \lambda]$ and, for all $\epsilon > 0$, the set $[\lambda - \epsilon < u \leq \lambda]$ (resp., $[\lambda \leq u < \lambda + \epsilon]$) is a neighborhood of M.

Definition 2.3. We say that $\lambda \in I\!R$ is a *singular value* of the topographic map of u if it corresponds to a zonal maximum, minimum, or it corresponds to a level where it begins or ends a maximal monotone section, i.e., there is a point $x \in \overline{\Omega}$ such that $\eta_+(x,\lambda) = \lambda$ or $\eta_-(x,\lambda) = \lambda$.

Definition 2.4. Let $\lambda \in I\!R$. Assume that (**H**) holds. The *signature* of the level set $[u \geq \lambda]$ (resp. $[u < \lambda]$) consists of a finite family of points $\{p_i : i = 1, ..., r\}$ (resp. $\{q_j : j = 1, ..., s\}$) such that each p_i (resp. q_j) is a point in a different connected component $X^{\lambda,i}$ (resp. $X_{\lambda,j}$) of $[u \geq \lambda]$ (resp. $[u < \lambda]$). The points p_i, q_j

are selected so that

$$u(p_i) = \sup_{x \in X^{\lambda,i}} u(x), \text{and} \quad u(q_j) = \inf_{x \in X_{\lambda,j}} u(x).$$

We denote the signature of $[u \geq \lambda]$ by $sig([u \geq \lambda])$, the signature of $[u < \lambda]$ by $sig([u < \lambda])$. We define $sig([u \geq \lambda], [u < \lambda]) = sig([u \geq \lambda]) \cup sig([u < \lambda])$.

To stress the marker point in the notation, we write X^{λ,p_i} instead of $X^{\lambda,i}$. As it is proved in [CSS02], under assumption (**H**), given $\lambda \in I\!R$, there is $\epsilon > 0$ such that $sig([u \geq \mu], [u < \mu])$ is constant for all $\mu \in (\lambda - \epsilon, \lambda]$.

Definition 2.5. We say that $\lambda \in I\!R$ is a *critical value* for u if there is a sequence $\mu_n \downarrow \lambda$ such that $sig([u \geq \mu_n], [u < \mu_n]) \neq sig([u \geq \lambda], [u < \lambda])$ for each $n = 1, 2, \ldots$.

Proposition 2.6. *Let $\lambda \in I\!R$. Assume that* (**H**) *holds. Then λ is a critical value of u if and only if λ is a singular value of the topographic map of u.*

The following result is proved in [CSS02] and permits us to compute the singular values of the topographic map of u by computing the critical values of u. Moreover, at the discrete level, a topological change occurs when going from level λ to level $\lambda - 1$ if and only if either *i)* $sig([u \geq \lambda]) \neq sig([u \geq \lambda - 1])$, or *ii)* $sig([u < \lambda]) \neq sig([u < \lambda - 1])$. This is equivalent to saying that *i)* two connected components of $[u \geq \lambda]$ merged at level $\lambda - 1$, a phenomenon which we call merging of upper connected components, or a new connected component of the upper level sets was born at level $\lambda - 1$, or *ii)* two connected components of $[u < \lambda - 1]$ merged at level λ, a phenomenon which we call merging of lower connected components, or a new connected component of the lower level sets was born at level λ. We call the first type of criticality of upper type, while the second is called of lower type. For a detailed description of the algorithm we refer to the reader to [CSS02]. Many different approaches exist to compute the basic Morse structure of images or 3D data. We have presented a simple computational approach and have studied its mathematical properties in [CSS02]. Let us also note that the singularities of the tree of shapes of an image as introduced in [Mon00], [BCM01], coincide with the notion of critical value defined here [CSS02].

3. Drainage structures

In the previous section we have developed an algorithm to compute the Morse structure of an image. This Morse structure consists on the maxima, minima, and the level lines where a topology change occurs (i.e., the boundaries of the maximal monotone sections). This can be considered as a reasonable *global* description of a DEM. There are also other structures which are of special interest due to its topographic significance in DEM data. These structures correspond mainly to the drainage structures (e.g., rivers and ravines). We will present an approach which is related to the one in [See89]. Strictly speaking, we do not compute the drainage structures but a version of them which is adapted to our purposes. In a simplistic way we can think of the drainage structures as the set of points for which there exists at least one direction in which the flow of water is accumulated or repealed. We can write down this definition mathematically by considering the set of points

$$\{\vec{x} : \exists \vec{v} \in I\!\!R^2,\ \varepsilon \geq 0 \ \text{ s. t. } u(\vec{x}) \leq u(\vec{x} + t\vec{v}) \ \forall t \in (-\varepsilon, \varepsilon), \\ \text{or} \quad u(\vec{x}) \geq u(\vec{x} + t\vec{v}) \ \forall t \in (-\varepsilon, \varepsilon)\}.$$

Observe that this set of points will contain in particular the maxima and minima of u. It has been shown in [CSS02] that this set of points is contained in an at most a countable family of curves. Intuitively, these curves contain the drainage structures (ridges and valleys), giving also information about boundaries of plateaus for example. In the discrete case we only consider 4 different directions (values of v) corresponding to 4 different profiles in the image. Concretely, we search maxima and minima of the vertical, horizontal and diagonal profiles. In [See89] only two directions were used, namely the horizontal and vertical ones. The problem is that a large number of extrema can appear due to low oscillations, mainly due to noise. In order to solve this problem we have to choose the most significant extrema, in our case those which have maximum *contrast.* Finally, if any computed extremum lies in a flat zone (i.e., it is a zonal extremum), we shall replace it by the boundary of the flat zone.

Figure 1 illustrates the whole geometric sampling process. From left to right and top to bottom we show the original DEM image, the level lines corresponding to its Morse structure, the curves corresponding to the extrema of the profiles, and the final sampling made by the union of the two previous structures. A thinning step has been performed in order to obtain one pixel width curves.

4. The interpolation and coding step

We have described two algorithms that compute important points and curves from an image, thereby providing the basic geometric description of DEM data. These algorithms can be considered as a non uniform geometric sampling of the image. The next step is to interpolate the missing data from our sampling. There exists several algorithms to interpolate data from curves and/or points. We have in particular tested three of them: the Laplacian model ($min \int \|\nabla u\|^2 dxdy$), the AMLE model ($min \lim_{p\to\infty} \int \|\nabla u\|^p dxdy$), and the thin plate model ($\int (\triangle(u))^2 dxdy$). We remark that, rigorously speaking, only the AMLE model can be used to interpolate values specified on points [CMS98]. In spite of this, we shall also use the Laplacian since there are many curves in the data and we may think of points as small regions. In order to evaluate these interpolation schemes we have chosen as a measure of goodness the entropy of the residual between the original image and the interpolated one. This is a natural choice since we want to minimize the number of bits used to encode the errors between the interpolated and the original images. After several tests we have decided to use the Laplacian interpolation since it reported the minimum entropy of the residual. In order to control the maximum

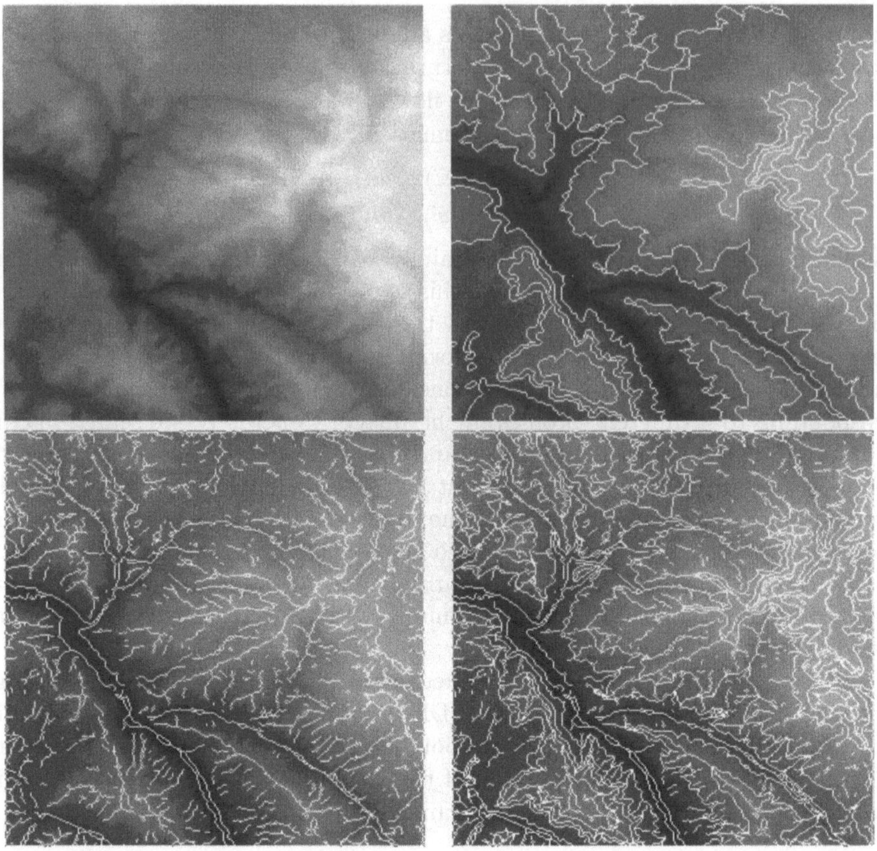

FIGURE 1. From left to right and up to down: the original DEM image, computed level lines (white), computed ridge/valley structure (white) and the whole image sampling (white).

(sup) error we simply store/encode the quantized error information (that is why the entropy of the residual was a natural measure of goodness for the interpolant). See the Appendix for more details on these interpolation processes.

Now, we need to consider how to encode both the initial curves (geometrically sampled data) and the residuals once a sup error e is specified. The geometry of the sampled curves and their gray levels are encoded separately. To encode the geometry we use a differential chain coding strategy. For the gray levels we use an ENO (Essentially Non Oscillatory) based encoding scheme [AD00] which also controls the sup error. Finally we compress both the geometry and the gray values of the curves using an arithmetic coder. Having these curves and the data on them, we can interpolate them by means of the Laplace equation to obtain the first estimate of the image. Finally, to control the maximum error, we need to store the

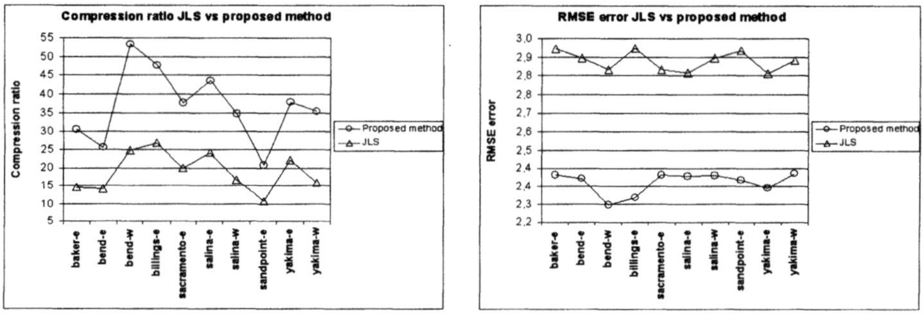

FIGURE 2. Left: Compression ratio of JLS vs the proposed method for a set of ten images compressed with $e = 5$. Right: Root mean square error ($RMSE$) for the two methods.

residuals r. Encoding the residuals can be simply done by quantizing them using $r^q = sign(r)\left\lfloor \frac{|r|}{e} \right\rfloor e$, and then coding the resulting r^q with an arithmetic coder.

The compression ratios using this approach were already satisfactory. We observed that the encoding of the geometry represented the main cost in bits. To further improve the encoding of the geometry we have adopted a simple multiscale approach. We compute and encode the curves and the residuals in a subsampled image and then zoom out the result and recompute new residuals. If required, before sampling, we may filter the given image. In order to avoid the well known oscillation problem of classical splines when performing the zoom out step we have used a *shape preserving spline* which respects the monotonicity of the original data (no new maxima or minima are created).

5. Compression results

We report results on a set of 10 DEM images of size 1200 by 1200 pixels with 8 bits per pixel. We also report the results of our algorithm when applied to the same set of images with a resolution of 16 bits per pixel. In order to compare our results, we use the JPEG-LS standard for lossless and controlled lossy image compression, being this one the only standard that allows a control on the maximal per pixel error [WSS99], and we also use JPEG-2000 [Ada01] in which, by reintroducing the errors, we are able to control the maximum error. In the following JPEG-LS is denoted by JLS, while ours is denoted by ME, standing for morphological encoding. We also report the $RMSE$ as frequently done for lossy image compression algorithms.

Figure 2 shows a more detailed study for the case $e = 5$. In the left plot we can see the compression ratio of both methods for the set of 10 images, the right plot corresponds to the $RMSE$ for both methods. The average CR for the case of

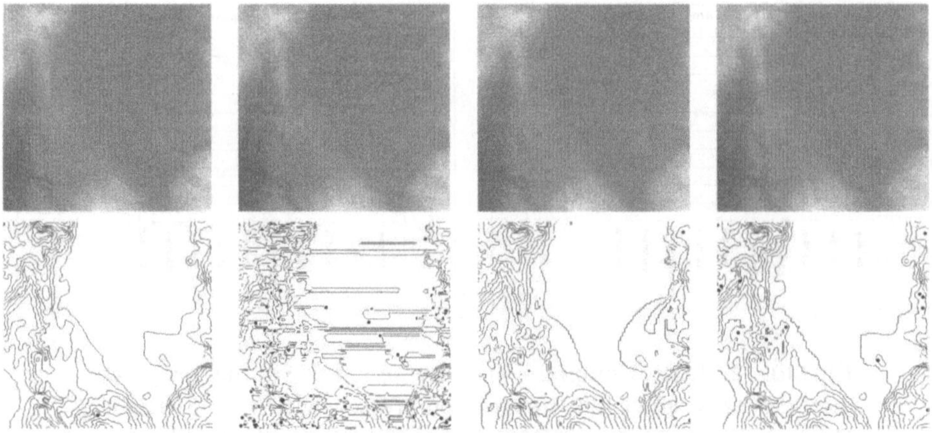

FIGURE 3. Up: from left to right a region of the original DEM image *baker-e* and after compressing it with JLS, ME and $JPEG-2000$ with $e = 5$. Bottom: level lines of the above images.

JLS and ME are **18,9872** and **36,7245** respectively. That is, our proposed scheme ME reaches almost twice the compression ratio of JLS. In addition, the average $RMSE$ (expressed in gray values, not meters) for the cases of JLS and ME is **2,9038** and **2,3782**, respectively.

We have also tested our algorithm using the full rank of the original images (16 bits) obtaining a compression ratio of $12,6203$ and a $RMSE$ of 2.5935 meters on average with a maximum error e of 5 meters.

We have compressed also the set of images at the same size as the one obtained with ME (for the 8 bit images only) using $JPEG-2000$. Although $JPEG-2000$ does not control the *sup* error, it is possible to control it afterwards by storing some additional values in an auxiliary file. For this set of images this additional cost is in average of 584 bytes, which is negligible, since it does not modify the compression ratio in practice. In addition, the average $RMSE$ is of 0.9326, which is much better than in the case of ME.

Figure 3 displays the original image and its compressed versions using JLS, ME and $JPEG-2000$. The figure shows both the gray scale images and its level sets. Note that the topographic structures are better preserved in the case of the ME and $JPEG-2000$ compression, while they are severely distorted in the case of JLS compression.

6. Conclusions

In this work we have presented some techniques to compute a basic geometric representation of images. This representation is given by the Morse and drainage structures of the image. As an application, this geometric image representation was used to derive a non uniform sampling strategy which, when combined with standard interpolation and coding techniques, provided a novel DEM compression algorithm. This algorithm produces compression ratios similar to $JPEG-2000$ and permits to control the maximal error in the decoded image, a property which is fundamental for most DEM applications.

Appendix

In this Appendix, we give a result which explains why the Laplacian and the AMLE interpolations are in some sense *shape preserving.* For the detailed proof we refer to [CSS02]. We shall consider an interpolation operator as a transformation E which associates to each open bounded set Ω and each function $\varphi \in C(\partial\Omega)$ a function $E(\varphi,\Omega) \in C(\overline{\Omega})$ such that $E(\varphi,\Omega)|_{\partial\Omega} = \varphi$. We shall say that the interpolation operator satisfies the stability principle if

$$E(E(\varphi,\Omega)|_{\partial\Omega'},\Omega') = E(\varphi,\Omega)|_{\Omega'}$$

for any open bounded set Ω, any $\varphi \in \partial\Omega$, and any open bounded set $\Omega' \subseteq \Omega$. Suppose that the interpolation operator E satisfies the stability property, we say that E satisfies the maximum principle if

$$\inf_{\partial\Omega'} \varphi \leq \inf_{\overline{\Omega'}} E(\varphi,\Omega') \leq \sup_{\overline{\Omega'}} E(\varphi,\Omega') \leq \sup_{\partial\Omega'} \varphi \tag{1}$$

for any open bounded set Ω, any open bounded set $\Omega' \subseteq \Omega$, and any $\varphi \in C(\partial\Omega')$. If E satisfies the maximum principle and $\varphi = \alpha$ in $\partial\Omega$, where $\alpha \in I\!R$, then $E(\varphi,\Omega) = \alpha$ in $\overline{\Omega}$, and the same is true in any open bounded set $\Omega' \subseteq \Omega$.

Theorem 6.1. *Let Ω_2 be an open simply connected set. Let Ω_1 be an open set whose boundary $\partial\Omega_1$ is connected and $\overline{\Omega_1} \subseteq \Omega_2$. Let $\Omega = \Omega_2 \setminus \overline{\Omega_1}$. Assume that $u|_{\partial\Omega_1} = \lambda$, $u|_{\partial\Omega_2} = \mu$ with $\lambda < \mu$ or $\lambda > \mu$. Let E be an interpolation operator satisfying the stability and the maximum principle. Then $E(u|_{\partial\Omega},\Omega)$ contains only a monotone section in the sense that $\overline{\Omega}$ is a monotone section of $E(u|_{\partial\Omega},\Omega)$. In particular, if $E(u|_{\partial\Omega},\Omega)$ is the $AMLE$ extension of the boundary data inside Ω, then $E(u|_{\partial\Omega},\Omega)$ contains only a monotone section.*

7. Acknowledgments

The first and second authors acknowledge partial support by the Departament d'Universitats, Recerca i Societat de la Informació de la Generalitat de Catalunya and by PNPGC project, reference BFM2000-0962-C02-01. G. Sapiro is partially supported by ONR-N00014-97-1-0509 and a NSF CAREER Award. F. Arándiga is partially supported by MCYT BFM 2001-2814.

References

[AD00] F. Aràndiga and R. Donat. *Nonlinear multiscale decompositions: the approach of A. Harten.* Numerical Algorithms, 23:175–216, 2000.

[Ada01] M. D. Adams. *The jpeg-2000 still image compression standard.* Available at http://www.ece.ubc.ca/ mdadams, 2001.

[BCM01] C. Ballester, V. Caselles, and P. Monasse. *The tree of shapes of an image.* Preprint CMLA, Cachan, Paris, 2001.

[BPS97] C. Bajaj, V. Pascucci, and D.R. Schikore. *The contour spectrum.* Proceedings Visualization '97, 167–173, 1997.

[CK98] J.L. Cox and D.B. Karron. *Digital morse theory with suggested applications.* Manuscript available at http://www.casi.net, 1998.

[CMS98] V. Caselles, J.M. Morel, and C. Sbert. *An axiomatic approach to image interpolation.* IEEE Transactions on Image Processing, 7(3):376–386, 1998.

[CSS02] V. Caselles, G. Sapiro, and A. Solé. *Morse description and morphological encoding of continuous data.* Preprint, 2002.

[FS96] W.R. Franklin and A. Said. *Lossy compression of elevation data.* 7th Int. Symposium on Spatial Data Handling, 1996.

[KK94] I.S. Kweon and T. Kanade. *Extracting topograpic terrain features from elevation maps.* CVGIP: Image Understanding, 59(2):171–182, 1994.

[Kro50] A.S. Kronrod. *On functions of two variables.* Uspehi Mathematical Sciences (NS), 35(5):24–134, 1950.

[Mon00] P. Monasse. *Représentation morphologique d'images numériques et application au recalage.* PhD thesis, Université de Paris-Dauphine, 2000.

[PS00] L. Garrido P. Salembier. *Binary partition tree as an efficient representation for image processing, segmentation, and information retrieval.* IEEE Transactions on Image Processing, 9(4):561–576, 2000.

[See89] W.W. Seemuller. *The extraction of ordered vector drainage networks from elevation data.* Comp. Vision Graphics and Image Processing, 47, 1989.

[SKK91] Y. Shinagawa, T.L. Kunii, and Y.L. Kergosien. *Surface coding based on morse theory.* IEEE Computer Graphics and Appl., 11(5):66–78, 1991.

[WSS99] M.J. Weinberger, G. Seroussi, and G. Sapiro. *From LOCO-I to the JPEG-LS standard.* Proceedings of the Int. Conf. on Image Processing, 4:68–72, 1999.

(A. Solé and V. Caselles) Dept. de Tecnologia, UPF, Pg. de Circumvalació, 8, 08003 Barcelona, Spain
E-mail address: {andreu.sole,vicent.caselles}@tecn.upf.es

(G. Sapiro) Dept. of Electrical and Computer Engineering, University of Minnesota, Minneapolis, MN 55455, USA
E-mail address: guille@ece.umn.edu

(F. Arándiga) Dept. Matemática Aplicada, Universitat de Valencia, C/ Dr. Moliner, s/n, 46100 Burjassot (València), Spain
E-mail address: arandiga@uv.es

International Series of Numerical Mathematics, Vol. 147, 313–327

Behavior of a Rigid Body in an Incompressible Viscous Fluid Near a Boundary

Victor N. Starovoitov

Abstract. The paper deals with the problem of the motion of a rigid body in a viscous incompressible fluid. It is investigated under which conditions the body gets to the wall with zero velocity or does not touch it at all. An example is constructed which shows that the weak solutions of the problem admit collisions of the body with the boundary of the flow domain. In this case, the velocity field satisfies the energy estimate only.

1. Introduction

During the last decade, there have appeared a lot of papers where the problem of the motion of rigid bodies in a fluid was studied (see [6]–[3] and references therein). Most of them deal with the situation where the bodies do not collide with the boundary of the flow and with other bodies. The possibility of collisions is one of the main difficulties of the problem. Up to now, it is not entirely understood how to state the problem properly, because the description of the collisions requires generally some additional relations. However, there are situations that admit a global solution of the problem without any collision laws. Namely, in [6, 7, 8] the global solvability of the problem is proved under the condition that the boundaries of the bodies and of the flow domain are curves of class C^2 in the 2-D case and spheres in 3-D case. It is shown there that such a smoothness yields zero relative velocity of a body when it touches another one. This fact implies that impulse forces do not appear by the collisions. Note that there remain some questions related to the extension of the solution to the time after the collision. In particular, the solution can be extended in such a way that the body will be immovable after it touches the wall (see [3]). This question, as well as questions of solvability of the problem, will not be considered in this paper. Our aim here is to study the behavior of a rigid body in a fluid when it is touching the wall or it is getting to the wall. Of course, all the results can be easily reformulated in the case of the interaction of two bodies.

2000 *Mathematics Subject Classification.* 35Q30, 76D99.
Key words and phrases. rigid body, viscous fluid, collision.

Let us recall the statement of the problem of the motion of a rigid body in a viscous incompressible fluid (*Problem A*) as it was made in [6, 7, 9, 8]. The tensor $D(\boldsymbol{u})$, which is called deformation rate tensor if $\boldsymbol{u}$ is a velocity vector field, plays an essential role in our considerations. This tensor is the symmetric part of the gradient of $\boldsymbol{u}$ and its components read as follows

$$D_{ij}(\boldsymbol{u}) = \frac{1}{2}\Big(\frac{\partial u_i}{\partial x_j} + \frac{\partial u_j}{\partial x_i}\Big).$$

The kernel of this tensor consists of the rigid vector fields. Assume that $\mathcal{S}$ is a connected domain in $\mathbb{R}^n$ and $\boldsymbol{u} : \mathcal{S} \to \mathbb{R}^n$ is a vector field. Then $D(\boldsymbol{u}) = 0$ in $\mathcal{S}$ if and only if there exist a vector $\boldsymbol{a} \in \mathbb{R}^n$ and a skew-symmetric tensor $\mathcal{Q} \in \mathbb{R}^n \times \mathbb{R}^n$ such that

$$\boldsymbol{u}(\boldsymbol{x}) = \boldsymbol{a} + \mathcal{Q}\boldsymbol{x} \qquad \text{for} \quad \boldsymbol{x} \in \mathcal{S}. \tag{1}$$

The proof of this fact can be found, for instance, in [10]. Velocity fields of the form (1) correspond to rigid motions. In particular, if $n = 3$, then there exists a vector $\boldsymbol{\omega}$ such that $\mathcal{Q}\boldsymbol{x} = \boldsymbol{\omega} \times \boldsymbol{x}$, where the sign $\times$ stands for the vector product, and the representation (1) has the form $\boldsymbol{u}(\boldsymbol{x}) = \boldsymbol{a} + \boldsymbol{\omega} \times \boldsymbol{x}$. Thus, it is possible to specify rigid bodies by the condition that the deformation rate tensor vanishes in the domains corresponding to the bodies.

Let Ω be a domain in $\mathbb{R}^n$ ($n = 2, 3$). We denote by $H^1_p(\Omega)$, $p \in [1, \infty]$, the space $\{\boldsymbol{u} \in \big(W^{1,p}_0(\Omega)\big)^n : \operatorname{div} \boldsymbol{u} = 0 \text{ in } \Omega\}$, and by $H_p(\Omega)$ the completion of $H^1_p(\Omega)$ in the norm of $L^p(\Omega)$. Given an open subset $\mathcal{S}$ of Ω, we introduce the following space

$$K_p(\mathcal{S}, \Omega) = \{\boldsymbol{u} \in H^1_p(\Omega) : D(\boldsymbol{u}) = 0 \text{ in } \mathcal{S}\}.$$

We omit the index p in the designation of the spaces if $p = 2$, i.e., $H^1(\Omega) = H^1_2(\Omega)$, $H(\Omega) = H_2(\Omega)$, and $K(\mathcal{S}, \Omega) = K_2(\mathcal{S}, \Omega)$. The dual space $H^1(\Omega)^*$ with respect to the pivot space $H(\Omega)$ will be denoted by $H^{-1}(\Omega)$. The space $K(\mathcal{S}, \Omega)$ was introduced and investigated in [6, 7] (see also [9]).

The motion of a rigid body can be described, for instance, by introducing a set-valued function $\mathcal{S} : [0, T] \to 2^{\Omega}$, $T \in (0, \infty)$. We assume that $\mathcal{S}(t)$ is an open subset of Ω for all $t \in [0, T]$. Since the body moves without deformation, there exists a family of preserving orientation isometries $\mathcal{M}_{s,t} : \mathbb{R}^n \to \mathbb{R}^n \big(s, t \in [0, T]\big)$ such that $\mathcal{S}(t) = \mathcal{M}_{s,t}\mathcal{S}(s)$. The characteristic function of $\mathcal{S}$ will be denoted also by $\mathcal{S}$:

$$\mathcal{S}(\boldsymbol{x}, t) = \begin{cases} 1, & \boldsymbol{x} \in \mathcal{S}(t), \\ 0, & \boldsymbol{x} \in \Omega \setminus \mathcal{S}(t). \end{cases}$$

Finally, we denote by $L^q(0, T; K_p(\mathcal{S}, \Omega))$ the space of functions $\boldsymbol{u} \in L^q(0, T; H^1_p(\Omega))$ such that $\boldsymbol{u}(\cdot, t) \in K_p(\mathcal{S}(t), \Omega)$ for almost all $t \in [0, T]$.

Definition 1.1. We say that a couple of functions

$$\boldsymbol{u} \in L^\infty(0, T; H(\Omega)) \cap L^2(0, T; K(\mathcal{S}, \Omega)),$$
$$\mathcal{S} \in C^{0,1/q}(0, T; L^q(\Omega)), \quad 1 \le q < \infty,$$

and a family of preserving orientation isometries $\mathcal{M}_{s,t} : \mathbb{R}^n \to \mathbb{R}^n \big(s, t \in [0, T]\big)$ form a *weak solution* of Problem A if $\mathcal{S}$ is a characteristic function,

$$\mathcal{S}(t) = \mathcal{M}_{s,t}\mathcal{S}(s) \quad \text{for all} \quad s, t \in [0, T], \tag{2}$$

and the integral equalities

$$\int_{\Omega_T} \Big(\boldsymbol{u} \cdot \boldsymbol{w}_t + \big(\boldsymbol{u} \otimes \boldsymbol{u} - D(\boldsymbol{u})\big) : D(\boldsymbol{w}) + \boldsymbol{g} \cdot \boldsymbol{w}\Big)\, d\boldsymbol{x}dt = -\int_{\Omega} \boldsymbol{u}_0 \cdot \boldsymbol{w}_0 \, d\boldsymbol{x}, \tag{3}$$

$$\int_{\Omega_T} \mathcal{S}\,\big(\eta_t + \boldsymbol{u} \cdot \nabla \eta\big)\, d\boldsymbol{x}dt = -\int_{\Omega} \mathcal{S}_0 \eta_0 \, d\boldsymbol{x} \tag{4}$$

hold for all functions $\boldsymbol{w} \in W^{1,2}(\Omega_T) \cap L^\infty(0, T; K(\mathcal{S}, \Omega))$ and $\eta \in C^1(\Omega_T)$ such that $\boldsymbol{w}|_{t=T} = 0$, $\eta|_{t=T} = 0$.

Here, $\Omega_T = \Omega \times [0, T]$, $\boldsymbol{g}$ is the external force vector field, $f_0 = f|_{t=0}$ with f standing for $\boldsymbol{u}$, $\boldsymbol{w}$, $\mathcal{S}$, and η.

This definition deals with the case of one rigid body in a fluid. Besides, we have supposed, for simplicity, that the densities of the fluid and of the body are identically equal to 1.

Theorem 1.2. *Let $\Omega \subset \mathbb{R}^n$, $\mathcal{S}_0 \subset \Omega$, $\boldsymbol{u}_0 \in H(\Omega)$, $\boldsymbol{g} \in L^2(0, T; H^{-1}(\Omega))$, and*
a) Ω and $\mathcal{S}_0$ are connected domains of the class C^2 if $n = 2$,
b) Ω and $\mathcal{S}_0$ are balls if $n = 3$.
Then for any arbitrary $T \in (0, \infty)$ there exists a weak solution to Problem A. Moreover, the isometry $\mathcal{M}_{s,t}$ is Lipschitz-continuous with respect to s and t.

This result was proved in [6] and [7] (see also [9] and [8]).

In the next sections, we investigate the properties of the solution of Problem A in the situation when the body $\mathcal{S}$ is in a neighborhood of the boundary of the flow domain Ω. The most interesting question is related to the contacts of the body with the boundary. The motion of the body during and before the contact is studied in Sections 2 and 3, respectively. Generally speaking, all the results below make sense if the contact is possible. Theorem 3.2 gives sufficient conditions which yield the impossibility of the collisions. On the other hand, an example constructed in Section 4 shows that velocity fields satisfying the energy estimate admit the collisions.

2. Behavior of the body during the contact

As it follows from Theorem 1.2, $\mathcal{S}(t)$ is a continuous function (for instance, in the Hausdorff metric). Suppose that $\overline{\mathcal{S}(t)} \cap \partial\Omega \neq \varnothing$ for all t from some time interval $J \subset [0, T]$. If the velocity vector field $\boldsymbol{u}$ is a function from the space $L^2(0, T; K(\mathcal{S}, \Omega))$, then $\boldsymbol{u}(t) \in K(\mathcal{S}(t), \Omega)$ for almost all $t \in J$. Thus, in order to describe the motion of the body in this situation, we have to investigate functions from the space $K(\mathcal{S}, \Omega)$ (or $K_p(\mathcal{S}, \Omega)$ in a more general case) with $\overline{\mathcal{S}} \cap \partial\Omega \neq \varnothing$.

At first, we consider the case $n = 2$.

Theorem 2.1. *Let Ω, $\mathcal{S} \subset \mathbb{R}^2$ be connected domains of class $C^{1,\alpha}$, $\alpha \in (0,1]$, such that $\mathcal{S} \subset \Omega$ and* $\operatorname{dist}(\mathcal{S}, \partial\Omega) = 0$. *If $\boldsymbol{u} \in K_p(\mathcal{S}, \Omega)$ with $p \geq \frac{2+\alpha}{2\alpha}$, then $\boldsymbol{u}(\boldsymbol{x}) = 0$ as $\boldsymbol{x} \in \mathcal{S}$.*

Proof. Assume that $\boldsymbol{u}$ is extended by zero to $\mathbb{R}^2 \setminus \Omega$. Let M be a point from the set $\overline{\mathcal{S}} \cap \partial\Omega$, which is non-empty by assumption.

Let us introduce a new orthogonal coordinate system (ξ_1, ξ_2) with the origin at M. The axis ξ_1 is tangent to $\partial\Omega$ and the axis ξ_2 is directed towards the domain $\mathcal{S}$. Since Ω and $\mathcal{S}$ are domains of class $C^{1,\alpha}$, there exist positive constants k and ρ_0 such that the sets $\pi^+ = \{\boldsymbol{\xi} \in \mathbb{R}^2 : |\xi_1| < \rho_0,\, k\,|\xi_1|^{1+\alpha} < \xi_2 < k\,\rho_0^{1+\alpha}\}$ and $\pi^- = \{\boldsymbol{\xi} \in \mathbb{R}^2 : |\xi_1| < \rho_0,\, -k\,\rho_0^{1+\alpha} < \xi_2 < -k\,|\xi_1|^{1+\alpha}\}$ belong to $\mathcal{S}$ and $R^2 \setminus \Omega$, respectively (see Fig. 1).

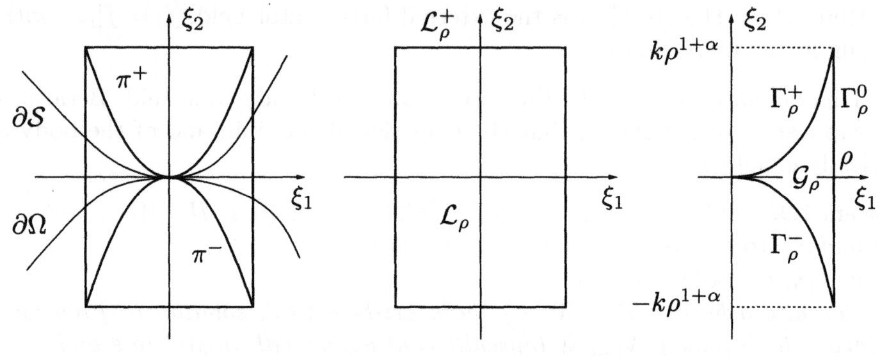

FIGURE 1. Touching of the body $\mathcal{S}$ and the boundary $\partial\Omega$. The domains π^+, π^-, $\mathcal{L}_\rho$ and $\mathcal{G}_\rho$.

Since $D(\boldsymbol{u}) = 0$ in $\mathcal{S}$, there exists constants $\boldsymbol{a} \in \mathbb{R}^2$ and $\omega \in \mathbb{R}$ such that $\boldsymbol{u}(\boldsymbol{x}) = \boldsymbol{a} + \omega\,(\boldsymbol{x} - \boldsymbol{x}_M)^\perp$ for $\boldsymbol{x} \in \mathcal{S}$. Here, $\boldsymbol{x}_M$ is the radius-vector of the point M and $\boldsymbol{x}^\perp$ is the vector with the components $(x_2, -x_1)$. In the variables $\boldsymbol{\xi}$, this representation can be written as $\boldsymbol{u}(\boldsymbol{\xi}) = \boldsymbol{a} + \omega\,\boldsymbol{\xi}^\perp$.

Let us show that $\boldsymbol{a} = 0$. Consider the rectangular domain $\mathcal{L}_\rho = \{\boldsymbol{\xi} \in \mathbb{R}^2 : |\xi_1| < \rho,\, -k\,\rho^{1+\alpha} < \xi_2 < k\,\rho^{1+\alpha}\}$, $\rho \in (0, \rho_0)$. We denote by $\mathcal{L}_\rho^+$ its upper side, i.e., $\mathcal{L}_\rho^+ = \{\boldsymbol{\xi} \in \overline{\mathcal{L}_\rho} : \xi_2 = k\,\rho^{1+\alpha}\}$. For all $p \in [1, \infty]$, we have

$$\|\boldsymbol{u}\|_{L^p(\mathcal{L}_\rho^+)} \leq \big(2k\,\rho^{1+\alpha}\big)^{(p-1)/p}\, \|\nabla \boldsymbol{u}\|_{L^p(\mathcal{L}_\rho)}.$$

On the other hand,

$$\begin{aligned}
\Big(\int_{\mathcal{L}_\rho^+} |\boldsymbol{u}(\boldsymbol{\xi})|^p\, ds\Big)^{1/p} &= \Big(\int_{\mathcal{L}_\rho^+} |\boldsymbol{a} + \omega\,\boldsymbol{\xi}^\perp|^p\, ds\Big)^{1/p} \\
&\geq \Big(\int_{\mathcal{L}_\rho^+} |\boldsymbol{a}|^p\, ds\Big)^{1/p} - |\omega| \Big(\int_{\mathcal{L}_\rho^+} |\boldsymbol{\xi}|^p\, ds\Big)^{1/p} \\
&\geq |\boldsymbol{a}|(2\rho)^{1/p} - |\omega|(2\rho)^{1/p}\Big(k\rho^{1+\alpha} + (p+1)^{-1/p}\rho\Big).
\end{aligned}$$

Therefore,

$$|\boldsymbol{a}| \leq C\rho^{\frac{(1+\alpha)(p-1)-1}{p}} \|\boldsymbol{u}\|_{W^{1,p}(\mathcal{L}_\rho)} + C\rho^{1+\alpha} + C\rho, \tag{5}$$

where C is a constant. In the assertion of the theorem, we supposed that $\alpha \in (0,1]$ and $p \geq \frac{2+\alpha}{2\alpha}$. As it follows from these conditions, $(1+\alpha)(p-1)-1 \geq 0$. Since $\boldsymbol{u} \in H^1_p(\Omega)$, the passage to the limit as $\rho \to 0$ in (5) yields $\boldsymbol{a} = 0$.

Thus, $\boldsymbol{u}(\boldsymbol{\xi}) = \omega\boldsymbol{\xi}^\perp$ for $\boldsymbol{\xi} \in \mathcal{S}$. In order to prove the theorem, we have to establish that $\omega = 0$. Let us consider the domain $\mathcal{G}_\rho = \{\boldsymbol{\xi} \in \mathbb{R}^2 : 0 < \xi_1 < \rho,\ -k\,\xi_1^{1+\alpha} < \xi_2 < k\,\xi_1^{1+\alpha}\}$, $\rho \in (0, \rho_0)$. Since $\operatorname{div}\boldsymbol{u} = 0$, we have

$$\int_{\partial\mathcal{G}_\rho} \boldsymbol{u}\cdot\boldsymbol{n}\,ds = 0,$$

where $\boldsymbol{n}$ is the outward normal to $\partial\mathcal{G}_\rho$. The boundary $\partial\mathcal{G}_\rho$ of the domain $\mathcal{G}_\rho$ consists of three parts:

$$\partial\mathcal{G}_\rho = \Gamma^+_\rho \cup \Gamma^-_\rho \cup \Gamma^0_\rho,$$

where $\Gamma^\pm_\rho = \partial\mathcal{G}_\rho \cap \partial\pi^\pm$ and $\Gamma^0_\rho = \{\boldsymbol{\xi} : \xi_1 = \rho,\ \xi_2 \in [-k\rho^{1+\alpha}, k\rho^{1+\alpha}]\}$.

We supposed that $\boldsymbol{u} = 0$ in $\mathbb{R}^2 \setminus \Omega$. This condition implies

$$\int_{\Gamma^-_\rho} \boldsymbol{u}\cdot\boldsymbol{n}\,ds = 0.$$

It is not difficult to calculate that

$$\int_{\Gamma^+_\rho} \boldsymbol{u}\cdot\boldsymbol{n}\,ds = \frac{1}{2}\,\omega\big(\rho^2 + (k\rho^{1+\alpha})^2\big).$$

Therefore,

$$\frac{1}{2}\,\omega\big(\rho^2 + (k\rho^{1+\alpha})^2\big) = -\int_{\Gamma^0_\rho} \boldsymbol{u}\cdot\boldsymbol{n}\,ds = -\int_{-k\rho^{1+\alpha}}^{k\rho^{1+\alpha}} u_1(\rho,\xi_2)\,d\xi_2,$$

where u_1 is the projection of the vector $\boldsymbol{u}$ on the axis ξ_1. By integrating this equality with respect to ρ from 0 to some $\gamma \in [0, \rho_0]$, we obtain

$$\frac{1}{2}\,\omega\int_0^\gamma \big(\rho^2 + (k\rho^{1+\alpha})^2\big)\,d\rho = -\int_{\mathcal{G}_\gamma} u_1(\boldsymbol{\xi})\,d\boldsymbol{\xi}.$$

The integral in the left-hand side of the last equality can be estimated from below as follows

$$\frac{1}{2}\int_0^\gamma \big(\rho^2 + (k\rho^{1+\alpha})^2\big)\,d\rho \geq \frac{1}{6}\,\gamma^3.$$

On the other hand, due to the Hölder and to the Poincaré inequalities,

$$\Big|\int_{\mathcal{G}_\gamma} u_1(\boldsymbol{\xi})\,d\boldsymbol{\xi}\Big| \leq |\mathcal{G}_\gamma|^{(p-1)/p}\|\boldsymbol{u}\|_{L^p(\mathcal{G}_\gamma)} \leq C\,|\mathcal{G}_\gamma|^{(p-1)/p}\,\gamma^{1+\alpha}\|\boldsymbol{u}\|_{W^{1,p}(\mathcal{L}_\gamma)},$$

where C is a constant. Since $|\mathcal{G}_\gamma| < k\gamma^{2+\alpha}$, we have

$$|\omega| \leq C\,\gamma^{(2\alpha p-2-\alpha)/p}\,\|\boldsymbol{u}\|_{W^{1,p}(\mathcal{L}_\gamma)}.$$

By passing to the limit in this inequality as $\gamma \to 0$, we find that $\omega = 0$.

Thus, $\boldsymbol{u} = 0$ in $\mathcal{S}$ and the theorem is proved. □

In the three-dimensional case, a body touching the boundary can move. Its possible motion is a rotation around the axis which is orthogonal to the boundary at the point of touching.

Theorem 2.2. *Let $\Omega, \mathcal{S} \subset \mathbb{R}^3$ be connected domains of class $C^{1,\alpha}$, $\alpha \in (0,1]$, such that $\mathcal{S} \subset \Omega$ and $\operatorname{dist}(\mathcal{S}, \partial\Omega) = 0$. If $\boldsymbol{u} \in K_p(\mathcal{S}, \Omega)$ with $p \geq \frac{3+\alpha}{2\alpha}$, then $\boldsymbol{u}(\boldsymbol{x}) = \boldsymbol{\omega} \times (\boldsymbol{x} - \boldsymbol{x}_M)$ as $\boldsymbol{x} \in \mathcal{S}$, where $\boldsymbol{x}_M$ is the radius-vector of a point $M \in \partial\mathcal{S} \cap \partial\Omega$ and $\boldsymbol{\omega}$ is a constant vector in $\mathbb{R}^3$ which is orthogonal to $\partial\Omega$ at the point M. In particular, if the set $\partial\mathcal{S} \cap \partial\Omega$ consists of more than one point, then $\boldsymbol{u}(\boldsymbol{x}) = 0$ as $\boldsymbol{x} \in \mathcal{S}$.*

Proof. The function $\boldsymbol{u}$ is assumed to be extended by zero to $\mathbb{R}^3 \setminus \Omega$. Let M be a point from the set $\partial\mathcal{S} \cap \partial\Omega$. We introduce a new coordinate system $\boldsymbol{\xi} = (\xi_1, \xi_2, \xi_3)$ with the origin at the point M. The axes ξ_1 and ξ_2 are tangent to $\partial\Omega$ and the axis ξ_3 is orthogonal to them. We shall also use the cylindrical coordinates (r, θ, z) associated with $\boldsymbol{\xi}$: $\xi_1 = r\cos\theta$, $\xi_2 = r\sin\theta$, $\xi_3 = z$.

As Ω and $\mathcal{S}$ are domains of class $C^{1,\alpha}$, there exist positive constants k and ρ_0 such that the sets $\pi^+ = \{\boldsymbol{\xi} \in \mathbb{R}^3 : r < \rho_0,\ k\,r^{1+\alpha} < z < k\,\rho_0^{1+\alpha}\}$ and $\pi^- = \{\boldsymbol{\xi} \in \mathbb{R}^3 : r < \rho_0,\ -k\,r^{1+\alpha} > z > -k\,\rho_0^{1+\alpha}\}$ are contained in $\mathcal{S}$ and $\mathbb{R}^3 \setminus \Omega$, respectively. Since $D(\boldsymbol{u}) = 0$ in $\mathcal{S}$, there exist vectors $\boldsymbol{a}$ and $\boldsymbol{\omega}$ such that $\boldsymbol{u}(\boldsymbol{x}) = \boldsymbol{a} + \boldsymbol{\omega} \times (\boldsymbol{x} - \boldsymbol{x}_M)$ for $\boldsymbol{x} \in \mathcal{S}$. In the variables $\boldsymbol{\xi}$, this representation reads as $\boldsymbol{u}(\boldsymbol{\xi}) = \boldsymbol{a} + \boldsymbol{\omega} \times \boldsymbol{\xi}$. Exactly the same arguments as in the proof of Theorem 2.1 imply that $\boldsymbol{a} = 0$. Thus, $\boldsymbol{u}(\boldsymbol{\xi}) = \boldsymbol{\omega} \times \boldsymbol{\xi}$.

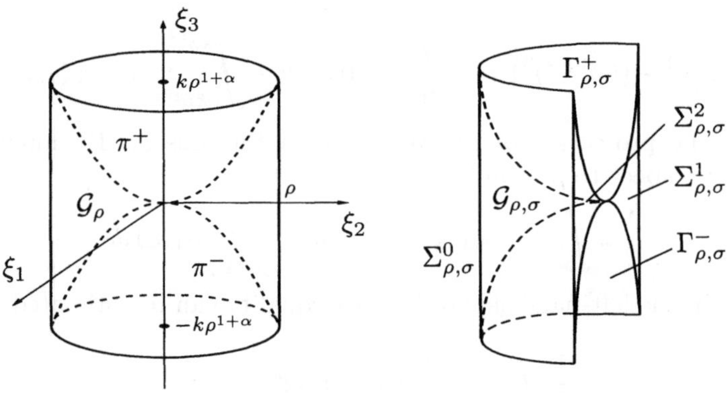

FIGURE 2. The domains π^+, π^-, $\mathcal{G}_\rho$ and $\mathcal{G}_{\rho,\sigma}$.

Let us investigate the vector $\boldsymbol{\omega}$. Consider the domains $\mathcal{G}_\rho = \{\boldsymbol{\xi} \in \mathbb{R}^3 : r \in [0,\rho),\ \theta \in [0, 2\pi],\ -k\,r^{1+\alpha} < z < k\,r^{1+\alpha}\}$ and $\mathcal{G}_{\rho,\sigma} = \{\boldsymbol{\xi} \in \mathcal{G}_\rho : \sigma < \theta < \sigma + \pi\}$ (see Fig. 2). Since $\operatorname{div} \boldsymbol{u} = 0$, we have

$$\int_{\partial\mathcal{G}_{\rho,\sigma}} \boldsymbol{u} \cdot \boldsymbol{n}\, ds = 0, \tag{6}$$

where $\boldsymbol{n}$ is the outward normal to $\partial\mathcal{G}_{\rho,\sigma}$. Let us divide $\partial\mathcal{G}_{\rho,\sigma}$ into five parts:

$$\partial\mathcal{G}_{\rho,\sigma} = \Gamma^{+}_{\rho,\sigma} \cup \Gamma^{-}_{\rho,\sigma} \cup \Sigma^{0}_{\rho,\sigma} \cup \Sigma^{1}_{\rho,\sigma} \cup \Sigma^{2}_{\rho,\sigma},$$

where

$$\Gamma^{\pm}_{\rho,\sigma} = \{\boldsymbol{\xi} \in \partial\mathcal{G}_{\rho,\sigma} \,:\, z = \pm k\, r^{1+\alpha}\}, \quad \Sigma^{0}_{\rho,\sigma} = \{\boldsymbol{\xi} \in \partial\mathcal{G}_{\rho,\sigma} \,:\, r = \rho\},$$

$$\Sigma^{1}_{\rho,\sigma} = \{\boldsymbol{\xi} \in \partial\mathcal{G}_{\rho,\sigma} \,:\, \theta = \sigma\}, \quad \Sigma^{2}_{\rho,\sigma} = \{\boldsymbol{\xi} \in \partial\mathcal{G}_{\rho,\sigma} \,:\, \theta = \sigma + \pi\}.$$

Note that

$$\int_{\Gamma^{-}_{\rho,\sigma}} \boldsymbol{u}\cdot\boldsymbol{n}\, ds = 0,$$

and

$$\int_{\Gamma^{+}_{\rho,\sigma}} \boldsymbol{u}\cdot\boldsymbol{n}\, ds = \boldsymbol{\omega}\cdot\int_{\Gamma^{+}_{\rho,\sigma}} \boldsymbol{\xi}\times\boldsymbol{n}\, ds = m(\rho)\,\boldsymbol{\omega}\cdot\boldsymbol{\tau}_{\sigma},$$

where

$$m(\rho) = \frac{2}{3}\,\rho^3 + k^2\,\frac{2\alpha+2}{2\alpha+3}\,\rho^{2\alpha+3},$$

and $\boldsymbol{\tau}_{\sigma} = (\cos\sigma, \sin\sigma, 0)$ in the coordinate system $\boldsymbol{\xi}$.

Thus, (6) yields the estimate

$$m(\rho)\,|\boldsymbol{\omega}\cdot\boldsymbol{\tau}_{\sigma}| \le \sum_{i=0}^{2}\int_{\Sigma^{i}_{\rho,\sigma}} |\boldsymbol{u}|\, ds. \tag{7}$$

It is not difficult to see that

$$\int_0^{\delta}\int_{\Sigma^{0}_{\rho,\sigma}} |\boldsymbol{u}|\, ds d\rho = \int_{\mathcal{G}_{\delta,\sigma}} |\boldsymbol{u}|\, d\boldsymbol{\xi} \le \int_{\mathcal{G}_{\delta}} |\boldsymbol{u}|\, d\boldsymbol{\xi} \le C\,\delta^{\frac{p-1}{p}(\alpha+3)}\Big(\int_{\mathcal{G}_{\delta}} |\boldsymbol{u}|^p\, d\boldsymbol{\xi}\Big)^{1/p} \tag{8}$$

for $\delta > 0$. We used here the Hölder inequality and the estimate $|\mathcal{G}_{\delta}| \le C\,\delta^{3+\alpha}$.

If $\zeta_j(r,\sigma,z)$, $j = 1,2$, denotes the value of the function $|\boldsymbol{u}(r,\theta,z)|$ as $\theta = \sigma + \pi(j-1)$, then

$$\begin{aligned}\int_0^{\delta}\int_{\Sigma^{j}_{\rho,\sigma}} |\boldsymbol{u}|\, ds d\rho &= \int_0^{\delta}\int_0^{\rho}\int_{-k\, r^{1+\alpha}}^{k\, r^{1+\alpha}} \zeta_j(r,\sigma,z)\, dz dr d\rho \le \delta\int_0^{\delta}\int_{-k\, r^{1+\alpha}}^{k\, r^{1+\alpha}} \zeta_j(r,\sigma,z)\, dz dr \\ &\le \delta\Big(\int_0^{\delta}\int_{-k\, r^{1+\alpha}}^{k\, r^{1+\alpha}} r^{-\frac{1}{p-1}}\, dz dr\Big)^{\frac{p-1}{p}}\Big(\int_0^{\delta}\int_{-k\, r^{1+\alpha}}^{k\, r^{1+\alpha}} r\zeta_j^p\, dz dr\Big)^{1/p} \\ &\le C\,\delta^{\frac{p-1}{p}(\alpha+3)}\Big(\int_0^{\delta}\int_{-k\, r^{1+\alpha}}^{k\, r^{1+\alpha}} r\zeta_j^p\, dz dr\Big)^{1/p}.\end{aligned}$$

By integrating this inequality with respect to σ from 0 to 2π, we deduce the estimate

$$\int_0^{2\pi}\int_0^{\delta}\int_{\Sigma^{j}_{\rho,\sigma}} |\boldsymbol{u}|\, ds d\rho d\sigma \le C\,\delta^{\frac{p-1}{p}(\alpha+3)}\Big(\int_{\mathcal{G}_{\delta}} |\boldsymbol{u}|^p\, d\boldsymbol{\xi}\Big)^{1/p} \qquad \text{for} \;\; j = 1,2. \tag{9}$$

Therefore, (7), (8) and (9) imply

$$\int_0^\delta m(\rho)\,d\rho \int_0^{2\pi} |\boldsymbol{\omega}\cdot\boldsymbol{\tau}_\sigma|\,d\sigma \le C\,\delta^{\frac{p-1}{p}(\alpha+3)}\Big(\int_{\mathcal{G}_\delta} |\boldsymbol{u}|^p\,d\boldsymbol{\xi}\Big)^{1/p}.$$

Furthermore, by employing the Poincaré inequality and the estimate $m(\rho) \ge \frac{2}{3}\rho^3$, we obtain

$$\int_0^{2\pi} |\boldsymbol{\omega}\cdot\boldsymbol{\tau}_\sigma|\,d\sigma \le C\,\delta^{\frac{2\alpha p-\alpha-3}{p}}\Big(\int_{co(\mathcal{G}_\delta)} |\nabla\boldsymbol{u}|^p\,d\boldsymbol{\xi}\Big)^{1/p},$$

where $co(\mathcal{G}_\delta)$ is the smallest cylinder containing the domain $\mathcal{G}_\delta$ (its convex hull). Since $\boldsymbol{u} \in H^1_p(\Omega)$ and $|co(\mathcal{G}_\delta)| \to 0$ as $\delta \to 0$, we have

$$\int_0^{2\pi} |\boldsymbol{\omega}\cdot\boldsymbol{\tau}_\sigma|\,d\sigma = 0.$$

That is, $\boldsymbol{\omega}\cdot\boldsymbol{\tau}_\sigma = 0$ for all $\sigma \in [0, 2\pi]$. This fact implies that the vector $\boldsymbol{\omega}$ is directed along the axis ξ_3.

The theorem is proved. □

As it follows from Theorem 1.2, the weak solution of Problem A satisfies the energy estimate. This means that $\boldsymbol{u} \in K(\mathcal{S};\Omega)$ for almost all $t \in [0,T]$. Therefore, we have to put $p = 2$ in Theorems 2.1 and 2.2. In this case, the assertions of these theorems hold if $\alpha \ge 2/3$ and $\alpha = 1$ for the 2-D and 3-D problems, respectively.

3. Behavior of the body before the contact

In this section, we investigate the motion of the body when it is getting to the boundary of the flow domain Ω. Suppose that the velocity vector field $\boldsymbol{u}$ is rigid ($D(\boldsymbol{u}) = 0$) in $\mathcal{S}(t)$ for almost all t and $\mathcal{S}(t) = \mathcal{M}_{0,t}\mathcal{S}_0$, where $\mathcal{S}_0$ is the domain occupied by the body at some initial time and $\mathcal{M}_{s,t}$ is the family of isometries associated to $\boldsymbol{u}$, defined in Definition 1.1. Therefore, $\mathcal{S}(t) \subset \Omega$ for all t. We denote by $h(t)$ the distance between $\mathcal{S}(t)$ and $\partial\Omega$.

Theorem 3.1. *Let Ω and $\mathcal{S}_0$ be domains in $\mathbb{R}^n$ ($n = 2,3$) of class $C^{1,\alpha}$, $\alpha \in (0,1]$, and $\mathcal{S}_0 \subset \Omega$. If $\boldsymbol{u} \in L^\infty(0,T;L^2(\Omega)) \cap L^1(0,T;K_p(\mathcal{S},\Omega))$, $p \in [1,\infty]$, then h is Lipschitz-continuous with respect to t and there exists a constant C such that*

$$\Big|\frac{dh(t)}{dt}\Big| \le C\,h^\beta(t)\|\boldsymbol{u}(t)\|_{H^1_p(\Omega)} \tag{10}$$

for almost all $t \in [0,T]$ with $\beta = 2 - \frac{1}{1+\alpha}\big(1 + \frac{n-1}{p}\big) - \frac{1}{p}$.

Proof. We prove the theorem in the case $n = 3$. The two-dimensional case can be managed in the same way. At first, we show that h is a Lipschitz-continuous function. As $D(\boldsymbol{u}) = 0$ in $\mathcal{S}$, there exist vector fields $\boldsymbol{a}$ and $\boldsymbol{\omega}$ such that $\boldsymbol{u}(\boldsymbol{x},t) = \boldsymbol{a}(t) + \boldsymbol{\omega}(t) \times (\boldsymbol{x} - \boldsymbol{x}_*(t))$ as $\boldsymbol{x} \in \mathcal{S}(t)$. Here, $\boldsymbol{x}_*(t)$ is the mass center of $\mathcal{S}$,

$\boldsymbol{a}(t) = d\boldsymbol{x}_*(t)/dt$ and $\boldsymbol{\omega}(t)$ are the translational and angular velocity of the body $\mathcal{S}(t)$, respectively. It is not difficult to calculate that

$$\int_{\mathcal{S}} \left(|\boldsymbol{a}|^2 + |\boldsymbol{\omega}|^2|\boldsymbol{x} - \boldsymbol{x}_*|^2\right) d\boldsymbol{x} \leq \int_{\mathcal{S}} |\boldsymbol{u}|^2 \, d\boldsymbol{x}.$$

Since $\boldsymbol{u} \in L^\infty(0,T;L^2(\Omega))$, the functions $\boldsymbol{a}$ and $\boldsymbol{\omega}$ are bounded. This fact implies that the mapping $\mathcal{M}_{0,t}$ and, as a consequence, the function $h(t)$ are Lipschitz-continuous with respect to t.

Let us turn to the proof of (10). Let $P(t) \in \partial\mathcal{S}(t)$ and $Q(t) \in \partial\Omega$ be points such that $h(t) = \operatorname{dist}(P(t), Q(t))$. It may occur that the points P and Q are not uniquely defined. In this case, we take an arbitrary pair.

As in the proof of Theorem 2.2, we introduce a new orthogonal system of coordinates $\boldsymbol{\xi} = (\xi_1, \xi_2, \xi_3)$ with the origin at the point $Q(t)$. The axis ξ_3 is directed from $Q(t)$ to $P(t)$ and the axes ξ_1 and ξ_2 are tangential to $\partial\Omega$. Denote by $\hat{\boldsymbol{\xi}}$ the two-dimensional vector (ξ_1, ξ_2). Since $\mathcal{S}(t)$ and Ω are domains of class $C^{1,\alpha}$, there exist positive constants k and ρ_0 such that the sets $\pi_+ = \{\boldsymbol{\xi} \in \mathbb{R}^3 \,:\, |\hat{\boldsymbol{\xi}}| < \rho_0,\ h(t) + k\,|\hat{\boldsymbol{\xi}}|^{1+\alpha} < \xi_3 < h(t) + k\,\rho_0^{1+\alpha}\}$ and $\pi_- = \{\boldsymbol{\xi} \in \mathbb{R}^3 \,:\, |\hat{\boldsymbol{\xi}}| < \rho_0,\ -k\,|\hat{\boldsymbol{\xi}}|^{1+\alpha} > \xi_3 > -k\,\rho_0^{1+\alpha}\}$ belong to $\mathcal{S}(t)$ and $\mathbb{R}^3 \setminus \Omega$, respectively. Let us consider the domain $\mathcal{G}_\rho = \{\boldsymbol{\xi} \in \mathbb{R}^3 \,:\, |\hat{\boldsymbol{\xi}}| < \rho,\ -k\,|\hat{\boldsymbol{\xi}}|^{1+\alpha} < \xi_3 < h(t) + k\,|\hat{\boldsymbol{\xi}}|^{1+\alpha}\}$. The condition $\operatorname{div}\boldsymbol{u} = 0$ implies that

$$\int_{\partial\mathcal{G}_\rho} \boldsymbol{u} \cdot \boldsymbol{n} \, ds = 0,$$

where $\boldsymbol{n}$ is the outward normal to $\partial\mathcal{G}_\rho$. Notice that $\partial\mathcal{G}_\rho = \Gamma_\rho^+ \cup \Gamma_\rho^- \cup \Gamma_\rho^0$, where $\Gamma_\rho^\pm = \partial\mathcal{G}_\rho \cap \partial\pi^\pm$.

We consider the function $\boldsymbol{u}$ to be extended by zero to $\mathbb{R}^3 \setminus \Omega$. That is,

$$\int_{\Gamma_\rho^-} \boldsymbol{u} \cdot \boldsymbol{n} \, ds = 0.$$

If N is a point on $\partial\mathcal{S}(t)$, then we denote by $\boldsymbol{u}_N$ and by $\boldsymbol{n}_N$ the velocity and the outward normal to $\partial\mathcal{S}(t)$ N. Since $\boldsymbol{u}$ is a rigid velocity vector field, it is not difficult to calculate that

$$\int_{\Gamma_\rho^+} \boldsymbol{u} \cdot \boldsymbol{n} \, ds = \pi\rho^2(\boldsymbol{u}_P \cdot \boldsymbol{n}_P).$$

Therefore,

$$\pi\rho^2|\boldsymbol{u}_P \cdot \boldsymbol{n}_P| \leq \int_{\Gamma_\rho^0} |\boldsymbol{u}| \, ds.$$

By integrating this inequality with respect to ρ from 0 to some $\sigma \in (0, \rho_0)$, we obtain

$$\frac{\pi}{3}\sigma^3|\boldsymbol{u}_P \cdot \boldsymbol{n}_P| \leq \int_{\mathcal{G}_\sigma} |\boldsymbol{u}| \, d\boldsymbol{\xi}. \tag{11}$$

The integral in the right-hand side of (11) can be estimated as follows

$$\int_{\mathcal{G}_\sigma} |\boldsymbol{u}|\, d\boldsymbol{\xi} \le |\mathcal{G}_\sigma|^{\frac{p-1}{p}}\, \|\boldsymbol{u}\|_{L^p(\mathcal{G}_\sigma)} \le C\,(h + 2k\,\sigma^{1+\alpha})\, |\mathcal{G}_\sigma|^{\frac{p-1}{p}}\, \|\nabla \boldsymbol{u}\|_{L^p(\Omega)},$$

where C is a constant. Since $|\mathcal{G}_\sigma| \le C\,\sigma^2(h + 2k\,\sigma^{1+\alpha})$, we get

$$|\boldsymbol{u}_P \cdot \boldsymbol{n}_P| \le C\,\sigma^{-1-\frac{2}{p}}\,\big(h(t) + 2k\,\sigma^{1+\alpha}\big)^{2-\frac{1}{p}}\, z(t),$$

where $z(t) = \|\boldsymbol{u}(t)\|_{H^1_p(\Omega)}$.

Let us take $\sigma = c\,h^{1/(1+\alpha)}$, where c is a constant such that the condition $\sigma \in (0, \rho_0)$ holds. It is possible to put $c = \rho_0\, \mathrm{diam}(\Omega)^{-\frac{1}{1+\alpha}}$. Thus,

$$|\boldsymbol{u}_P \cdot \boldsymbol{n}_P| \le C\, h^\beta(t)\, z(t), \tag{12}$$

where $\beta = -\frac{1}{1+\alpha}\left(1 + \frac{2}{p}\right) + 2 - \frac{1}{p}$.

Let us define the function $\eta : [0,T] \times \partial\Omega \to \mathbb{R}$ such that $\boldsymbol{y} = \boldsymbol{x} + \eta(t, \boldsymbol{x})\, \boldsymbol{\nu}_x \in \partial\mathcal{S}(t)$, where $\boldsymbol{\nu}_x$ is the internal normal to $\partial\Omega$ at $\boldsymbol{x}$. Since $\partial\Omega$ and $\partial\mathcal{S}(t)$ are surfaces of class $C^{1+\alpha}$, the function $\eta(t, \boldsymbol{x})$ is well defined for $\boldsymbol{x}$ belonging to a neighborhood $U(t) \subset \partial\Omega$ of the point $Q(t)$. It is not difficult to see that

$$\frac{\partial \eta(t, \boldsymbol{x})}{\partial t} = (\boldsymbol{u}_y \cdot \boldsymbol{n}_y)(\boldsymbol{n}_y \cdot \boldsymbol{\nu}_x)$$

for almost all $t \in [0,T]$ and all $\boldsymbol{x} \in U(t)$, where $\boldsymbol{n}_y$ is the outward normal to $\partial\mathcal{S}(t)$ at the point $\boldsymbol{y} = \boldsymbol{x} + \eta(t, \boldsymbol{x})\, \boldsymbol{\nu}_x$. In particular, if $\boldsymbol{x}$ is the radius vector of the point $Q(t)$, then $\boldsymbol{n}_y \cdot \boldsymbol{\nu}_x = -1$ and

$$\frac{\partial \eta(t, \boldsymbol{x})}{\partial t} = -\boldsymbol{u}_P \cdot \boldsymbol{n}_P. \tag{13}$$

Let us show that

$$\frac{dh(t)}{dt} = \frac{\partial \eta}{\partial t}(t, Q(t)) \tag{14}$$

for almost all $t \in [0,T]$. Indeed, for all $t \in [0,T]$ and $\tau > 0$, we have

$$h(t+\tau) - h(t) = \min_{\boldsymbol{x} \in \partial\Omega} \eta(t+\tau, \boldsymbol{x}) - \min_{\boldsymbol{x} \in \partial\Omega} \eta(t, \boldsymbol{x}) \le \eta(t+\tau, Q(t)) - \eta(t, Q(t)),$$

$$h(t) - h(t-\tau) \ge \eta(t, Q(t)) - \eta(t-\tau, Q(t)).$$

Since the functions h and η are Lipschitz-continuous with respect to t, the last two inequalities imply (14).

Thus, (10) follows from (12), (13), and (14). The theorem is proved. □

The theorem we have just proved enables us to investigate the motion of the body near the boundary of the flow domain.

Theorem 3.2. *Let the conditions of Theorem* 3.1 *be satisfied and, besides, let* $\boldsymbol{u} \in L^q(0,T; K_p(\mathcal{S}, \Omega))$ *with* $p, q \in [1, \infty]$. *The following assertions hold true*
a) if $\beta < 1$ *and* $h(t_*) = 0$ *for some* $t_* \in [0,T]$, *then* $\lim\limits_{t \to t_*} h(t)\, |t - t_*|^{-\gamma} = 0$, *where* $\gamma = \frac{q-1}{q}\, \frac{1}{1-\beta}$;
b) if $\beta \ge 1$ *and* $h(t_*) > 0$ *for some* $t_* \in [0,T]$, *then* $h(t) > 0$ *for all* $t \in [0,T]$.

Proof. If $s, t \in [0,T]$, by integrating (10) from s to t we obtain

$$F_\beta\big(h(s)\big) - C\Big|\int_s^t z(\lambda)\,d\lambda\Big| \le F_\beta\big(h(t)\big) \le F_\beta\big(h(s)\big) + C\Big|\int_s^t z(\lambda)\,d\lambda\Big|, \tag{15}$$

where $z(t) = \|\boldsymbol{u}(t)\|_{H^1_p(\Omega)}$ and

$$F_\beta(h) = \begin{cases} \frac{1}{1-\beta}\, h^{1-\beta}, & \beta \ne 1, \\ \log h, & \beta = 1. \end{cases}$$

Assertion (a) of the theorem follows from (15) by applying the Hölder inequality and the fact that $z \in L^q(0,T)$.

Assertion (b) is also a simple consequence of (15). Namely, if $\beta = 1$, then

$$h(t) \ge h(t_*)\, \exp\Big(- C\,\Big|\int_{t_*}^t z(\lambda)\,d\lambda\Big|\Big)$$

for all $t \in [0,T]$, and the required assertion follows. The case $\beta > 1$ is similar. □

It is interesting to apply Theorem 3.2 to the weak solutions of Problem A. These solutions satisfy the energy estimate only. That is, $\boldsymbol{u} \in L^\infty(0,T; L^2(\Omega)) \cap L^2(0,T; K(\mathcal{S},\Omega))$ and $p = q = 2$. A body gets to a wall with a zero speed if $\gamma \ge 1$. This condition is satisfied if $\alpha \ge \frac{1}{2}$ in the two-dimensional case ($n = 2$) and if $\alpha = 1$ in the case $n = 3$.

Another interesting question is related to the possibility of collisions. According to Theorem 3.2(b), collisions are impossible if $\beta \ge 1$. For $\alpha = 1$, this will be the case if $p \ge n+1$. Thus, strong solutions of Problem A do not admit collisions.

4. An example of collisions

Theorem 3.2(a) describes the motion of the body when it collides with the boundary of the flow domain. But this theorem does not guarantee that collisions occur. In this section, we construct an example which shows that collisions are possible.

We consider the two-dimensional case ($n = 2$). Our task is to find a vector velocity field $\boldsymbol{u}$ that satisfies the energy estimate and solves Problem A with some external force field $\boldsymbol{g}$.

Suppose that $\Omega = B_R(\mathbf{0})$ and $\mathcal{S}(t) = B_r(\boldsymbol{b}(t))$, $r < R$, where $B_\rho(\boldsymbol{b}) = \{\boldsymbol{x} \in \mathbb{R}^2 : |\boldsymbol{x} - \boldsymbol{b}| < \rho\}$, $\mathbf{0} = (0,0)$ and $\boldsymbol{b}(t) = (b(t), 0)$. The function $b(t)$ is such that $\mathcal{S}(t) \subset \Omega$ for all t, i.e., $|b(t)| \le R - r$. Sometimes, it will be more convenient to use the function $\sigma(t) = b(t)(R-r)^{-1}$. Therefore, $\sigma \in [0,1)$, and $\sigma(t) \to 1$ if $\operatorname{dist}(\mathcal{S}(t), \partial\Omega) \to 0$. Let $\boldsymbol{x}$ and $\boldsymbol{\xi}$ be two coordinate systems in $\mathbb{R}^2$. Define a mapping $\boldsymbol{F}(\cdot,t) : \Omega \to \Omega$ as follows

$$\begin{aligned} x_1 &= F_1(\xi_1,\xi_2,t) = \xi_1 + \sigma(t)\Big(R - \sqrt{\xi_1^2+\xi_2^2}\Big), \\ x_2 &= F_2(\xi_1,\xi_2,t) = \xi_2. \end{aligned}$$

It is not difficult to verify that $\boldsymbol{F}(\Omega,t)=\Omega$ and $\boldsymbol{F}\big(B_r(\mathbf{0}),t\big)=\mathcal{S}(t)$. Besides, if $\boldsymbol{\xi}\in\partial\Omega$, then $\boldsymbol{F}(\boldsymbol{\xi},t)=\boldsymbol{\xi}$.

The inverse mapping $\boldsymbol{G}$ looks as follows

$$\xi_1 = G_1(x_1,x_2,t) = \frac{1}{1-\sigma^2}\Big(x_1-\sigma R+\sigma\sqrt{x_2^2(1-\sigma^2)+(\sigma R-x_1)^2}\,\Big),$$
$$\xi_2 = G_2(x_1,x_2,t) = x_2.$$

Thus, $\boldsymbol{G}(\Omega,t)=\Omega$ and $\boldsymbol{G}\big(\mathcal{S}(t),t\big)=B_r(\mathbf{0})$.

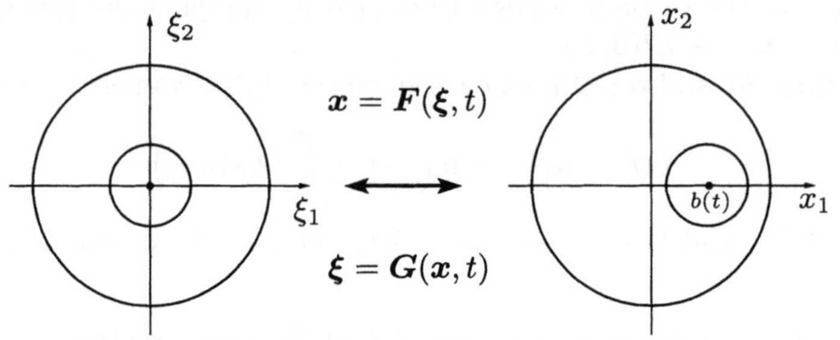

FIGURE 3. The mappings $\boldsymbol{F}$ and $\boldsymbol{G}$.

Note that we do not consider the mapping $\boldsymbol{F}$ as a Lagrangian flow. It simply maps the domain $\Omega\setminus\mathcal{S}(t)$ to a more symmetric one. The symmetric domain corresponds to the value of b being equal to zero. In this case Ω and $\mathcal{S}$ are concentric disks centered at the origin.

Let us introduce the polar coordinates (ρ,θ) associated with $\boldsymbol{\xi}=(\xi_1,\xi_2)$: $\xi_1=\rho\cos\theta,\quad \xi_2=\rho\sin\theta$. The Jacobians of the mappings $\boldsymbol{F}$ and $\boldsymbol{G}$ have a simple form in these coordinates:

$$J_{\scriptscriptstyle F}(\boldsymbol{\xi},t)=\det\Big(\frac{\partial\boldsymbol{F}}{\partial\boldsymbol{\xi}}\Big)=1-\sigma(t)\cos\theta,$$

$$J_{\scriptscriptstyle G}\big|_{\boldsymbol{x}=\boldsymbol{F}(\boldsymbol{\xi},t)}=\det\Big(\frac{\partial\boldsymbol{G}}{\partial\boldsymbol{x}}\Big)\Big|_{\boldsymbol{x}=\boldsymbol{F}(\boldsymbol{\xi},t)}=J_{\scriptscriptstyle F}^{-1}=\frac{1}{1-\sigma(t)\cos\theta}.$$

Now we want to define the stream function $\psi(\boldsymbol{x},t)$ of a flow which would correspond to the considered situation. Namely, the velocity $\boldsymbol{u}$ of the fluid should be such that

$$\boldsymbol{u}=0 \quad\text{on}\quad \partial\Omega \qquad\text{and}\qquad D(\boldsymbol{u})=0 \quad\text{in}\quad \mathcal{S}. \tag{16}$$

Let $\lambda:[0,\infty)\to[0,1]$ be a smooth function such that $\lambda(\rho)=1$ for $\rho<r$ and $\lambda'(r)=\lambda(R)=\lambda'(R)=0$. It is possible to take

$$\lambda(\rho)=(R-r)^{-3}(\rho-R)^2(2\rho-3r+R)$$

If, in particular, $R=3$ and $r=1$, then $\lambda(\rho)=\frac{1}{4}\rho(\rho-3)^2$.

Define the stream function as follows

$$\psi(\boldsymbol{x},t)=\dot{b}(t)\,x_2\,\varphi(\boldsymbol{G}(\boldsymbol{x},t)).$$

where $\varphi(\boldsymbol{\xi}) = \lambda(\sqrt{\xi_1^2+\xi_2^2}\,)$ and $\dot{b} = db/dt$. The level lines of the function ψ are schematically shown in Fig. 4.

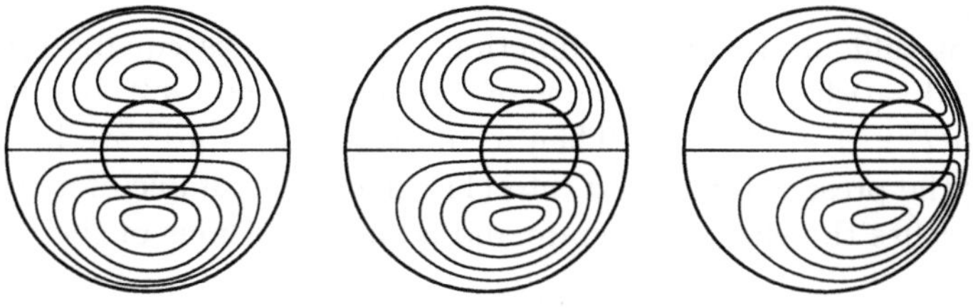

FIGURE 4. The level lines of the stream function ψ for various values of σ.

The corresponding velocity vector field $\boldsymbol{u}(\boldsymbol{x},t) = \nabla_x^{\perp}\psi(\boldsymbol{x},t)$ has the components

$$\begin{aligned} u_1(\boldsymbol{x},t) &= \frac{\partial\psi}{\partial x_2} = \dot{b}(t)\,\varphi(\boldsymbol{G}(\boldsymbol{x},t)) + \dot{b}(t)\,x_2\,\boldsymbol{G}_{x_2}\cdot\nabla_\xi\varphi(\boldsymbol{\xi})|_{\boldsymbol{\xi}=\boldsymbol{G}(\boldsymbol{x},t)}, \\ u_2(\boldsymbol{x},t) &= -\frac{\partial\psi}{\partial x_1} = -\dot{b}(t)\,x_2\,\boldsymbol{G}_{x_1}\cdot\nabla_\xi\varphi(\boldsymbol{\xi})|_{\boldsymbol{\xi}=\boldsymbol{G}(\boldsymbol{x},t)}. \end{aligned} \tag{17}$$

Note that $\boldsymbol{u}$ satisfies (16).

We have to check that $\boldsymbol{u} \in L^\infty(0,T;L^2(\Omega)) \cap L^2(0,T;H^1(\Omega))$. The norm of $\boldsymbol{u}$ in $L^2(\Omega)$ can be calculated explicitly:

$$\|\boldsymbol{u}\|^2_{L^2(\Omega)} = \|u_1\|^2_{L^2(\Omega)} + \|u_2\|^2_{L^2(\Omega)} = \mu_1(R,r)\,\nu_1(\sigma)\,|\dot{\sigma}|^2,$$

where $\mu_1(R,r) = \int_r^R \rho^3\,(\lambda'(\rho))^2\,d\rho < \infty$ and

$$\nu_1(\sigma) = \int_0^{2\pi} \frac{\sin^2\theta}{1-\sigma\cos\theta}\,d\theta = 2\pi\,\frac{1-\sqrt{1-\sigma^2}}{\sigma^2} = 2\pi\,\frac{1}{1+\sqrt{1-\sigma^2}}.$$

It is not difficult to see that $\nu_1(\sigma) \in [\pi, 2\pi]$. Thus, the following assertions holds true.

Lemma 4.1. *The function $\boldsymbol{u}$ belongs to the space $L^\infty(0,T;L^2(\Omega))$ if and only if $\dot{\sigma} \in L^\infty(0,T)$.*

In order to prove that $\boldsymbol{u} \in L^2(0,T;H^1(\Omega))$, it is enough to estimate $\Delta\psi$ in $L^2(\Omega_T)$. After some computations, we obtain

$$\|\Delta\psi\|^2_{L^2(\Omega)} \le \mu_2(R,r)\nu_2(\sigma)\,|\dot{\sigma}|^2,$$

where $\nu_2(\sigma) = (1-\sigma)^{-3/2}$ and μ_2 is a bounded positive constant depending on R and r. This implies the following lemma.

Lemma 4.2. *The function $\boldsymbol{u}$ belongs to the space $L^2(0,T;H^1(\Omega))$ if and only if $(1-\sigma)^{-3/2}|\dot{\sigma}|^2 \in L^1(0,T)$.*

Notice that the distance between the body and the boundary reads as $h(t) = (R-r)(1-\sigma(t))$ and the condition in Lemma 4.2 has the form $h^{-3/4}|\dot{h}| \in L^2(0,T)$. This condition is in accordance with (10). Therefore, the estimate in Theorem 2.2 is precise.

It is not difficult to find a function $\sigma(t)$ which is equal to zero at some $t_* \in [0,T]$ and satisfies the conditions of Lemmas 4.1 and 4.2. Really, let us take $T = 1$ and $\sigma(t) = 1 - (t - t_*)^4$. Then $\dot{\sigma}$ and $(1-\sigma)^{-3/2}|\dot{\sigma}|^2$ are bounded. Thus, substituting the present function σ into (17) gives an example of a vector velocity field that satisfies the energy estimate and admits collisions of the body with the boundary.

However, we cannot say yet that $\boldsymbol{u}$ is a weak solution of Problem A. Let us consider the time derivative of $\boldsymbol{u}$. There exists a positive constant $\mu_3(R,r)$ such that

$$\|\psi_t\|_{L^2(\Omega)} \leq \mu_3(R,r)\big(|\ddot{\sigma}| + |\dot{\sigma}|^2\big).$$

This estimate yields the following lemma.

Lemma 4.3. *If $\ddot{\sigma} \in L^2(0,T)$, then $\boldsymbol{u}_t \in L^2(0,T;H^{-1}(\Omega))$.*

Notice that $\sigma(t) = 1 - (t - t_*)^4$ satisfies the condition of this lemma. By substituting the function $\boldsymbol{u}$ corresponding to this σ into (3), we easily deduce the result below.

Theorem 4.4. *If $\sigma(t)$ is a function such that $\ddot{\sigma} \in L^2(0,T)$ and $(1-\sigma)^{-3/2}|\dot{\sigma}|^2 \in L^1(0,T)$, then there exists an external force vector field $\boldsymbol{g} \in L^2(0,T;H^{-1}(\Omega))$ such that the velocity vector field $\boldsymbol{u}$ given by (17) is a weak solution of Problem A.*

References

[1] C. Conca, J.A. San Martín, M. Tucsnak, *Existence of solutions for the equations modelling the motion of a rigid body in a viscous fluid*, Comm. Partial Differential Equations **25**, No. 5-6 (2000), 1019–1042.

[2] B. Desjardins, M. J. Esteban, *Existence of weak solutions for the motion of rigid bodies in a viscous fluid*, Arch. Ration. Mech. Anal. **146**, No. 1 (1999), 59–71.

[3] E. Feireisl, *On the motion of rigid bodies in a viscous incompressible fluid*, to appear.

[4] G.P. Galdi, *On the steady self-propelled motion of a body in a viscous incompressible fluid*, Arch. Ration. Mech. Anal. **148**, No.1 (1999), 53–88.

[5] M.D. Gunzburger, H.-C. Lee, G. Seregin, *Global existence of weak solutions for viscous incompressible flows around a moving rigid body in three dimensions*, J. Math. Fluid Mech. **2**, No. 3 (2000), 219–266.

[6] K.-H. Hoffmann, V.N. Starovoitov, *On a motion of a solid body in a viscous fluid. Two–dimensional case*, Adv. Math. Sci. Appl. **9**, No. 2 (1999), 633-648.

[7] K.-H. Hoffmann, V.N. Starovoitov, *Zur Bewegung einer Kugel in einer zähen Flüssigkeit*, Documenta Mathematica **5** (2000), 15–21.

[8] J.A. San Martín, V.N. Starovoitov, M. Tucsnak, *Global weak solutions for the two-dimensional motion of several rigid bodies in an incompressible viscous fluid,* Arch. Ration. Mech. Anal. **161**, No. 2 (2002), 113–147.

[9] V.N. Starovoitov, *Non–regular problems of hydrodynamics,* Novosibirsk State University, 2000, Doctor of Sciences Thesis, 230 p. (in Russian)

[10] R. Temam, *Problèmes mathématiques en plasticité,* Méthodes Mathématiques de l'Informatique, 12. Publié avec le concours du C.N.R.S. Paris: Gauthier-Villars. Bordas. VII, 1983.

Center of Advanced European
Studies and Research,
Friedensplatz 16
D-53111 Bonn, Germany
E-mail address: starovoitov@caesar.de

Lavrentyev Institute of Hydrodynamics,
630090 Novosibirsk, Russia

International Series of Numerical Mathematics, Vol. 147, 329–341

Crystal Growth, Coarsening and the Convective Cahn–Hilliard Equation

Stephen J. Watson

Abstract. The coarsening dynamics of a faceted vicinal crystalline surface growing into its melt by attachment kinetics is considered. The convective Cahn–Hilliard equation ($\mathcal{CCH}$) is derived as a small amplitude expansion of such surface evolutions restricted to 1-D morphologies, with the local surface slope serving as the order parameter. A summary of the *sharp interface* theory for $\mathcal{CCH}$, that follows from a matched asymptotic analysis, is also presented [26]. It takes the form of a nearest neighbor interaction between two non-symmetrically related phase boundaries (*kink* and *anti-kink*). The resulting coarsening dynamical system $\mathcal{CDS}$ for the phase boundaries exhibits novel coarsening mechanisms. In particular, binary coalescence of phase boundaries is impossible. Also, ternary coalescence occurs only through two kinks meeting an anti-kink resulting in a kink (*kink-ternary*); the alternative of two anti-kinks meeting a kink is impossible. This behavior stands in marked contrast to the Cahn–Hilliard equation $\mathcal{CH}$ where binary coalescence of phase boundaries is generic. Numerical simulations of $\mathcal{CCH}$ are presented to validate the predictions of the sharp interface theory $\mathcal{CDS}$.

[1] S. J. Watson, F. Otto, S. H. Davis, *Coarsening dynamics of the convective Cahn–Hilliard equation*, Physica D **178** (3-4) (2003) 127–148.

1. Introduction

First-order phase transformation under local thermodynamic equilibrium results in phases which are identified through minimization of appropriate thermodynamic potentials; e.g. the Maxwell rule for a vapor-liquid mixture. However, if equilibrium is not present then *kinetic effects* may present themselves, resulting in new phase selection rules; e.g., *solute trapping* during rapid solidification of a binary alloy.

The *phase ordering* of an initially homogeneous unstable state will generally proceed in two distinct stages:

Spinodal decomposition: A primary instability, such as surface tension, will drive segregation (nucleation) into a sub-optimal fine-grained structure of pure phases separated by diffuse but narrow phase boundaries.

Coarsening: The *morphological* (characteristic) length scale $\mathcal{L}_{\mathcal{M}}(t)$ grows in time as coalescing phase boundaries annihilate.

We use the term *coarsening dynamics* of a phase ordering process in a broad sense which includes, i) the detailed study of the temporal evolution of the phase boundaries, ii) the mechanisms by which coarsening occurs, as well as iii) gross features like the temporal change in the *morphological length scale* $\mathcal{L}_{\mathcal{M}}(t)$. The coarsening dynamics of a non-equilibrium (*driven*) phase ordering system can be profoundly different from its equilibrium counterpart, as we shall see.

A thermodynamically unstable crystal surface with strongly anisotropic surface energy in contact with its melt will *spinodally decompose* into a faceted surface [5, 22, 17] ; here the local normal to the surface serves as the phase variable (*order parameter*). Now, the coarsening dynamics of a faceted crystal surface in thermal equilibrium with its melt (*thermally annealed surface*) is known experimentally to change dramatically if, instead, the crystal is subject to net growth [20, 11, 21]. In particular, growth accelerates the rate at which the *morphological* length scale of the surface, $\mathcal{L}_{\mathcal{M}}$, increases. In addition, 2D surface morphologies are radically altered [20, 18, 11]. We coin the term *annealing-to-growth transition* in reference to the above effects.

We consider the *annealing-to-growth* transition in the setting where *attachment kinetics* is the dominant mass transfer mechanism, and restrict our attention to one-dimensional surface morphologies (hill/valley structures). We derive the *convective Cahn–Hilliard* equation ($\mathcal{CCH}$) via a *small amplitude* expansion of the physical model associated with net growth of the faceted surface. In scaled form ($\mathcal{CCH}$) reads,

$$q_t - \varepsilon q q_x = (W'(q) - q_{xx})_{xx}\,, \tag{1}$$

where the local surface slope $q(x,t)$ serves as the order parameter, subscripts denote the partial derivative with respect to time t and space x respectively, and $'$ denotes the q-derivative. The *effective free energy* $W(q)$ takes the form of a symmetric *double well* with minima at $q = \pm 1$, thereby capturing the anisotropy of the crystal surface energy. Also, the dimensionless small parameter ε multiplying the convective term qq_X is a measure of the growth strength.

In the absence of net growth, $\varepsilon = 0$, eqn. (1) reduces to the well studied Cahn–Hilliard equation ($\mathcal{CH}$)

$$q_t = (W'(q) - q_{xx})_{xx}\,. \tag{2}$$

This equation originally arose in modelling the *spinodal decomposition* of binary alloys under isothermal conditions [4]; there q is related to the *phase fraction*. Here it serves as a model for the evolution of a thermally annealed 1-D faceted (hill/valley) crystal surface [22, 17]. Among the properties of the $\mathcal{CH}$ theory are the order parameter (phase) selection $q = \pm 1$ through the bi-tangent construction on $W(q)$ [3]; a thermodynamic equilibrium construction. The coarsening is driven by the non-local free energy [15]

$$\int (W(q) + \frac{1}{2}{q_x}^2)\ dx,$$

and exhibits the following features [14, 1, 6, 23]:

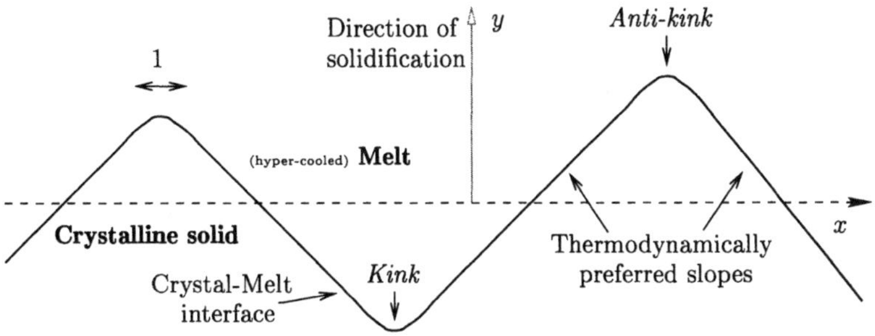

FIGURE 1. Schematic representation of the growing faceted surface; the facet angles have been exaggerated

- Binary coalescence of phase boundaries is the generic coarsening event.
- $\mathcal{L}_{\mathcal{M}}(t) \sim \ln t$

The coarsening dynamics of $\mathcal{CH}$ serves as the archetypal example of a thermodynamic equilibrium phase ordering process [2].

Remark 1.1. *The bi-tangent construction on $W(q)$ may be viewed as the "small slope" version of the Wulff construction on the surface energy $\hat{\gamma}(\theta)$, which determines thermodynamically compatible facet orientations* [7].

The coarsening dynamics of $\mathcal{CCH}$ $(0 < \varepsilon \ll 1)$ [1] stands in marked contrast with its equilibrium counterpart $\mathcal{CH}$. This is an expression of the the experimentally observed *annealing-to-growth* transition alluded to earlier. First, the bi-tangent construction for phase selection is broken by $\mathcal{CCH}$ [16]. This departure from thermodynamic equilibrium marks $\mathcal{CCH}$ as a *driven* phase ordering system [10]. Two non-symmetrically related phase boundaries consequently emerge, which we refer to as kink (hill corner) and anti-kink (valley corner); see figs. 1 and 2. In turn, the kinks and anti-kinks interact through a convection-diffusion mechanism yielding, to leading order, a nearest neighbors *Coarsening Dynamical System* ($\mathcal{CDS}$) representation for the phase ordering process [26]. The $\mathcal{CDS}$ theory predicts the following surprising and novel features for the coarsening dynamics of $\mathcal{CCH}$ [26]:

i. $\mathcal{L}_{\mathcal{M}} \sim t^{1/2}$ provided $1 \ll \mathcal{L}_{\mathcal{M}} \ll \frac{1}{\varepsilon}$
ii. Binary coalescence of phase boundaries is impossible.
iii. Ternary coalescence of phase boundaries may only occur through a *kink-ternary*; two kinks meet an anti-kink resulting in a kink.

Furthermore, a *crossover* from power law to logarithmically slow coarsening occurs as $\mathcal{L}_{\mathcal{M}}$ crosses the (dimensionless) *Peclet length scale* $\mathcal{L}_{\mathcal{P}} \sim \frac{1}{\varepsilon}$ [26]. All of the above

[1]See Remark 1.2

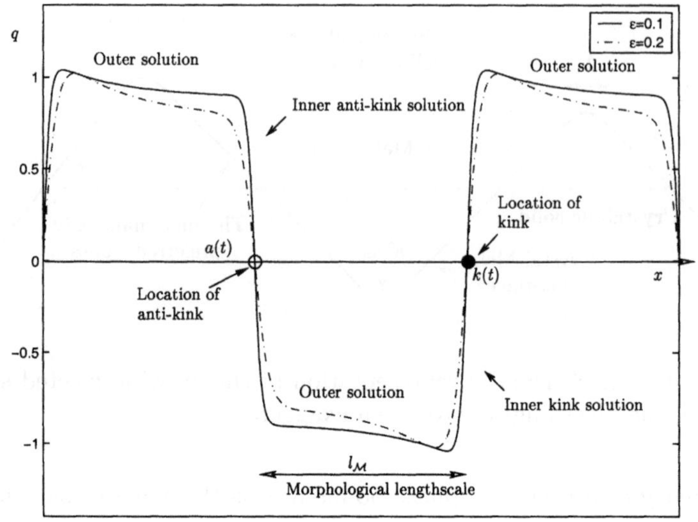

FIGURE 2. Structure of the matched asymptotic composite solution of a periodic profile for $\mathcal{CCH}$ [26]. Here, • and ∘ denote the location of the kink and the anti-kink, respectively.

are confirmed by direct numerical simulations on $\mathcal{CCH}$, which have been carried out in [26, 12].

Remark 1.2. *When $\varepsilon \sim 1$, the morphologies of solutions do not coarsen in time, but rather display periodic patterns, while for $\varepsilon \gg 1$ solutions become* rough [12]. *This is related to the fact that formally the convective Cahn–Hilliard approaches the Kuramoto–Sivashinsky (KS) equation as $\varepsilon \to \infty$, and solutions of the (KS) equation are known to display spatio-temporal chaos.*

The remainder of this paper is structured as follows. First, in Section 2 we describe the geometrically exact equation of motion for attachment kinetics controlled crystal growth. A small amplitude expansion of this equation about a *vicinal surface* is then presented in Sec. 3, from which we derive the convective Cahn–Hilliard equation ($\mathcal{CCH}$). In Section 4 we present a synopsis of the *Coarsening Dynamical System $\mathcal{CDS}$* associated with $\mathcal{CCH}$. We conclude in Sec. 5 with direct numerical simulations of $\mathcal{CCH}$ and $\mathcal{CDS}$.

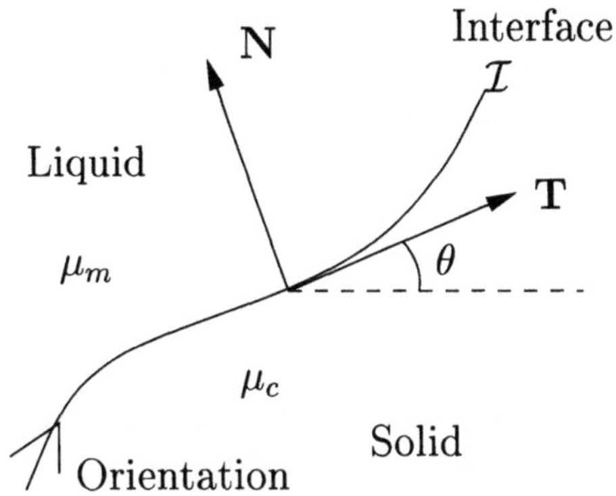

FIGURE 3. Sign conventions for a liquid/solid interface $\mathcal{I}$

2. Attachment kinetics controlled crystal growth

Here we describe a physical model for the growth of a 2D crystal with anisotropic surface energy, where *attachment kinetics* [8] is the dominant mass transport mechanism. The resulting geometric evolution equation for the crystal surface [9, 13] takes the form of a (*driven*) anisotropic Willmore flow [27].

Let $\mathcal{C} \subset \mathbb{R}^2$ denote the boundary between a crystal and its melt. We orient the curve as in fig. 2 and let θ denote the (oriented) polar angle of the local tangent plane, s the arc length, and $\kappa = \frac{d\theta}{ds}$ the (signed) curvature. We consider the surface energy $\mathcal{S}$ of $\mathcal{I}$ to take the form [9]

$$\mathcal{S} = \mathcal{E} \int_{\mathcal{I}} \left(\hat{\gamma}(\theta) + \frac{1}{2}\alpha^2\kappa^2 \right) \, \mathrm{d}s \;, \tag{3}$$

where $\hat{\gamma} : [0, 2\pi] \to \mathbb{R}$ is a given dimensionless function of the local angle θ, while $\mathcal{E}$ and α are dimensional constants which set scales of energy per unit length and length, respectively. Surface energy anisotropy is captured by the dependence of $\hat{\gamma}$ on the local surface orientation θ. The curvature term models an *edge energy* which is concentrated in the region where two facets meet, with α setting the length scale of the associated rounded edge. It is argued that such edge energies arise from the presence of crystalline *steps* on the surface, and the constant α is related to the lattice constant d [24, 19].

The *surface chemical potential* $\mu_{\mathcal{S}}$ is defined by the variational derivative of $\mathcal{S}$; namely,

$$\mu_{\mathcal{S}} := \frac{\delta \mathcal{S}}{\delta \mathcal{I}} = \mathcal{E} \left[-\hat{s}(\theta)\; \kappa + \alpha^2 \left(\kappa_{ss} + \frac{1}{2}\kappa^3 \right) \right], \tag{4}$$

where the surface stiffness $\hat{s}(\theta) := \hat{\gamma}(\theta) + \hat{\gamma}''(\theta)$. Note that μ_S, which has the dimensions of energy per unit area, expresses solely the local gain or loss of surface energy as the interface evolves.

A crystal surface with orientation θ_0 is *thermodynamically unstable* if $\hat{s}(\theta_0) < 0$, and *stable* if $\hat{s}(\theta_0) > 0$. Unstable orientations are missing in the equilibrium crystal shape (Wulff shape). Also, a pair of stable facets which meet are globally stable if the corresponding facet angles satisfy the Wulff construction [7].

We neglect the coupling of the surface evolution to the dissipation of heat by imposing constant bulk chemical potentials at the solid and melt sides of the interface; this assumption is valid, for example, in the limit of infinite heat conductivity. Furthermore, we assume that the mass density of the crystal and melt coincide, so that there is no induced fluid flow during solidification and melting. Now, on a point of terminology, bulk chemical potentials are usually given in units of energy per unit mass. Here, because of our uniform density assumption, we may conveniently take μ_c and μ_m to denote the energy per unit area of the bulk crystal and its melt respectively.

The *interfacial chemical potential* $\mu_{\mathcal{I}}$, which accounts for changes in both surface and bulk energies upon variation of the interface $\mathcal{I}$, is given by

$$\mu_{\mathcal{I}} = \mu_S + \mu_c - \mu_m. \tag{5}$$

We assume that isotropic *attachment kinetics* (solidification/melting) is the dominant mass transfer mechanism, i.e.,

$$\mathcal{B}\, V_n = -\mu_{\mathcal{I}}, \tag{6}$$

where V_n is the normal velocity of the interface, and $\mathcal{B} > 0$ is the (constant) kinetic coefficient with dimensions ($[\mathcal{B}] = MT^{-1}L^{-1}$). Taking all of the above together, we arrive at the following geometric evolution equation for the interface [9, 13]:

$$\mathcal{B}\, V_n = \mathcal{E}\left[\left(\hat{\gamma}(\theta) + \hat{\gamma}_{\theta\theta}(\theta)\right)\kappa - \alpha^2\left(\kappa_{ss} + \kappa^3\right)\right] + \left(\mu_l - \mu_s\right) \tag{7}$$

Note that the planar front $\theta \equiv 0$ moving with normal velocity $V_n = \frac{\mu_m - \mu_c}{\mathcal{B}}$ is a solution to (6). Switching to a frame of reference co-moving with this front, the evolution equation for the interface takes the form

$$\mathcal{B}\, V_n = \mathcal{E}\left[\left(\hat{\gamma}(\theta) + \hat{\gamma}_{\theta\theta}(\theta)\right)\kappa - \alpha^2\left(\kappa_{ss} + \kappa^3\right)\right] + (\mu_l - \mu_s)(1 - \cos\theta) \tag{8}$$

3. Long-wave approximation

We assume that $\theta \equiv 0$ is a *vicinal* surface, with a surface stiffness for neighboring orientations of the specific form [2]

$$\hat{s}(\theta) = 3\left(\frac{\theta}{\eta}\right)^2 - 1 =: \tilde{s}\left(\frac{\theta}{\eta}\right), \tag{9}$$

where $0 < \eta \ll 1$. This guarantees that the planar front $\theta \equiv 0$ is thermodynamically unstable, and, furthermore, that there is a globally stable pair of facets with

[2]see remark 3.1

$\theta \sim \pm\eta$. In addition, we assume that the surface energy scale $\mathcal{E}$ appearing in (3) scales with η:

$$\mathcal{E} = \eta E_0, \tag{10}$$

where $E_0 > 0$.

Remark 3.1. *The term* vicinal *refers to an unstable surface that is in the vicinity of stable orientations. Our specific parabolic choice of surface stiffness is simply for convenience, but could be replaced by a general function with similar structure.*

We now proceed to derive the convective Cahn–Hilliard $\mathcal{CCH}$ as a *small amplitude* expansion of equation (8), subject to the constraints on the surface energy given by (9) and (10). In this sense, $\mathcal{CCH}$ serves as a model for the *spinodal decomposition* and subsequent coarsening of a growing faceted surface.

First, we re-scale (8) with respect to the length scale α, and the time scale $\frac{\alpha\mathcal{B}}{\mathcal{E}}$; i.e., $s \to \frac{s}{\alpha}$ and $t \to \frac{\mathcal{E}}{\alpha\mathcal{B}}t$. We arrive at the dimensionless form

$$V_n = \hat{s}(\theta)\kappa - \left(\kappa_{ss} + \frac{1}{2}\kappa^3\right) + \tilde{\varepsilon}(1 - \cos\theta), \tag{11}$$

where the dimensionless parameter

$$\varepsilon := \frac{\alpha(\mu_m - \mu_c)}{\eta E_0} \tag{12}$$

is a measure of the growth strength.

Consider now the evolution of a free surface in the vicinity of the unstable planar front $\theta \equiv 0$. Letting $y = h(x,t)$ be the graph of the free surface, we assume O(1) variation in x,

$$\frac{\partial}{\partial x} \sim \mathrm{O}(1). \tag{13}$$

together with the small amplitude expansion

$$h = \eta h_1 + \eta^2 h_2 + \cdots . \tag{14}$$

Note that (13) and (14) together imply the small slope condition

$$h_x = \mathrm{O}(\eta). \tag{15}$$

It now follows that the normal velocity V_n has the expansion

$$V_n = \eta\frac{\partial h_1}{\partial t} + \mathrm{O}(\eta^2).$$

Also, arc-length derivatives coincide with x-derivatives to leading order:

$$\frac{\partial}{\partial s} = \frac{\partial}{\partial x} + \mathrm{O}(\eta^2),$$

In addition, the angle θ and the local slope h_x coincide to leading order,

$$\theta = h_x + \mathrm{O}(\eta^3).$$

Since $\kappa = d\theta/ds$, we also conclude

$$\kappa = \eta\frac{\partial^2 h_1}{\partial x^2} + \mathrm{O}(\eta^2), \qquad \kappa_{ss} = \eta\frac{\partial^4 h_1}{\partial x^4} + \mathrm{O}(\eta^2), \qquad \kappa^3 = \mathrm{O}(\eta^3).$$

Last, we find

$$(1 - cos\theta) = \eta^2 \left(\frac{\partial h_1}{\partial x}\right)^2 + \mathrm{O}(\eta^3),$$

and, upon recalling (9),

$$\hat{s}(\theta) = \tilde{s}\left(\frac{\partial h_1}{\partial x}\right) + \mathrm{O}(\eta).$$

We expand (11) in powers of η, and then gather the (leading) terms of order $\mathrm{O}(\eta)$, which yields

$$\frac{\partial h_1}{\partial t} - \varepsilon\frac{1}{2}\left(\frac{\partial h_1}{\partial x}\right)^2 = \tilde{s}\left(\frac{\partial h_1}{\partial x}\right)\frac{\partial^2 h_1}{\partial x^2} - \frac{\partial^4 h_1}{\partial x^4} \tag{16}$$

where

$$\varepsilon = \frac{(\mu_m - \mu_c)\alpha}{E_0}.$$

Now, differentiating (16) with respect to x and setting $q = h_{1x}$ (the re-scaled local slope), we arrive at the convective Cahn–Hilliard equation ($\mathcal{CCH}$)

$$q_t - \varepsilon q q_x = \left(W'(q) - q_{xx}\right)_{xx}. \tag{17}$$

Here, for the particular choice of $\hat{s}$ given by (9), the *effective free energy* is

$$W(q) := \int\int \tilde{s}(q)\mathrm{d}q^2 = \frac{1}{4}q^4 - \frac{1}{2}q^2.$$

Remark 3.2. *For a thermally annealed surface we have $\mu_m = \mu_c$, in which case $\varepsilon = 0$ and* (17) *reduces to the Cahn–Hilliard equation ($\mathcal{CH}$).*

4. The coarsening dynamical system $\mathcal{CDS}$

The coarsening dynamics of $\mathcal{CCH}$ ($0 < \varepsilon \ll 1$) and $\mathcal{CH}$ ($\varepsilon = 0$) stand in marked contrast with one another. The origin of this distinction rests in the *driven* (non-equilibrium) nature of the phase ordering of $\mathcal{CCH}$. We summarize here the key *morphological* features of the coarsening theory for $\mathcal{CCH}$; a complete treatment appears in [26].

First, the presence of convection induces two non-symmetrically related phase boundaries (*kink* and *anti-kink*) [16]. In turn, the kinks and the anti-kinks interact with their nearest neighbors through a convective-diffusive mechanism which arises from their associated non-equilibrium offsets. Fig. 2 shows a uniformly valid matched asymptotic expansion of a periodic (though unstable) profile for $\mathcal{CCH}$. It clearly exhibits the kink/anti-kink asymmetry, as well as the convective-diffusive balance which occurs in the *outer region* between them.

Definition 4.1 (Index convention). *Given an alternating array of kink/anti-kink locations on the line, we adopt the (arbitrary) ordering convention of letting kinks have odd indices, which in turn gives anti-kinks even indices.*

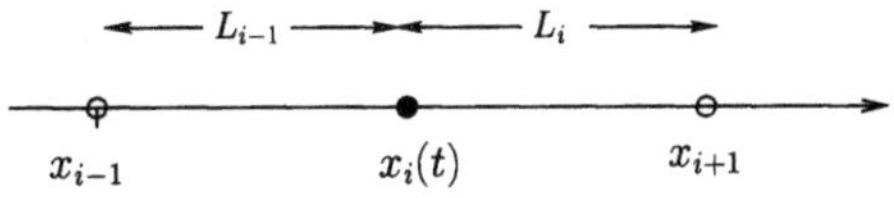

FIGURE 4. Here $x_i(t)$ denotes the location of the i'th phase boundary, $L_i(t) := x_{i+1} - x_i$ is the distance between the i'th and $i+1$'th phase boundary, and • and ∘ denote kink and anti-kink, respectively; given our indexing convention , in this figure i is understood to be odd.

Let $x_i(t)$ denote the location of the i'th phase boundary at time t, and set $L_i(t) := x_{i+1} - x_i$ to be the distance between the i'th and $i+1$'th phase boundary. A matched asymptotic analysis [26] shows that the motion of phase boundaries is governed, to leading order, by the nearest neighbor dynamical system

$$\frac{dx_i}{dt} = (-1)^{i+1}\left[J(L_i) - J(L_{i-1})\right], \tag{18}$$

where the (effective) *flux function* $J(\mathcal{L})$ is given by

$$J(\mathcal{L}) := \frac{\varepsilon}{2\sqrt{2}} \frac{1}{\exp(\mathcal{L}/2) - 1} \tag{19}$$

This theory is valid for all morphological (characteristic) length scales $\mathcal{L}_{\mathcal{M}} \gg 1$; i.e., whenever the system enters the *interfacial regime.*

Remark 4.2. *The skew-symmetry between the evolution equations for kink and anti-kink reflects the broken mirror symmetry $x \to -x$ of the convective Cahn–Hilliard equation.*

The dynamical system (18) *coarsens* in time as kinks and anti-kinks meet and then annihilate one another. By coarsening, we mean that the average separation between phase boundaries grows in time. We refer to the dynamical system (18) subject to such coarsening events as $\mathcal{CDS}$ (*coarsening dynamical system*).

5. Numerical simulation of $\mathcal{CDS}$ and $\mathcal{CCH}$

A direct numerical simulation of $\mathcal{CCH}$ with $\varepsilon = 0.1$ is shown in fig. 5; a pseudo-spectral explicit in time method on a periodic domain. The initial condition involved a small random (Gaussian) perturbation of the unstable homogeneous state $q = 0$. One initially observes spinodal decomposition with a wavelength consistent with the dominant linear instability of the system. An interfacial regime subsequently emerges, and coarsening then proceeds solely via a unique type of ternary event whereby two kinks meet an anti-kink resulting in a kink; the *kink-ternary.* On

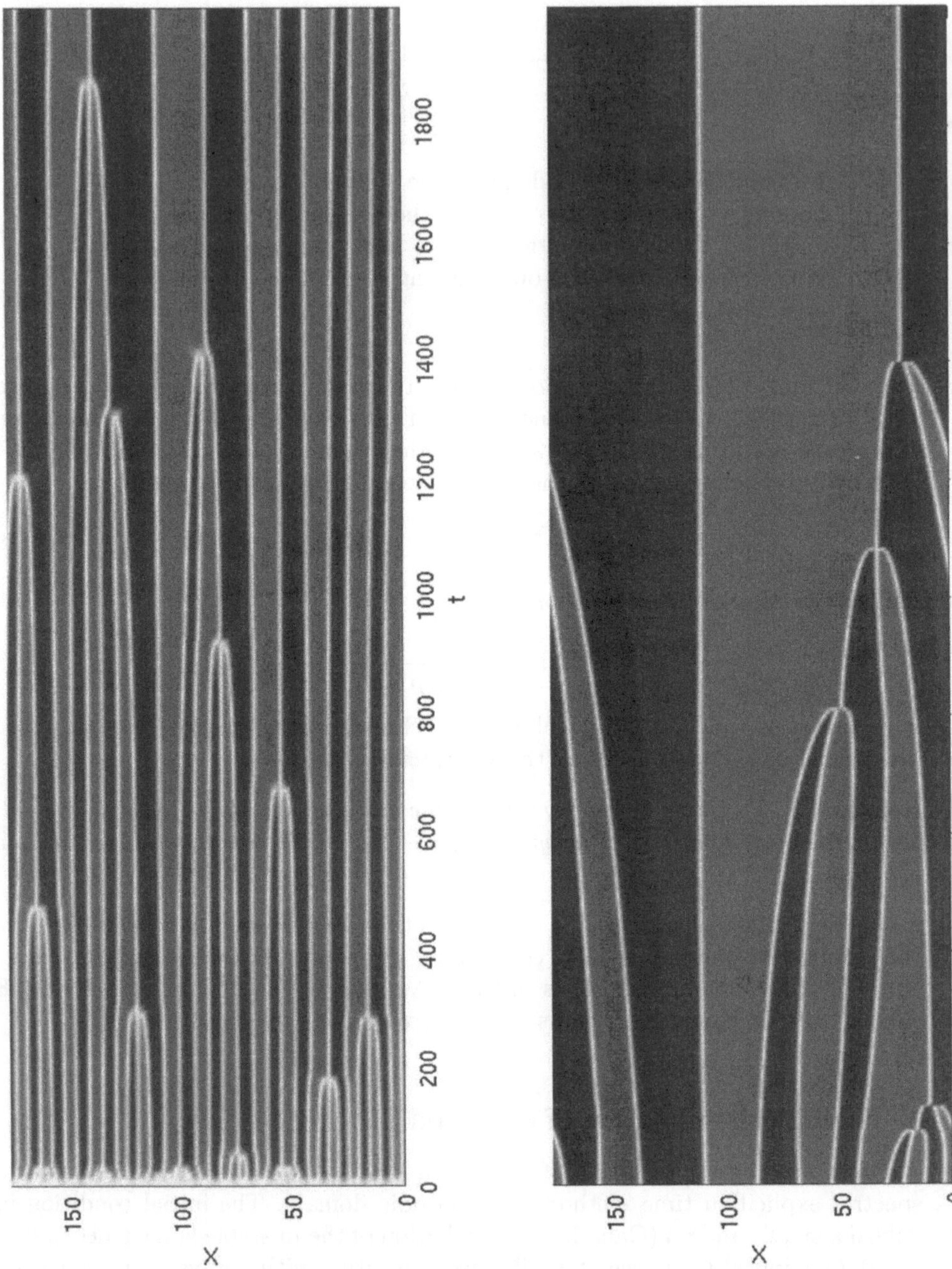

FIGURE 5. Simulation of the $\mathcal{CCH}$ evolution equation for $\varepsilon = 0.1$. The color scheme is gray for $q = 1$ through white, $q = 0$, to black for $q = -1$; courtesy A. A. Golovin.

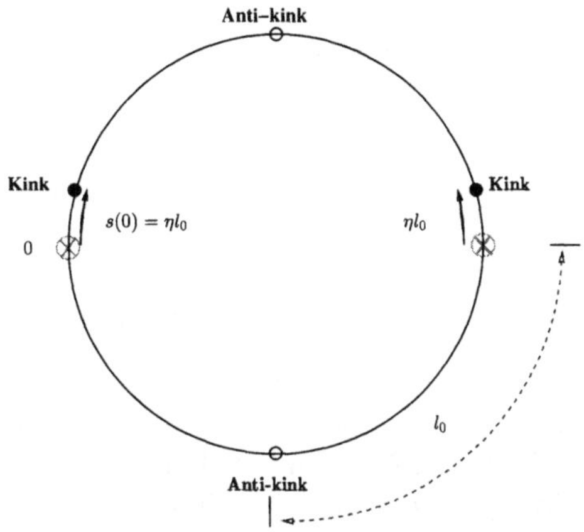

FIGURE 6. The initial location of the kinks • and anti-kinks ○ for the comparative study of $\mathcal{CCH}$ and $\mathcal{CDS}$; here ⊗ denotes the location of the kink prior to the small offset η from periodicity in the most unstable direction [26].

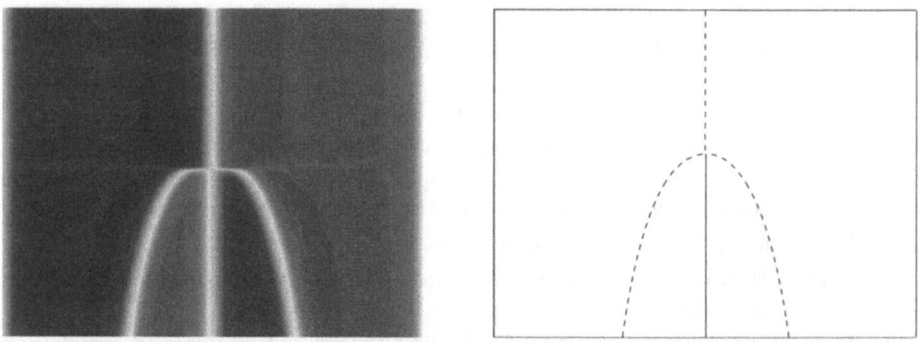

FIGURE 7. Numerical simulation of the $\mathcal{CCH}$ and $\mathcal{CDS}$ on a periodic domain with initial data as in fig. 6; courtesy of Scott Norris. The space scale (horizontal axis) and time scale (vertical axis) are left unspecified since the match displayed is generic.

the other hand, binary coalescence of phase boundaries is the generic coarsening event of the Cahn–Hilliard evolution [14, 6, 23].

Now, consider the simple case of a periodic domain of four equi-spaced phase boundaries perturbed in the manner exhibited in fig. 6. The direct numerical simulations of $\mathcal{CCH}$ and $\mathcal{CDS}$ are presented for comparison in fig 7. One notes excellent agreement up to the point where the phase boundaries meet. Here, the finite width of phase boundaries plays a role that is not captured by our sharp interface theory. However, the time scale over which they differ markedly is only a small fraction of the phase boundary lifetimes. Whence, we expect the coarsening rate of $\mathcal{CDS}$ to faithfully represent that of $\mathcal{CCH}$.

Remark 5.1. *A linear stability analysis of an equi-spaced kink/anti-kink array has a dominant instability in which kinks are* pairwise pinched *toward the intervening anti-kinks, in the manner exhibited in Fig. 6. This justifies the choice of initial data for the comparative study of $\mathcal{CDS}$ and $\mathcal{CCH}$ shown in Fig. 7.*

Acknowledgment: The author gratefully acknowledges the financial support of the SFB 611 at the University of Bonn (June-July, 2003) and also the partial support of the NSF grant # DMR-0102794. The author would also like to thank Scott Norris for kindly providing the comparative simulations for $\mathcal{CDS}$ and $\mathcal{CCH}$.

References

[1] N. Alikakos, P. Bates, G. Fusco, *Slow motion for the Cahn–Hilliard equation in one space dimension*, J. Diff. Eq. **90** 1, 81–135 (1991).

[2] A. J. Bray, *Theory of phase ordering kinetics*, Advances in Physics **43** (3), (1997), 357–459.

[3] J. W. Cahn, W. C. Carter, *Crystal shapes and phase equilibria: A common mathematical basis*, Metall. Mater. Trans. A **27** (6), (1996), 1431–1440.

[4] J. W. Cahn, J. E. Hilliard, *Free energy of nonuniform systems. I. Interfacial free energy*, J. Chem. Phys. **28**, (1958), 258.

[5] J. W. Cahn, D. W. Hoffman, *A vector thermodynamics for anisotropic surfaces — II. Curved and faceted surfaces*, Acta. Metall. **22**, (1974), 1205.

[6] H. Calisto, M. Clerc, R. Rojas, E. Tirapegui, *Bubbles interactions in the Cahn–Hilliard equation*, Phys. Rev. Lett. **85** (18), (2000), 3805–3808.

[7] A. A. Chernov, *Stability of faceted faces*, J. Cryst. Growth **24/25**, (1974), 11.

[8] Stephen H. Davis, Theory of Solidification, *Cambridge Monographs on Mechanics*, Cambridge University Press, 2001.

[9] A. Di Carlo, M. E. Gurtin, P. Podio-Guidugli, *A regularised equation for anisotropic motion-by curvature*, SIAM J. Math. Anal., **52**, 1111–1119 (1992)

[10] C. L. Emmott, A. J. Bray, *Coarsening dynamics of a one-dimensional driven Cahn–Hilliard system*, Phys. Rev. E **54**, (1996), 4568.

[11] A. A. Golovin, S. H. Davis, A. A. Nepomnyashchy, *A convective Cahn–Hilliard model for the formation of facets and corners in crystal growth*, Physica D **118**, (1998), 202–230.

[12] A. A. Golovin, A. A. Nepomnyashchy, S. H. Davis, M. A. Zaks, *Convective Cahn–Hilliard models: from coarsening to roughening*, Phys. Rev. Lett. **86** (8), (2001), 1550–1553.

[13] M. E. Gurtin, Thermomechanics of Evolving Phase Boundaries in the Plane, Clarendon Press, Oxford, 1993.

[14] T. Kawakatsu, T. Munakata, *Kink dynamics in a one-dimensional conserved TDGL system*, Prog. Theor. Phys. **74**, (1985), 11–19.

[15] R. V. Kohn, F. Otto, *Upper bounds for coarsening rates*, Comm. Math. Phys. **229**, (2002), 375–395.

[16] K. Leung, *Theory of morphological instability in driven systems*, J. Stat. Phys. **61**, (1990), 345.

[17] F. Liu, H. Metiu, *Dynamics of phase-separation of crystal surfaces*, Phys. Rev. B **48**, (1993), 5808.

[18] D. Moldovan, L. Golubovic, *Interfacial coarsening dynamics in epitaxial growth with slope selection*, Phys. Rev. E **61** (6), (2000), 6190–6214.

[19] P. Nozieres, *Shape and growth in Crystals*, in: C. Godreche (Eds.), Solids far from equilibrium, Cambridege University Press, 1992, 1–154.

[20] M. Siegert, *Coarsening dynamics of crystalline thin films*, Phys. Rev. Lett., **81** (25), (1998) 5481.

[21] P. Smilauer, M Rost, J. Krug, *Fast coarsening in unstable epitaxy with desorption*, Phys. Rev. E **59** (6), (1999), 6263.

[22] J. Stewart, N. Goldenfeld, *Spinodal decomposition of a crystal surface*, Phys. Rev. A **46**, (1992), 6505.

[23] X. Sun, M. J. Ward, *Dynamics and coarsening of interfaces for the viscous Cahn–Hilliard equation in one spatial dimension*, Studies in Applied Mathematics, **105** (2000), 203–234.

[24] W. A. Tiller, *The Science of Crystallization: Microscopic Interfacial Phenomena*, Cambridge University Press, Cambridge, UK (1991).

[25] S. J. Watson, *The spectrum and generalized eigenvector basis of alternating circulant matrices* (submitted).

[26] S. J. Watson, F. Otto, S. H. Davis, *Coarsening dynamics of the convective Cahn–Hilliard equation*, Physica D **178** (3-4) (2003) 127–148.

[27] T. Willmore, *Riemannian Geometry*, Oxford University Press, New York 1993.

Engineering Sciences and Applied Mathematics,
Northwestern University,
2145 Sheridan Road,
Evanston IL,
USA
E-mail address: s-watson@northwestern.edu

List of Participants

Toyohiko Aiki *(Gifu University)*
Dmitri Alexandrov *(Urals State University)*
Grégoire Allaire *(Ecole Polytechnique, Palaiseau)*
Luigi Ambrosio *(Scuola Normale Superiore, Pisa)*
Fuensanta Andreu *(Universidad de Valencia)*
Daniele Andreucci *(Università La Sapienza di Roma)*
Masayasu Aso *(Chiba University)*
Eberhard Bänsch *(Weierstrass Institute and Freie Universität Berlin)*
Fabio Bagagiolo *(Università di Trento)*
Tudor Barbu *(Romanian Academy of Sciences)*
Peter Bates *(Michigan State University)*
Giovanni Bellettini *(Università di Roma Tor Vergata)*
Michal Benes *(Czech Technical University)*
Marco L. Bernardi *(Università di Pavia)*
Giorgio Bertotti *(IEN Galileo Ferraris)*
Elena Bonetti *(Università di Pavia)*
Giovanna Bonfanti *(Università di Brescia)*
Folkmar A. Bornemann *(Munich University of Technology)*
Nikolai Botkin *(Stiftung Caesar, Bonn)*
Martin Brokate *(TU München)*
Martin Burger *(Industrial Mathematics Institute, University of Linz)*
Erik Burman *(Ecole Polytechnique Fédérale de Lausanne)*
Vincenzo Capasso *(Università degli Studi di Milano)*
Antonin Chambolle *(Université de Paris-Dauphine)*
Zhiming Chen *(Institute of Computational Mathematics, Beijing)*
Pierluigi Colli *(Università di Pavia)*
Cristina Costantini *(Università di Chieti)*
Elaine Crooks *(Oxford University)*
Pierre Degond *(CNRS and University Paul Sabatier)*
Irina Denissova *(Russian Academy of Sciences)*
Jeffrey J. Derby *(University of Minnesota)*
Emmanuele DiBenedetto *(Vanderbilt University)*
Zhengzhong Ding *(Hangzhou University of Commerce)*
Nicolas Dirr *(MPI for Mathematics in the Sciences, Leipzig)*
Wolfgang Dreyer *(WIAS, Berlin)*
Frank Duderstadt *(WIAS, Berlin)*
Gerhard Dziuk *(University of Freiburg)*
Charles M. Elliott *(University of Sussex)*
Ramon Escobedo *(Università di Milano)*
Antonio Fasano *(Università di Firenze)*
Francesca Fierro *(Università degli Studi di Milano)*
Stefano Finzi Vita *(Università La Sapienza di Roma)*

Michel Frémond *(Laboratoire Lagrange, LCPC Paris)*
J. Michael Fried *(University of Freiburg)*
Avner Friedman *(Ohio State University)*
Takeshi Fukao *(Chiba University)*
Giorgio Fusco *(Università di L'Aquila)*
Lorenzo Fusi *(Università di Firenze)*
Leszek Gasinski *(Jagiellonian University)*
Stefania Gatti *(Politecnico di Milano)*
Luca Gerardo-Giorda *(Università di Trento)*
Ugo Gianazza *(Università di Pavia)*
Roberto Gianni *(Università La Sapienza di Roma)*
Gianni Gilardi *(Università di Pavia)*
Martin E. Glicksman *(Rensselaer Polytechnic Institute)*
Jan Goncerzewicz *(Wroclaw University of Technology)*
Maria T. A. Gonzales *(Cuban Society of Mathematics)*
Maurizio Grasselli *(Politecnico di Milano)*
Pierre Gremaud *(North Carolina State University)*
Giovanna Guidoboni *(Università di Ferrara)*
Danielle Hilhorst *(Université Paris-Sud)*
Karl-Heinz Hoffmann *(Stiftung Caesar, Bonn)*
Sam Howison *(Oxford University)*
Tiberiu Ioana *(Babes-Bolyai University)*
Francoise Issard-Roch *(Université Paris Sud)*
Kazuo Ito *(University of Hannover)*
Akio Ito *(School of Engineering, Kinki University)*
Bum Ja Jin *(Università di Ferrara)*
Atsushi Kadoya *(Hiroshima Shudo University)*
Nobuyuki Kenmochi *(Chiba University)*
Daniel Kessler *(University of Maryland)*
Tero Kilpelainen *(University of Jyvaskyla)*
Robert Kohn *(Courant Institute, New York)*
Yoshihito Kohsaka *(University of Tohoku)*
Ralph Kornhuber *(Freie Universität Berlin)*
Tetsuya Koyama *(Hiroshima Institute of Technology)*
Oliver Kriessl *(University of Freiburg)*
Omar Lakkis *(Università degli Studi di Milano)*
Anne Langlois *(Ecole des Mines et Institut Elie Cartan, Nancy)*
Maria B. Lerena *(Universidad Autonoma de Madrid)*
Chu-Pin Lo *(Dahan Institute of Technology)*
Stephan Luckhaus *(University of Leipzig)*
Fabio Luterotti *(Università di Brescia)*
Abdeslem Lyaghfouri *(King Fahd University of Petroleum and Minerals)*
Enrico Magenes *(IMATI-CNR, Pavia)*
Daniela Mansutti *(IAC-CNR, Roma)*
Paola Manucci *(Università di Padova)*
Simon Masnou *(Université de Paris 6)*
Jose M. Mazon *(Universidad de Valencia)*

Juergen Mehnert *(University of Freiburg)*
Anvarbek Meirmanov *(Universidade da Beira Interior)*
Michael Metzger *(University of Erlangen-Nürnberg)*
Alessandra Micheletti *(Università degli Studi di Milano)*
Karol Mikula *(Slovak University of Technology)*
Masayasu Mimura *(Hiroshima University)*
Emil Minchev *(Chiba University)*
José S. Moll Cebolla *(Universitade de Valencia)*
Maria Giovanna Mora *(Max-Planck Institut, Leipzig)*
Maria Morandi Cecchi *(Università di Padova)*
Pedro Morin *(Instituto de Matemàtica Aplicada Litoral-Santa Fe)*
Georg Müller *(University Erlangen-Nürnberg)*
Oliver Müller *(University of Freiburg, Germany)*
Stefan Müller *(Max Planck Institut, Leipzig)*
Maria del Carmen Muniz *(Universidad de Santiago de Compostela)*
Cornel Murea *(Université de Haute Alsace)*
Kazuaki Nakane *(Osaka Institute of Techonogy)*
Matteo Negri *(Max Planck Institut, Leipzig)*
Barbara Niethammer *(Institut für Angewandte Mathematik, Bonn)*
Ricardo Nochetto *(University of Maryland)*
Igbida Noureddine *(University de Picardie Jule Vernes, Amiens)*
Matteo Novaga *(Università di Pisa)*
John R. Ockendon *(Oxford University)*
Takanobu Okazaki *(Chiba University)*
Daniel T. Onofrei *(WPI Worcester Politechnic Institute)*
Felix Otto *(University of Bonn)*
Maurizio Paolini *(Università Cattolica di Brescia)*
Oliver Penrose *(Heriot-Watt University, Edinburgh)*
Nelly Point *(Conservatoire National des Arts et Metiers, Paris)*
Errico Presutti *(Università Tor Vergata di Roma)*
Mario Primicerio *(Università di Firenze)*
Alfio Quarteroni *(Politecnico di Milano - EPF Lausanne)*
Vincenzo Recupero *(Università di Trento)*
Fernando Reitich *(University of Minnesota)*
Riccardo Ricci *(Università di Firenze)*
Elisabetta Rocca *(Università di Pavia)*
Jean R. Roche *(Université Henri Poincaré)*
Josè-Francisco Rodrigues *(CMAF/Universidade de Lisboa)*
Riccarda Rossi *(Università di Pavia)*
Tomas Roubicek *(Charles University, Prague)*
Maria R. Russo *(Università di Padova)*
Witold Rudnicki *(University of Warsaw)*
Raluca Rusu *(University of Freiburg)*
Claudia Salani *(Università degli Studi di Milano)*
Elena Sartori *(Politecnico di Milano)*
Naoki Sato *(Nagaoka College of Tecnology)*
Giuseppe Savaré *(Università di Pavia)*

Jean-Francois Scheid *(Université Nancy 1, France)*
Giulio Schimperna *(Università di Pavia)*
Alfred Schmidt *(Universität Bremen)*
Ken Shirakawa *(Tokyo Denki University)*
Andrés F. Solé Martinez *(Universitat Pompeu Fabra)*
Jurgen Sprekels *(WIAS, Berlin)*
Victor Starovoitov *(Center of Advanced European Studies and Research, Bonn)*
Ulisse Stefanelli *(IMATI-CNR, Pavia)*
Vanessa Styles *(University of Sussex)*
Tohru Tsujikawa *(Miyazaki University)*
Holavanahalli S. Udaykumar *(University of Iowa)*
José M. Urbano *(Universidade de Coimbra)*
Andreas Veeser *(Università degli Studi di Milano)*
Juan J. L. Velazquez *(Universidad Complutense de Madrid)*
Claudio Verdi *(Università degli Studi di Milano)*
Augusto Visintin *(Università di Trento)*
Stephen J. Watson *(Northwestern University)*
Adam Wheeler *(University of Southampton)*
Noriaki Yamazaki *(Muroran Institute of Technology)*
Fahuai Yi *(South China Normal University)*
Boris Zaltzman *(Ben-Gurion University of the Negev)*

ISNM

MIX
Papier aus verantwortungsvollen Quellen
Paper from responsible sources
FSC® C105338

If you have any concerns about our products,
you can contact us on
ProductSafety@springernature.com

In case Publisher is established outside the EU,
the EU authorized representative is:
Springer Nature Customer Service Center GmbH
Europaplatz 3, 69115 Heidelberg, Germany

Printed by Libri Plureos GmbH
in Hamburg, Germany